T0327654

DIFFERENCE AND DIFFERENTIAL EQUATIONS WITH APPLICATIONS IN QUEUEING THEORY

DIFFERENCE AND DIFFERENTIAL EQUATIONS WITH APPLICATIONS IN QUEUEING THEORY

Aliakbar Montazer Haghighi
Department of Mathematics
Prairie View A&M University
Prairie View, TX

Dimitar P. Mishev
Department of Mathematics
Prairie View A&M University
Prairie View, TX

Library of Congress Cataloging-in-Publication Data:
Haghighi, Aliakbar Montazer.
    Difference and differential equations with applications in queueing theory / Aliakbar
Montazer Haghighi Department of Mathematics, Prairie View A&M University, Prairie View,
Texas, Dimitar P. Mishev Department of Mathematics, Prairie View A&M University,
Prairie View, Texas.
        pages cm
    Includes bibliographical references and index.
    ISBN 978-1-118-39324-6 (hardback) 1. Queuing theory. 2. Difference equations.
3. Differential equations. I. Mishev, D. P. (Dimiter P.) II. Title.
    QA274.8.H338 2013
    519.8′2–dc23
                                        2013001302

10 9 8 7 6 5 4 3 2 1

CONTENTS

![] PREFACE

Topics in difference and differential equations with applications in queueing theory typically span five subject areas: (1) probability and statistics, (2) transforms, (3) differential equations, (4) difference equations, and (5) queueing theory. These are addressed in at least four separate textbooks and taught in four different courses in as many semesters. Due to this arrangement, students needing to take these courses have to wait to take some important and fundamental required courses until much later than should be necessary. Additionally, based on our long experience in teaching at the university level, we find that perhaps not all topics in one subject are necessary for a degree. Hence, perhaps we as faculty and administrators should rethink our traditional way of developing and offering courses. This is another reason for the content of this book, as of the previous one from the authors, to offer several related topics in one textbook. This gives the instructor the freedom to choose topics according to his or her desire to emphasize, yet cover enough of a subject for students to continue to the next course, if necessary.

The methodological content of this textbook is not exactly novel, as "mathematics for engineers" textbooks have reflected this method for long past. However, that type of textbook may cover some topics that an engineering student may already know. Now with this textbook the subject will be reinforced. The need for this practice has generally ignored some striking relations that exist between the seemingly separate areas of a subject, for instance, in statistical concepts such as the estimation of parameters of distributions used in queueing theory that are derived from differential–difference equations. These concepts commonly appear in queueing theory, for instance, in measures on effectiveness in queuing models.

All engineering and mathematics majors at colleges and universities take at least one course in ordinary differential equations, and some go further to take courses in partial differential equations. As mentioned earlier, there are many books on "mathematics for engineers" on the market, and one that contains some applications using Laplace and Fourier transforms. Some also have included topics of probability and statistics, as the one by these authors. However, there is a lack of applications of probability and statistics that use differential equations, although we did it in our book. Hence, we felt that there is an urgent need for a textbook that recognizes the corresponding relationships between the various areas and a matching cohesive course. Particularly, theories of queues and reliability are two of those topics, and this book is designed to achieve just that. Its five chapters, while retaining their individual integrity, flow from selected topics in probability and statistics to differential and difference equations to stochastic processes and queueing theory.

Chapter 1 establishes a strong foundation for what follows in Chapter 2 and beyond. Classical Fourier and Laplace transforms as well as Z-transforms and generating functions are included in Chapter 2. Partial differential equations are often used to construct models of the most basic theories underlying physics and engineering, such as the system of partial differential equations known as Maxwell's equations, from which one can derive the entire theory of electricity and magnetism, including light. In particular, elegant mathematics can be used to describe the vibrating circular membrane. However, our goal here is to develop the most basic ideas from the theory of partial differential equations and to apply them to the simplest models arising from physics and the queueing models. Detailed topics of ordinary and partial differential and difference equations are included in Chapter 3 and Chapter 4 that complete the necessary tools for Chapter 5, which discusses stochastic processes and queueing models. However, we have also included the power series method of solutions of differential equations, which can be applied to, for instance, Bessel's equation.

In our previous book, we required two semesters of calculus and a semester of ordinary differential equations for a reader to comprehend the contents of the book. In this book, however, knowledge of at least two semesters of calculus that includes some familiarity with terminology such as the gradient, divergence, and curl, and the integral theorems that relate them to each other, are needed. However, we discuss not only the topics in differential equation, but also the difference equations that have vast applications in the theory of signal processing, stochastic analysis, and queueing theory.

Few instructors teach the combined subject areas together due to the difficulties associated with handling such a rigorous course with such hefty materials. Instructors can easily solve this issue by teaching the class as a multi-instructor course.

We should note that throughout the book, we use boldface letters, Greek or Roman (lowercase or capital) for vectors and matrices. We shall write $P(n)$ or P_n to mean P as a function of a discrete parameter n. Thus, we want to make sure that students are well familiar with functions of discrete variables as well as continuous ones. For instance, a vibrating string can be regarded as a continuous object, yet if we look at a fine enough scale, the string is made up of molecules, suggesting a discrete model with a large number of variables. There are many cases in which a discrete model may actually provide a better description of the phenomenon under study than a continuous one.

We also want to make sure that students realize that solution of some problems requires the ability to carry out lengthy calculations with confidence.

Of course, all of these skills are necessary for a thorough understanding of the mathematical terminology that is an essential foundation for the sciences and engineering. We further note that subjects discussed in each chapter could be studied in isolation; however, their cohesiveness comes from a thorough understanding of applications, as discussed in this book.

We hope this book will be an interesting and useful one to both students and faculty in science, technology, engineering, and mathematics.

ALIAKBAR MONTAZER HAGHIGHI

DIMITAR P. MISHEV

Houston, Texas
April 2013

CHAPTER ONE

Probability and Statistics

The organization of this book is such that by the time reader gets to the last chapter, all necessary terminology and methods of solutions of standard mathematical background has been covered. Thus, we start the book with the basics of probability and statistics, although we could have placed the chapter in a later location. This is because some chapters are independent of the others.

In this chapter, the basics of probability and some important properties of the theory of probability, such as discrete and continuous random variables and distributions, as well as conditional probability, are covered.

After the presentation of the basics of probability, we will discuss statistics. Note that there is still a dispute as to whether statistics is a subject on its own or a branch of mathematics. Regardless, statistics deals with gathering, analyzing, and interpreting data. Statistics is an important concept that no science can do without. Statistics is divided in two parts: *descriptive statistics* and *inferential statistics*. Descriptive statistics includes some important basic terms that are widely used in our day-to-day lives. The latter is based on *probability theory*. To discuss this part of the statistics, we include point estimation, interval estimation, and hypothesis testing.

We will discuss one more topic related to both probability and statistics, which is extremely necessary for business and industry, namely *reliability of a system*. This concept is also needed in applications such as queueing networks, which will be discussed in the last chapter.

In this chapter, we cover as much probability and statistics as we will need in this book, except some parts that are added for the sake of completeness of the subject.

1.1. BASIC DEFINITIONS AND CONCEPTS OF PROBABILITY

Nowadays, it has been established in the scientific world that since quantities needed are not quite often predictable in advance, randomness should be

Difference and Differential Equations with Applications in Queueing Theory, First Edition.
Aliakbar Montazer Haghighi and Dimitar P. Mishev.
© 2013 John Wiley & Sons, Inc. Published 2013 by John Wiley & Sons, Inc.

accounted for in any realistic world phenomenon, and that is why we will consider random experiments in this book.

Determining probability, or chance, is to quantify the variability in the outcome or outcomes of a random experiment whose exact outcome or outcomes cannot be predicted by certainty. Satellite communication systems, such as radar, are built of electronic components such as transistors, integrated circuits, and diodes. However, as any engineer would testify, the components installed usually never function as the designer has anticipated. Thus, not only is the probability of failure to be considered, but the reliability of the system is also quite important, since the failure of the system may have not only economic losses but other damages as well. With probability theory, one may answer the question, "How reliable is the system?"

Definition 1.1.1. *Basics*

(a) Any result of performing an experiment is called an *outcome* of that experiment. A set of outcomes is called an *event*.

(b) If occurrences of outcomes are not certain or completely predictable, the experiment is called a *chance* or *random experiment*.

(c) In a random experiment, sets of outcomes that cannot be broken down into smaller sets are called *elementary* (or *simple* or *fundamental*) *events*.

(d) An elementary event is, usually, just a singleton (a set with a single element, such as $\{e\}$). Hence, a combination of elementary events is just an *event*.

(e) When any element (or outcome) of an event happens, we say that the *event occurred*.

(f) The *union* (set of all elements, with no repetition) of all events for a random experiment (or the set of all possible outcomes) is called the *sample space*.

(g) In "set" terminology, an *event* is a *subset* of the sample space. Two events A_1 and A_2 are called *mutually exclusive* if their intersection is the empty set, that is, they are disjoint subsets of the sample space.

(h) Let $A_1, A_2, \ldots, A_n$ be mutually exclusive events such that $A_1 \cup A_2 \cup \ldots \cup A_n = \Omega$. The set of $\{A_1, A_2, \ldots, A_n\}$ is then called a *partition* of the sample space Ω.

(i) For an experiment, a collection or a set of all individuals, objects, or measurements of interest is called a (statistical) *population*.

For instance, to determine the average grade of the differential equation course for all mathematics major students in four-year colleges and universities in Texas, the totality of students majoring mathematics in the colleges and universities in the Texas constitutes the population for the study.

Usually, studying the population may not be practically or economi-
cally feasible because it may be quite time consuming, too costly, and/
or impossible to identify all members of it. In such cases, sampling is
being used.

(j) A portion, subset, or a part of the population of interest (finite or
infinite number of them) is called a *sample.*

Of course, the sample must be *representative* of the entire population in order to make any prediction about the population.

(k) An element of the sample is called a *sample point.* By *quantification*
of the sample we mean changing the sample points to numbers.

(l) The *range* is the difference between the smallest and the largest sample
points.

(m) A sample selected such that each element or unit in the population
has the same chance to be selected is called a *random sample.*

(n) The *probability of an event A,* denoted by $P(A)$, is a number between
0 and 1 (inclusive) describing likelihood of the event A to occur.

(o) An event with probability 1 is called an *almost sure event.* An event
with probability 0 is called a *null* or an *impossible event.*

(p) For a sample space with n (finite) elements, if all elements or outcomes
have the same chance to occur, then we assign probability $1/n$ to each
member. In this case, the sample space is called *equiprobable.*

For instance, to choose a digit at random from 1 to 5, we mean
that every digit of $\{1, 2, 3, 4, 5\}$ has the same chance to be picked,
that is, all elementary events in $\{1\}$, $\{2\}$, $\{3\}$, $\{4\}$, and $\{5\}$ are equiprobable. In that case, we may associate probability $1/5$ to each digit
singleton.

(q) If a random experiment is repeated, then the chance of occurrence of
an outcome, intuitively, will be approximated by the ratio of occurrences of the outcome to the total number of repetitions of the experiment. This ratio is called the *relative frequency.*

Axioms of Probabilities of Events

We now state properties of probability of an event A through *axioms of probability.* The Russian mathematician Kolmogorov originated these axioms in
early part of the twentieth century. By an axiom, it is meant a statement that
cannot be proved or disproved. Although all probabilists accept the three
axioms of probability, there are axioms in mathematics that are still contro-
versial, such as the axiom of choice, and not accepted by some prominent
mathematicians.

Let Ω be the sample space, $\mathcal{B}$ the set function containing all possible events
drawn from Ω, and P denote the probability of an event. The triplet $(\Omega, \mathcal{B}, P)$
is then called the *probability space.* Later, after we define a random variable,
we will discuss this space more rigorously.

Axioms of Probability

Axiom A1. $0 \le P(A) \le 1$ for each event A in $\mathcal{B}$.

Axiom A2. $P(\Omega) = 1$.

Axiom A3. If A_1 and A_2 are *mutually exclusive* events in $\mathcal{B}$, then:

$$P(A_1 \cup A_2) = P(A_1) + P(A_2),$$

where mutually exclusive events are events that have no sample point in common, and the symbol $\cup$ means the union of two sets, that is, the set of all elements in both set without repetition.

Note that the axioms stated earlier are for events. Later, we will define another set of axioms of probability involving random variables.

If the occurrence of an event has influence on the occurrence of other events under consideration, then the probabilities of those events change.

Definition 1.1.2

Suppose $(\Omega, \mathcal{B}, P)$ is a probability space and B is an event (i.e., $B \in \mathcal{B}$) with positive probability, $P(B) > 0$. The *conditional probability of A given B*, denoted by $P(A|B)$, defined on $\mathcal{B}$, is then given by:

$$P(A|B) = \frac{P(AB)}{P(B)}, \text{ for any event } A \text{ in } \mathcal{B}, \text{ and for } P(B) > 0. \qquad (1.1.1)$$

If $P(B) = 0$, then $P(A|B)$ is not defined. Under the condition given, we will have a new triple, that is, a new probability space $(\Omega, \mathcal{B}, P(A|B))$. This space is called the *conditional probability space induced on* $(\Omega, \mathcal{B}, P)$, *given B*.

Definition 1.1.3

For any two events A and B with conditional probability $P(B \mid A)$ or $P(A \mid B)$, we have the *multiplicative law*, which states:

$$P(AB) = P(B|A)P(A) = P(A|B)P(B). \qquad (1.1.2)$$

We leave it as an exercise to show that for n events $A_1, A_2, \dots, A_n$, we have:

$$P(A_1 A_2 \dots A_n) = P(A_1) \times P(A_2|A_1) \times P(A_3|A_1 A_2) \times \cdots \times P(A_n|A_1 A_2 \dots A_{n-1}). \qquad (1.1.3)$$

Definition 1.1.4

We say that events A and B are *independent* if and only if:

$$P(AB) = P(A)P(B). \qquad (1.1.4)$$

It will be left as an exercise to show that if events A and B are independent and $P(B) > 0$, then:

$$P(A|B) = P(A). \qquad (1.1.5)$$

It can be shown that if $P(B) > 0$ and (1.1.5) is true, then A and B are independent. For proof, see Haghighi et al. (2011a, p. 139).

The concept of independence can be extended to a finite number of events.

Definition 1.1.5
Events $A_1, A_2, \ldots, A_n$ are *independent* if and only if the probability of the intersection of any subset of them is equal to the product of corresponding probabilities, that is, for every subset $\{i_1, \ldots, i_k\}$ of $\{1, \ldots, n\}$ we have:

$$P\{(A_{i_1} A_{i_2} \ldots A_{i_n})\} = P(A_{i_1}) \times P(A_{i_2}) \times \cdots \times P(A_{i_k}). \qquad (1.1.6)$$

As one of the very important applications of conditional probability, we state the following theorem, whose proof may be found in Haghighi et al. (2011a):

Theorem 1.1.1. *The Law of Total Probability*
Let $A_1, A_2, \ldots, A_n$ be a partition of the sample space Ω. For any given event B, we then have:

$$P(B) = \sum_{i=1}^{n} P(A_i) P(B|A_i). \qquad (1.1.7)$$

Theorem 1.1.1 leads us to another important application of conditional probability. Proof of this theorem may also found in Haghighi et al. (2011a).

Theorem 1.1.2. *Bayes' Formula*
Let $A_1, A_2, \ldots, A_n$ be a partition of the sample space Ω. If an event B occurs, the probability of any event A_j given an event B is:

$$P(A_j|B) = \frac{P(A_j)P(B|A_j)}{\sum_{i=1}^{n} P(A_i)P(B|A_i)}, \quad j = 1, 2, \ldots, n. \qquad (1.1.8)$$

Example 1.1.1
Suppose in a factory three machines A, B, and C produce the same type of products. The percent shares of these machines are 20, 50, and 30, respectively. It is observed that machines A, B, and C produce 1%, 4%, and 2% defective items, respectively. For the purpose of quality control, a produced item is chosen at random from the total items produced in a day. Two questions to answer:

1. What is the probability of the item being defective?
2. Given that the item chosen was defective, what is the probability that it was produced by machine B?

Answers
To answer the first question, we denote the event of defectiveness of the item chosen by D. By the law of total probability, we will then have:

$$P(D) = P(A)P(D|A) + P(B)P(D|B) + P(C)P(D|C)$$
$$= 0.20 \times 0.01 + 0.50 \times 0.04 + 0.30 \times 0.20$$
$$= 0.002 + 0.020 + 0.006 = 0.028.$$

Hence, the probability of the produced item chosen at random being defective is 2.8%.

To answer the second question, let the conditional probability in question be denoted by $P(B | D)$. By Bayes' formula and answer to the first question, we then have:

$$P(B|D) = \frac{P(B)P(D|B)}{P(D)} = \frac{0.50 \times 0.04}{0.028} = 0.714.$$

Thus, the probability that the defective item chosen be produced by machine C is 71.4%.

Example 1.1.2
Suppose there are three urns that contain black and white balls as follows:

$$\begin{cases} \text{Urn 1:} & 2\ blacks \\ \text{Urn 2:} & 2\ whites \\ \text{Urn 1:} & 1\ black \text{ and } 1\ white. \end{cases} \qquad (1.1.9)$$

A ball is drawn randomly and it is "white." Discuss possible probabilities.

Discussion
The sample space Ω is the set of all pairs $(\cdot,\cdot)$, where the first dot represents the urn number (1, 2, or 3) and the second represents the color (black or white). Let U_1, U_2 and U_3 denote events that drawing was chosen from, respectively. Assuming that urns are identical and balls have equal chances to be chosen, we will then have:

$$P(U_1) = P(U_2) = P(U_3) = \frac{1}{3}. \qquad (1.1.10)$$

Also, $U_1 = (1,\cdot)$, $U_2 = (2,\cdot)$, $U_3 = (3,\cdot)$.

Let W denote the event that a white ball was drawn, that is, $W = \{(\cdot, w)\}$. From (1.1.9), we have the following conditional probabilities:

$$P(W|U_1) = 0, \quad P(W|U_2) = 1, \quad P(U_3) = \frac{1}{2}. \tag{1.1.11}$$

From Bayes' rule, (1.1.9), (1.1.10), and (1.1.11), we have:

$$P(U_1|W) = \frac{P(W|U_1)P(U_1)}{P(W|U_1)P(U_1) + P(W|U_2)P(U_2) + P(W|U_3)P(U_3)}, \tag{1.1.12}$$

$$= 0. \tag{1.1.13}$$

Note that denominator of (1.1.12) is:

$$0 + (1)\left(\frac{1}{3}\right) + \left(\frac{1}{2}\right)\left(\frac{1}{3}\right) = \frac{1}{3} + \frac{1}{6} = \frac{1}{2}. \tag{1.1.14}$$

Using (1.1.14), we have:

$$P(U_2|W) = \frac{P(W|U_2)P(U_2)}{P(W|U_1)P(U_1) + P(W|U_2)P(U_2) + P(W|U_3)P(U_3)}$$

$$= \frac{(1)\left(\frac{1}{3}\right)}{\frac{1}{2}} = \frac{2}{3}. \tag{1.1.15}$$

Again, using (1.1.14), we have:

$$P(U_3|W) = \frac{P(W|U_3)P(U_3)}{P(W|U_1)P(U_1) + P(W|U_2)P(U_2) + P(W|U_3)P(U_3)}$$

$$= \frac{\left(\frac{1}{2}\right)\left(\frac{1}{3}\right)}{\frac{1}{2}} = \frac{1}{3}. \tag{1.1.16}$$

Now, observing from (1.1.13), (1.1.15), and (1.1.16), there is a better chance that the ball was drawn from the second urn. Hence, if we assume that the ball was drawn from the second urn, there is one white ball that remains in it. That is, we will have the three urns with 0, 1, and 1 white ball, respectively, in urns 1, 2, and 3.

1.2. DISCRETE RANDOM VARIABLES AND PROBABILITY DISTRIBUTION FUNCTIONS

As we have seen so far, elements of a sample space are not necessarily numbers. However, for convenience, we would rather have them so. This is done through what is called a *random variable*. In other words, a *random variable* quantifies the sample space. That is, a *random variable* assigns numerical (or set) labels to the sample points. Formally, we define a random variable as follows:

Definition 1.2.1
A *random variable* is a function (or a mapping) on the sample space.

We note that a random variable is really neither a variable (as known independent variable) nor random, but as mentioned, it is just a function. Also note that sometimes the range of a random variable may not be numbers. This is simply because we defined a random variable as a mapping. Thus, it maps elements of a set into some elements of another set. Elements of either set do not have to necessarily be numbers.

There are two main types of random variables, namely, *discrete* and *continuous*. We will discuss each in detail.

Definition 1.2.2
A *discrete random variable* is a function, say X, from a countable sample space, Ω (that could very well be a numerical set), into the set of real numbers.

Example 1.2.1
Suppose we are to select two digits from 1 to 6 such that the sum of the two numbers selected equals 7. Assume that repetition is not allowed. The sample space under consideration will then be $S = \{(1, 6), (2, 5), (3, 4), (4, 3), (5, 2), (6, 1)\}$, which is discrete. This set can also be described as $S = \{(i, j): i + j = 7, i, j = 1, 2, \ldots, 6\}$.

Now, the random variable X can be defined by $X((i, j)) = k, k = 1, 2, \ldots, 6$. That is, the range of X is the set $\{1, 2, 3, 4, 5, 6\}$ such that, for instance, $X((1, 6)) = 1$, $X((2, 5)) = 2$, $X((3, 4)) = 3$, $X((4, 3)) = 4$, $X((5, 2)) = 5$, and $X((6, 1)) = 6$. In other words, the discrete random variable X has quantified the set of ordered pairs S to a set of positive integers from 1 to 6.

Example 1.2.2
Toss a fair coin three times. Denoting heads by H and tails by T, the sample space will then contain eight triplets as $\Omega = \{HHH, HHT, HTH, HTT, THH, THT, TTH, TTT\}$. Each tossing will result in either heads or tails. Thus, we might define the random variable X to take values 1 and 0 for heads and tails, respectively, at the jth tossing. In other words,

$$X_j = \begin{cases} 1, & \text{if } j\text{th outcome is heads,} \\ 0, & \text{if } j\text{th outcome is tails.} \end{cases}$$

Hence, $P\{X_j = 0\} = 1/2$ and $P\{X_j = 1\} = 1/2$. Now from the sample space we see that the probability of the element HTH is:

$$P\{X_1 = 1, X_2 = 0, X_3 = 1\} = \frac{1}{8}. \tag{1.2.1}$$

In contrast, product of individual probabilities is:

$$P\{X_1 = 1\} \times P\{X_2 = 0\} \times P\{X_3 = 1\} = \frac{1}{2} \times \frac{1}{2} \times \frac{1}{2} = \frac{1}{8}. \qquad (1.2.2)$$

From (1.2.1) and (1.2.2), we see that $X_1, X_2,$ and X_3 are mutually independent.

Now suppose we define X and Y as the total number of heads and tails, respectively, after the third toss. The probability, then, of three heads and three tails is obviously zero, since these two events cannot occur at the same time, that is, $P\{X = 3, Y = 3\} = 0$. However, from the sample space probabilities of individual events are $P\{X = 3\} = 1/8$ and $P\{Y = 3\} = 1/8$. Thus, the product is:

$$P\{X = 3\} \times P\{Y = 3\} = \frac{1}{8} \times \frac{1}{8} = \frac{1}{64} \neq 0.$$

Hence, X and Y, in this case, are not independent.

One of the useful concepts using random variable is the *indicator function* (or *indicator random variable* that we will define in the next section.

Definition 1.2.3

Let A be an event from the sample space Ω. The random variable $I_A(\omega)$ for $\omega \in A$ defined as:

$$I_A(\omega) = \begin{cases} 1, & \text{if} \quad \omega \in A, \\ 0, & \text{if} \quad \omega \in A^c, \end{cases} \qquad (1.2.3)$$

is called the indicator function (or indicator random variable).

Note that for every $\omega \in \Omega$, $I_\Omega(\omega) = 1$ and $I_\phi(\omega) = 0$.

We leave it as an exercise for the reader to show the following properties of random variables:

(a) if X and Y are two discrete random variables, then $X \pm Y$ and XY are also random variables, and

(b) if $\{Y = 0\}$ is empty, X/Y is also a random variable.

The way probabilities of a random variable are distributed across the possible values of that random variable is generally referred to as the *probability distribution* of that random variable. The following is the formal definition.

Definition 1.2.4

Let X be a discrete random variable defined on a sample space Ω and x is a typical element of the range of X. Let p_x denote the probability that the random variable X takes the value x, that is,

$$p_x = P([X = x]) \quad \text{or} \quad p_x = P(X = x), \qquad (1.2.4)$$

where p_X is called the *probability mass function* (pmf) of X and also referred to as the *(discrete) probability density function* (pdf) of X.

Note that $\sum_x p_x = 1$, where x varies over all possible values for X.

Example 1.2.3

Suppose a machine is in either "good working condition" or "not good working condition." Let us denote "good working condition" by 1 and " not good working condition" by 0. The sample space of states of this machine will then be $\Omega = \{0, 1\}$. Using a random variable X, we define $P([X = 1])$ as the probability that the machine is in "good working condition" and $P([X = 0])$ as the probability that the machine is not in "good working condition." Now if $P([X = 0]) = 4/5$ and $P([X = 0]) = 1/5$, then we have a distribution for X.

Definition 1.2.5

Suppose X is a discrete random variable, and x is a real number from the interval $(-\infty, x]$. Let us define $F_X(x)$ as:

$$F_X(x) = P([X \le x]) = \sum_{n=-\infty}^{x} p_n, \qquad (1.2.5)$$

where p_n is defined as $P([X = n])$ or $P(X = n)$. $F_X(x)$ is then called the *cumulative distribution function* (cdf) for X.

Note that from the set of axioms of probability mentioned earlier, for all x, we have:

$$p_x \ge 0, \quad \text{and} \quad \sum_x p_x = 1. \qquad (1.2.6)$$

We now discuss selected important discrete probability distribution functions. Before that, we note that a random experiment is sometimes called a *trial*.

Definition 1.2.6

A *Bernoulli trial* is a trial with exactly two possible outcomes. The two possible outcomes of a Bernoulli trial are often referred to as *success* and *failure* denoted by s and f, respectively. If a Bernoulli trial is repeated independently n times with the same probabilities of success and failure on each trial, then the process is called *Bernoulli trials*.

Notes:

(1) From Definition 1.2.6, if the probability of s is p, $0 \le p \le 1$, then, by the second axiom of probability, the probability of f will be $q = 1 - p$.

(2) By its definition, in a Bernoulli trial, the sample space for each trial has two sample points.

Definition 1.2.7

Now, let X be a random variable taking values 1 and 0, corresponding to success and failure, respectively, of the possible outcome of a Bernoulli trial, with p $(p > 0)$ as the probability of success and q as probability of failure. We will then have:

$$P(X = k) = p^k q^{1-k}, \quad k = 0, 1. \tag{1.2.7}$$

Formula (1.2.7) is the probability distribution function (pmf) of the Bernoulli random variable X.

Note that (1.2.7) is because first of all, $p^k q^{1-k} > 0$, and second, $\sum_{k=0}^{1} p^k q^{1-k} = p + q = 1$.

Example 1.2.4

Suppose we test 6 different objects for strength, in which the probability of breakdown is 0.2. What is the probability that the third object test be successful is, that is, does not breakdown?

Answer

In this case, we have a sequence of six Bernoulli trials. Let us assume 1 for a success and 0 for a failure. We would then have a 6-tuple (001000) to symbolize our objective. Hence, the probability would be $(0.2)(0.2)(0.8)(0.2)(0.2)$ $(0.2) = 0.000256$.

Now suppose we repeat a Bernoulli trial independently finitely many times. We would then be interested in the probability of given number of times that one of the two possible outcomes occurs regardless of the order of their occurrences. Therefore, we will have the following definition:

Definition 1.2.8

Suppose X_n is the random variable representing the number of successes in n independent Bernoulli trials. Denote the *pmf* of X_n by $B_k = b(k; n, p)$. $B_k = b(k; n, p)$ is called the *binomial distribution function* with *parameters n and p* of the random variable X, where the parameters n, p and the number k refer to the number of independent trials, probability of *success* in each trial, and the number of successes in n trials, respectively. In this case, X is called the *binomial random variable*. The notation $X \sim b(k; n, p)$ is used to indicate that X is a binomial random variable with parameters n and p.

We leave it as an exercise to prove that:

$$B_k = \binom{n}{k} p^k q^{n-k}, \quad k = 0, 1, 2, \ldots, n, \tag{1.2.8}$$

where $q = 1 - p$.

Example 1.2.5
Suppose two identical machines run together, each to choose a digit from 1 to 9 randomly five times. We want to know what the probability that a sum of 6 or 9 appears k times ($k = 0, 1, 2, 3, 4, 5$) is.

Answer
To answer the question, note that we have five independent trials. The sample space in this case for one trial has 81 sample points and can be written in a matrix form as follows:

$$\begin{pmatrix} (1,1) & (1,2) & \cdots & (1,8) & (1,9) \\ (2,1) & (2,2) & \cdots & (2,8) & (2,9) \\ \vdots & \ddots & \ddots & \vdots & \vdots \\ (8,1) & (8,2) & \ddots & (8,8) & (8,9) \\ (9,1) & (9,2) & \cdots & (9,8) & (9,9) \end{pmatrix}.$$

There are 13 sample points, where the sum of the components is 6 or 9. They are:

$$(1,5), (2,4), (3,3), (4,2), (5,1), (1,8), (2,7), (3,6), (4,5), (5,4), (6,3), (7,2), (8,1).$$

Hence, the probability of getting a sum as 6 or 9 on one selection of both machines together (i.e., probability of a success) is $p = 13/81$. Now let X be the random variable representing the total times a sum as 6 or 9 is obtained in 5 trials. Thus, from (1.2.8), we have:

$$P([X = k]) = \binom{5}{k}\left(\frac{13}{81}\right)^k \left(\frac{68}{81}\right)^{5-k}, \quad k = 0, 1, 2, 3, 4, 5.$$

For instance, the probability that the sum as 6 or 9 does not appear at all will be $(68/81)^5 = 0.42$, that is, there is a $(100 - 42) = 58\%$ chance that we do get at least a sum as 6 or 9 during the five trials.

Based on a sequence of independent Bernoulli trials, we now define two other important discrete random variables. Consider a sequence of independent Bernoulli trials with probability of success in each trial as p, $0 \leq p \leq 1$. Suppose we are interested in the total number of trials required to have the rth success, r being a fixed positive integer. The answer is in the following definition:

Definition 1.2.9
Let X be a random variable with *pmf* as:

$$f(k; r, p) = \binom{r+k-1}{k} p^r q^k, \quad k = 0, 1, \ldots . \tag{1.2.9}$$

Formula (1.2.9) is then called a *negative binomial* (or *Pascal*) *probability distribution function* (or *binomial waiting time*). In particular, if $r = 1$ in (1.2.9), then we will have:

$$f(k; 1, p) = P(x = k + 1) = pq^k, \quad k = 0, 1, , \dots. \tag{1.2.10}$$

The pmf given by (1.2.10) is called a geometric probability distribution function.

Example 1.2.6

As an example, suppose a satellite company finds that 40% of call for services received need advanced technology service. Suppose also that on a particular crazy day, all tickets written are put in a pool and requests are drawn randomly for service. Finally, suppose that on that particular day there are four advance service personnel available. We want to find the probability that the fourth request for advanced technology service is found on the sixth ticket drawn from the pool.

Answer

In this problem, we have independent trials with $p = 0.4$ as probability of success, that is, in need of advanced technology service, on any trial. Let X represent the number of the tickets on which the fourth request in question is found. Thus,

$$P(X = 4) = \binom{6}{4}(0.4)^4 (0.6)^2 = 0.09216.$$

Example 1.2.7

We now want to derive (1.2.9) differently. Suppose treatment of a cancer patient may result in "response" or "no response." Let the probability of a response be p and for a no response be $1 - p$. Hence, the simple space in this case has two outcomes, simply, "response" and "no response." We now repeatedly treat other patients with the same medicine and observe the reactions. Suppose we are looking for the probability of the number of trials required to have exactly k "responses."

Answer

Denoting the sample space by S, $S = \{$response, no response$\}$. Let us define the random variable X on S to denote the number of trials needed to have exactly k responses. Let A be the event, in S, of observing $k - 1$ responses in the first $x - 1$ treatments. Let B be the event of observing a response at the xth treatment. Let also C be the event of treating x patients to obtain exactly k responses. Hence, $C = A \cap B$. The probability of C is:

$$P(C) = P(A \cap B) = P(A) \cdot P(B \mid A).$$

In contrast, $P(B \mid A) = p$ and:

$$P(A) = \binom{x-1}{k-1} p^{k-1}(1-p)^{x-k}.$$

Moreover, $P(X = x) = P(C)$. Hence:

$$P(X = x) = \binom{x-1}{k-1} p^k (1-p)^{x-k}, \quad x = k, k+1, \ldots. \tag{1.2.11}$$

We leave it as an exercise to show that (1.2.11) is equivalent to (1.2.9).

Definition 1.2.10
Let n represent a sample (sampling without replacement) from a finite population of size N that consists of two types of items n_1 of "defective," say, and n_2 of "nondefective," say, $n_1 + n_2 = n$. Suppose we are interested in the probability of selecting x "defective" items from the sample. n_1 must be at least as large as x. Hence, x must be less than or equal to the smallest of n and n_1. Thus,

$$p_x \equiv P(X = x) = \frac{\binom{n_1}{x} \times \binom{N-n_1}{n-x}}{\binom{N}{n}}, \quad x = 0, 1, 2, \ldots, \min(n, n_1), \tag{1.2.12}$$

defines the general form of *hypergeometric pmf* of the random variable X.

Notes:

i. If sampling would have been with replacement, distribution would have been binomial.

ii. p_x is the probability of waiting time for the occurrence of exactly x "defective" outcomes. We could think of this scenario as an urn containing N white and green balls. From the urn, we select a random sample (a sample selected such that each element has the same chance to be selected) of size n, one ball at a time without replacement. The sample consists of n_1 white and n_2 green balls, $n_1 + n_2 = n$. What is the probability of having x white balls drawn in a row? This model is called an *urn model*.

iii. If we let x_i equal to 1 if a defective item is selected and 0 if a nondefective item is selected, and let x be the total number of defectives selected, then $x = \sum_{i=1}^{n} x_i$. Now, if we consider selection of a defective item as a success, for instance, then we could also interoperate (1.2.12) as:

$$p_x = \frac{(\text{number of ways for } x \text{ successes}) \times (\text{number of ways for } n-x \text{ failures})}{\text{total number of ways to select}}.$$

$$\tag{1.2.13}$$

Example 1.2.8
Suppose we have 100 balls in a box and 10 of them are red. If we randomly take out 40 of them (without replacement), what is the probability that we will have at least 6 red balls?

Answer
In this example, which is a hypergeometric distribution, if we assume that all 40 balls are withdrawn at the same time, $N = 100$, $n = 40$, $n_1 = 10$, "defective," is replaced by "red," and $n_2 = 90$, and "nondefective" is replaced by "nonred". The question is to find the probability of selecting at least six red balls. To find the probabilities, we need to calculate the probabilities of 7, 8, 9, and 10 red balls and sum them all or calculate $p \equiv 1 - P\{\text{number of red balls}\} \leq 5$. To do this, we could use statistical software, called Stata, for instance. However, using (1.2.12) or (1.2.13), we have:

$$P\{\text{number of red balls}\} \leq 5 = \frac{\binom{10}{0} \times \binom{90}{40-0}}{\binom{100}{40}} + \frac{\binom{10}{1} \times \binom{90}{40-1}}{\binom{100}{40}}$$

$$+ \frac{\binom{10}{2} \times \binom{90}{40-2}}{\binom{100}{40}} + \frac{\binom{10}{3} \times \binom{90}{40-3}}{\binom{100}{40}}$$

$$+ \frac{\binom{10}{4} \times \binom{90}{40-4}}{\binom{100}{4}} + \frac{\binom{10}{5} \times \binom{90}{40-5}}{\binom{100}{40}} = 0.846.$$

Thus, $p = 1 - 0.846 = 0.154 = 15.4\%$.

We caution that if one uses Excel formula as: "=HYPGEOMDIST (6,40,10,100)," which works out to be 10%, it is not quite right.

As our final important discrete random variable, we define the Poisson probability distribution function.

Definition 1.2.11
A *Poisson random variable* is a nonnegative random variable X such that:

$$p_k = P([X = x]) = \frac{e^{-\lambda}\lambda^k}{k!}, \quad k = 0, 1, \ldots, \qquad (1.2.14)$$

where λ is a constant. Formula (1.2.14) is called a *Poisson probability distribution function* with parameter λ.

Example 1.2.9
Suppose that the number of telephone calls arriving to a switchboard of an institution every working day has a Poisson distribution with parameter 20. What is the probability that there will be:

(a) 20 calls in one day?
(b) at least 30 calls in one day?
(c) at most 30 calls in one day?

Answers
Using $\lambda = 20$ in (1.2.12) we will have:

(a) $P_{30} = P([X = 30]) = ((e^{-20})(20^{30})/30!) = 0.0083.$
(b) $P([X \geq 30]) = \sum_{k=30}^{\infty}(e^{-20}20^k/k!) = 0.0218.$
(c) $P([X \leq 30]) = 1 - P([X \geq 30]) + P([X = 30])$
    $= 1 - 0.0218 + 0.0083 = 0.9865.$

Let X be a binomial random variable with distribution function B_k and $\lambda = np$ be fixed. We will then leave it as an exercise to show that:

$$B_k = \lim_{\substack{n \to \infty, \\ p \to 0}} = \frac{\lambda^k e^{-k}}{k!}, \quad k = 0, 1, 2, \ldots . \tag{1.2.15}$$

1.3. MOMENTS OF A DISCRETE RANDOM VARIABLE

We now discuss some properties of a discrete distribution.

Definition 1.3.1
Suppose X is a discrete random variable defined on a sample space Ω with pmf of p_X. The *mathematical expectation* or simply *expectation* of X, or *expected value* of X, or the *mean* of X or *the first moment* of X, denoted by $E(X)$, is then defined as follows: If Ω is finite and the range of X is $\{x_1, x_2, \ldots, x_n\}$, then:

$$E(X) = \sum_{i=1}^{n} x_i p_{X_i}, \tag{1.3.1}$$

and if Ω is infinite and the range of X is $\{x_1, x_2, \ldots, x_n, \ldots\}$, then:

$$E(X) = \sum_{i=1}^{\infty} x_i p_{X_i}, \tag{1.3.2}$$

provided that the series converges. If Ω is finite and $p_{X_i}, i = 1, 2, \ldots, n$, is constant for all i's, say $1/n$, then the right-hand side of (1.3.1) will become $x_1 + x_2$

$+ \ldots + x_n/n$. This expression is denoted by $\bar{x}$ and is called *arithmetic average* of $x_1, x_2, \ldots, x_n$, that is,

$$\bar{x} = \frac{x_1 + x_2 + \cdots + x_n}{n}. \tag{1.3.3}$$

$p_{x_i}, i = 1, 2, \ldots, n$ in (1.3.1), (1.3.2), and (1.3.3) is called the *weight* for the values of the random variable X. Hence, in (1.3.1) and (1.3.2), the weights vary and $E(X)$ is called the *weighted average*, while in (1.3.3) the weights are the same and $\bar{x}$ is called the *arithmetic average* or simply the *average*.

We next state some properties of the first moment without proof. We leave the proof as exercises.

Properties of the First Moment

1. The expected value of the indicator function $I_A(\omega)$ defined in (1.2.3) is $P(A)$, that is,

$$E(I_A) = P(A). \tag{1.3.4}$$

2. If c is a constant, then $E(c) = c$.
3. If c, c_1, and c_2 are constants and X and Y are two random variables, then:

$$E(cX) = cE(X) \quad \text{and} \quad E(c_1 X + c_2 Y) = c_1 E(X) + c_2 E(Y). \tag{1.3.5}$$

4. If $X_1, X_2, \ldots, X_n$ are n random variables, then:

$$E(X_1 + X_2 + \cdots + X_n) = E(X_1) + E(X_1) + \cdots + E(X_n). \tag{1.3.6}$$

5. Let X_1 and X_2 be two independent random variables with marginal mass (density) functions p_{x_1} and p_{x_2}, respectively. If $E(X_1)$ and $E(X_2)$ exist, we will then have:

$$E(X_1 X_2) = E(X_1) E(X_2). \tag{1.3.7}$$

6. For a finite number of random variables, that is, if $X_1, X_2, \ldots, X_n$ are n independent random variables, then:

$$E(X_1 X_2 \ldots X_n) = E(X_1) E(X_2) \ldots E(X_n). \tag{1.3.8}$$

We now extend the concept of moments.

Definition 1.3.2

Let X be a discrete random variable and r a positive integer. $E(X^r)$ is then called the rth *moment* of X or *moment of order r* of X. In symbols:

$$E(X^r) = \sum_{k=1}^{\infty} x_k^r P(X = x_k). \tag{1.3.9}$$

Note that if $r = 1$, $E(X^r) = E(X)$, that is, the moment of first order or the first moment of X is just the expected value of X. The second moment, that is, $E(X^2)$ is also important, as we will see later.

Let us denote $E(X)$ by μ, that is, $E(X) \equiv \mu$. It is clear that if X is a random variable, so is $X - \mu$, where μ is a constant. However, since $E(X - \mu) = E(X) - E(\mu) = \mu - \mu = 0$, we can *center* X by choosing the new random variable $X - \mu$. This leads to the following definition.

Definition 1.3.3
The rth moment of the random variable $X - \mu$, denoted by $\mu_r(X)$ is defined by $E[(X - \mu)^r]$, and is called the rth *central moment* of X, that is,

$$\mu_r(X) = E(X - \mu)^r. \tag{1.3.10}$$

Note that the random variable $X - \mu$ measures the *deviation* of X from its mean. Thus, we have the next definition:

Definition 1.3.4
The *variance* of a random variable, X, denoted by $Var(X)$ or equivalently by $\sigma^2(X)$, or if there is no fear of confusion, just σ^2, is defined as the second central moment of X, that is,

$$\sigma^2(X) = E\left[(X - \mu)^2\right]. \tag{1.3.11}$$

The positive square root of the variance of a random variable X is called the *standard deviation* and is denoted by $\sigma(X)$.

It can easily be shown that if X is a random variable and μ is finite, then:

$$Var(X) = E(X^2) - \mu^2. \tag{1.3.12}$$

It can also be easily proven that if X is a random variable and c is a real number, then:

$$Var(X + c) = Var(X), \tag{1.3.13}$$

$$Var(cX) = c^2 Var(X). \tag{1.3.14}$$

Example 1.3.1
Consider the Indicator function defined in (1.2.3). That is,

$$I_A(\omega) = \begin{cases} 1, & \text{if } \omega \in A, \\ 0, & \text{if } \omega \in A^c. \end{cases}$$

The expected value of $I_A(\omega)$ is:

$$E(I_A(\omega)) = 1 \cdot P(A) + 0 \cdot [1 - P(A)] = P(A).$$

Example 1.3.2
Consider the Bernoulli random variable defined in Definition 1.2.7. Thus, the random variable X takes two values 1 and 0, for instance, for success and failure, respectively. The probability of success is assumed to be p. Thus, the expected value of X is:

$$E(X) = 1 \cdot p + 0 \cdot (1 - p) = p.$$

To find the variance, note that:

$$E(X^2) = (1^2 =)1 \cdot p + (0^2 =)0 \cdot (1 - p) = p.$$

Hence,

$$Var(X) = p - p^2 = p(1 - p).$$

Example 1.3.3
We want to find the mean and variance of the random variable X having binomial distribution defined in (1.2.8).

Answer
From (1.2.8), we have:

$$E(X) = \sum_{k=0}^{n} k \binom{n}{k} p^k (1-p)^{n-k} = \sum_{k=1}^{n} k \binom{n}{k} p^k (1-p)^{n-k}$$

$$= np \sum_{r=0}^{n-1} \binom{n-1}{r} p^r (1-p)^{n-r-1} = np.$$

We leave it as an exercise to show that the $Var(X) = np(1 - p)$.

Example 1.3.4
Consider the Poisson distribution defined by (1.2.14). We want to find the mean and variance of the random variable X having Poisson pmf as given in (1.2.14).

Answer

$$E(X) = \sum_{x=0}^{\infty} x \frac{e^{-\lambda} \lambda^x}{x!} = \lambda e^{-\lambda} \sum_{x=0}^{\infty} \frac{\lambda^{x-1}}{(x-1)!} = \lambda e^{-\lambda} e^{\lambda} = \lambda,$$

$$E(X^2) = \sum_{x=0}^{\infty} x^2 \frac{e^{-\lambda}\lambda^x}{x!} = e^{-\lambda} \sum_{x=0}^{\infty} x \frac{\lambda^x}{(x-1)!} = e^{-\lambda} \sum_{x=0}^{\infty} (x-1+1) \frac{e^{-\lambda}}{(x-1)!}$$

$$= e^{-\lambda} \left[\sum_{x=1}^{\infty} \frac{(x-1)\lambda^x}{(x-1)!} + \sum_{x=1}^{\infty} \frac{\lambda^x}{(x-1)!} \right]$$

$$= e^{-\lambda} \left[\lambda^2 \sum_{x=2}^{\infty} \frac{\lambda^{x-2}}{(x-2)!} + \lambda \sum_{x=1}^{\infty} \frac{\lambda^{x-1}}{(x-1)!} \right]$$

$$= e^{-\lambda} \left[\lambda^2 e^{\lambda} + \lambda e^{\lambda} \right] = \lambda^2 + \lambda,$$

$$Var(X) = \lambda^2 + \lambda - \lambda^2 = \lambda.$$

1.4. CONTINUOUS RANDOM VARIABLES

So far we have been discussing discrete random variables, discrete distribution functions, and some of their properties. We now discuss continuous cases.

Definition 1.4.1
When the values of outcomes of a random experiment are real numbers (not necessarily integers or rational), the sample space, Ω, is a called a *continuous sample space*, that is, Ω is the entire real number set $\mathbb{R}$ or a subset of it (i.e., an interval or a union of intervals).

The set consisting of all subsets of real numbers $\mathbb{R}$ is extremely large and it will be impossible to assign probabilities to all of them. It has been shown in the theory of probability that a smaller set, say $\mathcal{B}$, may be chosen that contains all events of our interest. In this case, $\mathcal{B}$ is, loosely, referred to as the *Borel field*. We now pause to discuss Borel field more rigorously.

Definition 1.4.2
A nonempty set-function $\mathcal{F}$ is called a *σ-algebra* if it is closed under complements, and under finite or countable unions, that is,

(i) $A_1, A_2 \in \mathcal{F}$, then $A_1 \cup A_2$ and $A_1^c \in \mathcal{F}$, and
(ii) $A_i \in \mathcal{F}, i \geq 1$, then $\bigcup_{i=1}^{\infty} A_i \in \mathcal{F}$.

Note that axioms (i) and (ii) imply that, $\mathcal{F}$ should be closed under finite and countable intersections as well.

Example 1.4.1
The power set of a set $X, \mathcal{F}$, is a σ-algebra.

Definition 1.4.3
Earlier in this chapter we defined "function." The way it was defined, it was a "point function" since values were assigned to each point of a set. A *set function F* assigns values to sets or regions of the space.

Definition 1.4.4

A *measure, μ*, on a set $\mathcal{F}$ is a set function that assigns a real number to each subset of $\mathcal{F}$, (which intuitively determines the size of the set $\mathcal{F}$) such that:

i. $\mu(\phi) = 0$, where ϕ is the empty set,
ii. $\mu(A) \geq 0, \forall A \in \mathcal{F}$, that is, nonnegative, and
iii. if $\{A_i, i \in \mathbb{Z}\} \in \mathcal{F}$ is a finite or countable sequence of mutually disjoint sets in $\mathcal{F}$, then $\mu(\bigcup_{i \in \mathbb{Z}} A_i) = \sum_{i \in \mathbb{Z}} \mu(A_i)$, the countably additive axiom, where $\mathbb{Z}$ is the set of integers.

Notes:

1. What the third axiom says is that the measure of a "large" subset (the union of subsets A_i's) that can be partitioned into a finite (or countable) number of "smaller" disjoint subsets is the sum of the measures of the "smaller" subsets.
2. Generally speaking, if we were to associate a size to each subset of a given set so that we are consistent yet satisfying the other axioms of a measure, only trivial examples like the counting measure would be available. To remove this barrier, a measure would be defined only on a subcollection of all subsets, the so called *measurable subsets*, which are required to form a σ-algebra. In other words, elements of the σ-algebra are called *measurable sets*. This means that countable unions, countable intersections, and complements of measurable subsets are measurable.
3. Existence of a nonmeasurable set involves axiom of choice.
4. Main applications of measures are in the foundations of the Lebesgue integral, in Kolmogorov's axiomatization of probability theory, including ergodic theory. (Andrey Kolmogorov was a Russian mathematician who was a pioneer of probability theory.)
5. It can be proven that a measure is *monotone*, that is, if A is a subset of B, then the measure of A is less than or equal to the measure of B.

Let us now consider the following example.

Example 1.4.2

Consider the life span of a patient with cancer who is under treatment. Hence, the duration of the patient's life is a positive real number. This number is actually an outcome of our treatment (experiment). Let us denote this outcome by ω. Thus, the sample space is the set of all real numbers (of course, in reality, truncated positive real line). Now, we could include the nonpositive part of the real line to our sample space as long as probabilities assigned to them are zeros. Thus, the sample space would become just the real line. While the treatment goes on, we might ask, what is the probability that the patient dies before a preassigned time, say ω_t?

It might also be of interest to know the time interval, say $(\omega_{t_1}, \omega_{t_2})$, in which the dose level of a medicine needs to show a reaction.

To answer questions of these types, we would have to consider intervals $(-\infty, \omega_t)$ and $(\omega_{t_1}, \omega_{t_2})$ as events. Thus, we need to consider a family of sets, called Borel sets.

Definition 1.4.5

Let Ω be a sample space. A family (or a collection) B of subsets of Ω satisfies the following axioms:

Axiom B1. $\Omega \in B$,

Axiom B2. If $A \in B$, then $A^c \in B$, that is, is closed under complement, and

Axiom B3. If $\{A_i, i \in \mathbb{Z}\}$ is a finite or countable family of subsets of Ω in B, then also $\bigcup_{i \in \mathbb{Z}} A_i \in B$, that is, B is closed under the union of at most countable many of its members, called the *class of events*. The class of events B satisfying Axioms B1–B3 is called a *Borel field* or a *σ-field* (reads as sigma-field, which is a σ-algebra).

Example 1.4.3

The family of events $B = \{\phi, \Omega\}$ satisfies axioms B1–B3 stated in Definition 1.4.5 and hence is a Borel field. In fact, this is the smallest family that satisfies the axioms. This family is called the *trivial Borel field*.

Example 1.4.4

Let A be a nonempty set such that $\phi \subset A \subset \Omega$. Thus, $\{\phi, A, A^c, \Omega\}$ is the smallest Borel field that contains A.

Definition 1.4.6

The smallest Borel field of subsets of the real line $\mathbb{R}$ that contains all intervals $(-\infty, \omega)$ is called *Borel sets*.

Notes:

1. A subset of the real line $\mathbb{R}$ is a Borel set if and only if it belongs to the Borel field mentioned in Definition 1.4.5.
2. Since intersection of σ-algebras is again a σ-algebra, the Borel sets are intersection of all σ-algebras containing the collection of open sets in T.
3. We may define Borel sets as follows: let (X, T) be a topological space. The smallest σ-algebra containing the open sets in T is then called the collection of Borel sets0 in X, if such a collection exists.

Definition 1.4.7

Let X be a set and B be a Borel set. The ordered pair (X, B) is then called a *measurable space*.

Definition 1.4.8
Let Ω be a sample space. Also let $P(\cdot)$ be a nonnegative function defined on a Borel set B. The function $P(\cdot)$ is then called a *probability measure*, if and only if the following axioms M1–M3 are satisfied:

Axiom M1. $P(\Omega) = 1$.
Axiom M2. $P(A) \geq 0, \forall A \in B$.
Axiom M3. If $\{A_i, i \in I\} \in B$ is a finite or countable sequence of mutually disjoint sets (i.e., $A_i \cap A_j = \phi$, for each $i \neq j$) in B, then $P(\bigcup_{i \in I} A_i) = \sum_{i \in I} P(A_i)$, *the countably additive axiom*.

Example 1.4.5
Let Ω be a countable set and B the set of all subsets of Ω. Next, let $P(A) = \sum_{\omega \in A} p(\omega)$, where $p(\omega) \geq 0$ and $\sum_{\omega \in \Omega} p(\omega) = 1$. The function $P(\cdot)$, then, is a probability measure.

Definition 1.4.9
The triplet (Ω, B, P), where Ω is a sample space, B is a Borel set, and $P : B \rightarrow [0, 1]$ is a probability measure, is called the *probability space*.
    Note that a probability space is a measure space with a probability measure.

Definition 1.4.10
A measure λ on the real line $\mathbb{R}$ such that $\lambda((a, b]) = b - a, \forall a < b$, and $\lambda(\mathbb{R}) = \infty$, is called a *Lebesgue measure*.

Example 1.4.6
The Lebesgue measure of the interval $[0, 1]$ is its length, that is, 1.
    Note that a particularly important example is the Lebesgue measure on a Euclidean space, which assigns the conventional length, area, and volume of Euclidean geometry to suitable subsets of the n-dimensional Euclidean space $\mathbb{R}^n$.
    We now return to the discussion of a continuous random variable.
    If $A_1, A_2, A_3, \ldots$ is a sequence of mutually exclusive events represented as intervals of $\mathbb{R}$ and $P(A_i), i = 1, 2, \ldots$, is the probability of the event $A_i, i = 1, 2, \ldots$, then, by the third axiom of probability, A3, we will have:

$$P\left(\bigcup_{i=1}^{\infty} A_i\right) = \sum_{i=1}^{\infty} P(A_i). \tag{1.4.1}$$

For a random variable, X, defined on a continuous sample space, Ω, the probability associated with the sample points for which the values of X falls on the interval $[a, b]$ is denoted by $P(a \leq X \leq b)$.

Definition 1.4.11

Suppose the function $f(x)$ is defined on the set of real numbers, $\mathbb{R}$, such that $f(x) \geq 0$, for all real x, and $\int_{-\infty}^{\infty} f(x)dx = 1$. Then, $f(x)$ is called a *continuous probability density function (pdf)* (or just *density function*) on $\mathbb{R}$ and it is denoted by, $f_X(x)$. If X is a random variable that its probability is described by a continuous pdf as:

$$P(a \leq X \leq b) = \int_a^b f_X(x)dx, \text{ for any interval } [a, b], \qquad (1.4.2)$$

then X is called a *continuous random variable*. The *probability distribution function* of X, denoted by $F_X(x)$, is defined as:

$$F_X(x) = P(X \leq x) = \int_{-\infty}^x f(t)dt. \qquad (1.4.3)$$

Notes:

1. There is a significant difference between discrete pdf and continuous pdf. For a discrete pdf, $f_X = P(X = x)$ is a probability, while with a continuous pdf, $f_X(x)$ is not a probability. The best we can say is that $f_X(x)dx \approx P(x \leq X \leq x + dx)$ for all infinitesimally small dx.
2. If there is no fear of confusion, we will suppress the subscript "X" from $f_X(x)$ and $F_X(x)$ and write $f(x)$ and $F(x)$, respectively.

As it can be seen from (1.4.3), distribution function can be described as the area under the graph of the density function.

Note from (1.4.2) and (1.4.3) that if $a = b = x$, then:

$$P(x \leq X \leq x) = P(X = x) = \int_x^x f(t)dt = 0. \qquad (1.4.4)$$

What (1.4.4) says is that if X is a continuous random variable, then the probability of any given point is zero. That is, for a continuous random variable to have a positive probability, we have to choose an interval.

Notes:

(a) From (1.4.4), we will have:

$$P(a \leq X \leq b) = P(a < X \leq b) = P(a \leq X < b) = P(a < X < b). \quad (1.4.5)$$

(b) Since by the fundamental theorem of integral calculus, $f_x(x) = dF_x(x)/dx$, the density function of a continuous random variable can be obtained as the derivative of the distribution function, that is, $F_X'(x) = f_X(x)$.

Conversely, the cumulative distribution function can be recovered from the probability density function with (1.4.3).

(c) $F_X(x)$ collects all the probabilities of values of X up to and including x. Thus, it is the *cumulative distribution function (cdf)* of X.

(d) For $x_1 < x$, the intervals $(-\infty, x_1]$ and $(x_1, x]$ are disjoint and their union is $(-\infty, x]$. Hence, from (1.4.2) and (1.4.3), if $a \le b$, then:

$$P(a \le X \le b) = P(X \le b) - P(X \le a) = F_X(b) - F_X(a). \tag{1.4.6}$$

(e) It can easily be verified that $F_X(-\infty) = 0$ and $F_X(\infty) = 0$ (why?).

(f) As it can be seen from (1.4.3), the concept and definition of cdf applies to both discrete and continuous random variables. If the random variable is discrete, $F_X(x)$ is the sum of f_x's. However, if the random variable is continuous, then the sum becomes a limit and eventually an integral of the density function. The most obvious difference between the cdf for continuous and discrete random variables is that F_X is a continuous function if X is continuous, while it is a step function if X is discrete.

1.5. MOMENTS OF A CONTINUOUS RANDOM VARIABLE

As part of properties of continuous random variables, we now discuss continuous moments. Before doing that, we note that in an integral when the variable of integration, X, is replaced with a function, say $F(x)$, the integral becomes a Stieltjes integral that looks as $\int_a^b dF(x)$, found by Stieltjes in late nineteenth century. If $F(x)$ is a continuous function and its derivative is denoted by $f(x)$, then the *Lebesgue–Stieltjes integral* becomes $\int_a^b dF(x) = \int_a^b f(x)dx$. Henri Lebesgue was a French mathematician (1875–1941) and Thomas Joannes Stieltjes was a Dutch astronomer and mathematician (1856–1899).

Definition 1.5.1
Let X be a continuous random variable defined on the probability space (Ω, B, F) with pdf $f_X(x)$. The *mathematical expectation* or simply *expectation* of X, or *expected value* of X, or the *mean* of X or *the first moment* of X, denoted by $E(X)$, is then defined as the Lebesgue–Stieltjes integral:

$$E(X) = \int_\Omega X dF = \int_{-\infty}^\infty xf(x)dx, \tag{1.5.1}$$

provided that the integral exists. Formula (1.5.1), for a case for an arbitrary (continuous measurable) function of X, say $g(X)$, where X is a bounded random variable with continuous pdf $f_X(x)$, will be:

$$E(g(X)) = \int_\Omega g(X)dF = \int_{-\infty}^\infty g(x)f(x)dx, \tag{1.5.2}$$

provided that the integral converges absolutely.

Definition 1.5.2
The kth *moments* of the continuous random variable X with pdf $f_X(x)$, denoted by $E[X^k]$, $k = 1, 2, \ldots$, is defined as:

$$E(X^k) = \int_{-\infty}^{\infty} x^k f(x)\,dx, \tag{1.5.3}$$

provided that the integral exists, that is,

$$E(|X|^k) = \int_{-\infty}^{\infty} |x|^k f(x)\,dx < +\infty. \tag{1.5.4}$$

In other words, from (1.5.3) and (1.5.4), the kth of X exists if and only if the kth *absolute moment* of X, $E(|X|^k)$, is finite.

Notes:

(1) It can be shown that if the kth, $k = 1, 2, \ldots$ moment of a random variable (discrete or continuous) exists, then do all the lower order moments.

(2) Among standard known continuous distributions, *Cauchy probability distribution* with *density function*:

$$f_X(x) = \frac{1}{\pi(1+x^2)}, \tag{1.5.5}$$

is the only one that its kth, $k = 1, 2, \ldots$, moments do not exist for even values of k and exist for odd values of k in the sense that the Cauchy principal values of the integral exist and are equal to zero.

Several examples will be given in the next section.

1.6. CONTINUOUS PROBABILITY DISTRIBUTION FUNCTIONS

As in the discrete case, we now list a selected number of continuous probability distributions that we may be using in this book.

Definition 1.6.1
A continuous random variable X that has the probability density function:

$$f_X(x) = \begin{cases} \dfrac{1}{b-a}, & \text{if } a \leq b, \\ 0, & \text{elsewhere.} \end{cases} \tag{1.6.1}$$

has a *uniform distribution* over an interval $[a, b]$.

It is left for the reader to show that (1.6.1) defines a probability density function and that the *uniform distribution function of X* is given by:

$$F_X(x) = \begin{cases} 0, & \text{if } x < a, \\ \dfrac{x-a}{b-a}, & \text{if } a \le x \le b, \\ 1, & \text{if } x > b. \end{cases} \qquad (1.6.2)$$

Note that since the graphs of the uniform density and distribution functions have rectangular shapes, they are sometimes referred to as the *rectangular density functions* and *rectangular distribution functions*, respectively.

Example 1.6.1
Suppose X is distributed uniformly over $[0, 10]$. We want to find $P(3 < X \le 7)$. Thus:

$$P(3 < X \le 7) = \frac{1}{10} \int_3^7 dx = \frac{1}{10}(7-3) = 0.4.$$

Example 1.6.2
Suppose a counter registers events according to a Poisson distribution with rate 4. The counter begins at 8:00 A.M. and registers 1 event in 30 minutes. What is the probability that the event occurred by 8:20 A.M.?

Answer
We will restate the problem symbolically and then substitute the values of the parameters to answer the question. We have a Poisson distribution with rate λ and registration rate of 1 per τ minutes. We are to find the pdf of the occurrence at time t. Hence, we let $N(t)$ be the number of events registered from start to time t. We also let T_1 be the time of occurrence of the first event. Next, using properties of the conditional probability and the Poisson distribution, the cdf is:

$$\begin{aligned} F(t) &= P\{T_1 \le t | N(\tau) = 1, 0 \le t \le \tau\} \\ &= P\{N(t) = 1 | N(\tau) = 1\} \\ &= \frac{P\{N(t) = 1 \text{ and } N(\tau) = 1\}}{P\{N(\tau) = 1\}} \\ &= \frac{P\{N(\tau) = 1 | N(0) = 1\} P\{N(t) = 1\}}{P\{N(\tau) = 1\}} \\ &= \frac{P\{N(\tau - t) = 1 | N(0) = 1\} P\{N(t) = 1\}}{P\{N(\tau) = 1\}} \\ &= \frac{P\{N(\tau - t) = 0 | N(0) = 0\} P\{N(t) = 1\}}{P\{N(\tau) = 1\}} \\ &= \frac{e^{-(\tau - t)\lambda} \lambda t e^{-\lambda t}}{\lambda \tau e^{-\lambda \tau}} \\ &= \frac{t}{\tau}, \quad 0 < t < \tau. \end{aligned}$$

Therefore, T_1 is a uniform random variable in $(0, \tau)$. Hence, the first count happened before 8:20 A.M. with probability $(20/30) = 66.67\%$.

Definition 1.6.2
A continuous random variable X with pdf

$$f_X(t) = \begin{cases} \mu e^{-\mu t}, & t \geq 0, \\ 0, & \text{elsewhere,} \end{cases} \tag{1.6.3}$$

and cdf

$$F_X(t) = \begin{cases} 1 - e^{-\mu t}, & t \geq 0, \\ 0, & \text{elsewhere,} \end{cases} \tag{1.6.4}$$

is called a *negative exponential* (or *exponential*) *random variable*. Relation (1.6.3) and relation (1.6.4) are called *exponential density function* and *exponential distribution function*, respectively. μ is called the parameter for the pdf and cdf. See Figure 1.6.1 and Figure 1.6.2.

We note that the expected value of exponential distribution is the reciprocal of its parameter. This is because from (1.6.3) we have:

$$E(X) = \int_0^\infty \mu t e^{-\mu t} dt = \frac{1}{\mu}.$$

See also Figure 1.6.1 and Figure 1.6.2.

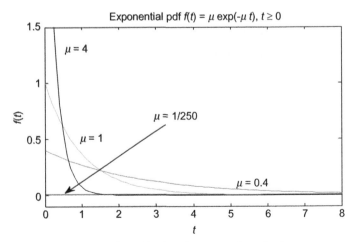

Figure 1.6.1. Exponential pdf with different values for its parameters.

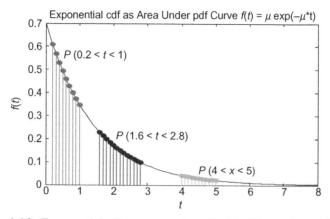

Figure 1.6.2. Exponential cdf as area under pdf with different intervals of t.

Example 1.6.3
Suppose it is known that the lifetime of a light bulb has an exponential distribution with parameter 1/250. We want to find the probability that the bulb works (a) for more than 300 hours and (b) for more than 300 if it has already worked 200 hours.

Answer
Let X be the random variable representing the lifetime of a bulb. From (1.6.3) and (1.6.4), we then have:

$$f_X(t) = \begin{cases} \dfrac{1}{250} e^{-\frac{1}{250}t}, & t \geq 0, \\ 0, & \text{elsewhere,} \end{cases} \quad \text{and} \quad F_X(t) = \begin{cases} 1 - e^{-\frac{1}{250}t}, & x \geq 0, \\ 0, & t < 0. \end{cases}$$

Therefore,

(a) $P(X > 300) = e^{-1.2} = 0.3012$, and

(b) $P(X > 300 | X > 200) = \dfrac{P(X > 300 \text{ and } X > 200)}{P(X > 200)}$

$$= \frac{P(X > 300)}{P(X > 200)} = \frac{e^{-1.2}}{e^{-0.8}} = e^{-.4} = 0.6703.$$

See Figure 1.6.3.

Definition 1.6.3
A continuous random variable, X, with probability density function $f(x)$ defined as:

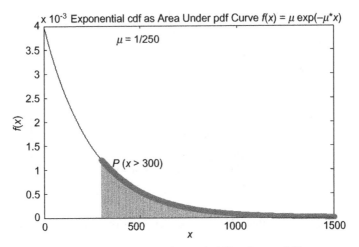

Figure 1.6.3. Exponential probability for x > 300.

$$f_X(x; \mu, \alpha) = \begin{cases} \dfrac{\mu^\alpha x^{\alpha-1}}{\Gamma(\alpha)} e^{-\mu x}, & x > 0, \alpha \text{ is real and} > 0, \\ 0, & x \le 0, \end{cases} \tag{1.6.5}$$

where $\Gamma(\alpha)$ is defined by:

$$\Gamma(\alpha) = \int_0^\infty x^{\alpha-1} e^{-x} dx, \tag{1.6.6}$$

where μ is a positive number, is called a *gamma random variable with parameters μ and α.* The corresponding distribution called *gamma distribution function* will, therefore, be:

$$F_X(x; \mu, \alpha) = \begin{cases} \dfrac{1}{\Gamma(\alpha)} \int_0^x \mu^\alpha u^{\alpha-1} e^{-\mu u} du, & \text{if } x \ge 0, \\ 0, & \text{if } x < 0. \end{cases} \tag{1.6.7}$$

Definition 1.6.4
In (1.6.4), if μ is a nonnegative integer, say k, then the distribution obtained is called the *Erlang distribution of order k*, denoted by $E_k(\mu; x)$, that is,

$$E_k(\mu; x) = \begin{cases} \dfrac{1}{\Gamma(k)} \int_0^{\mu x} u^{k-1} e^{-u} du, & \text{if } x \ge 0, \\ 0, & \text{if } x < 0. \end{cases} \tag{1.6.8}$$

The pdf in this case, denoted by $e_k(\mu; x)$, will be:

$$e_k(\mu; x) = \mu^k x^{k-1} e^{-\mu x}, \quad x \ge 0. \tag{1.6.9}$$

Notes:

(a) We leave it as an exercise to show that $f_X(x; \mu, \alpha)$ given by (1.6.4), indeed, defines a probability density function.

(b) The parameter μ in (1.6.8) is called the *scale parameter*, since values other than 1 either stretch or compress the pdf in the x-direction.

(c) In (1.6.4), if $\alpha = 1$, we will obtain the exponential density function with parameter μ defined by (1.6.4).

(d) $\Gamma(\alpha)$ is a positive function of α.

(e) If α is a natural number, say $\alpha = n$, then we leave it as an exercise to show that:

$$\Gamma(n) = (n-1)!, \quad n = 1, 2, \ldots, \tag{1.6.10}$$

where $n!$ is defined by:

$$n! = n(n-1)(n-2)\ldots(2)(1). \tag{1.6.11}$$

(f) We leave it as an exercise to show that from (1.6.5) and (1.6.10), one obtains:

$$0! = 1. \tag{1.6.12}$$

(g) Because of (1.6.11), the gamma function defined in (1.6.5) is called the *generalized factorial*.

(h) We leave it as an exercise to show that using double integration and polar coordinates, we obtain:

$$\int_0^\infty e^{-x^2} dx = \sqrt{2\pi}. \tag{1.6.13}$$

(i) We leave it as an exercise to show that using (1.6.13), one can obtain the following:

$$\Gamma\left(\frac{1}{2}\right) = \sqrt{\pi}\,\Gamma\left(\frac{1}{2}\right) = \sqrt{\pi}. \tag{1.6.14}$$

(j) Using (1.6.4), we denote the integral $\int_0^x u^{\alpha-1} e^{-u} du$, $x > 0$, by $\Gamma(\alpha, x)$, that is,

$$\Gamma(\alpha, x) = \int_0^x u^{\alpha-1} e^{-u} du, \quad x > 0. \tag{1.6.15}$$

The integral in (1.6.15), which is not an elementary integral, is called the *incomplete gamma function*.

(k) The parameter α in (1.6.4) could be a complex number whose real part must be positive for the integral to converge. There are tables available for values of $\Gamma(\alpha, x)$, defined by (1.6.15). As x approaches infinity, the integral in (1.6.5) becomes the gamma function defined by (1.6.4).

(1) If $k = 1$, then (1.6.6) reduces to the exponential distribution function (1.6.2). In other words, exponential distribution function is a special case of gamma and Erlang distributions.

Definition 1.6.5

In (1.6.4), if $\alpha = r/2$, where r is a positive integer, and if $\mu = 1/2$, then the random variable X is called the *chi-square* random variable with r degrees of freedom, denoted by $X^2(r)$. The pdf and cdf in this case with shape parameter r are:

$$f(x) = \frac{1}{\Gamma\left(\dfrac{r}{2}\right)} 2^{\frac{r}{2}} x^{\frac{r}{2}-1} e^{-\frac{x}{2}}, \quad 0 \le x < \infty \tag{1.6.16}$$

and

$$F(x) = \int_0^x \frac{1}{\Gamma\left(\dfrac{r}{2}\right) 2^{\frac{r}{2}}} v^{\frac{r}{2}-1} e^{\frac{-v}{2}} dv, \tag{1.6.17}$$

respectively, where $\Gamma(r/2)$ is the gamma function with parameter $r/2$ defined in (1.6.4).

Notes:

(a) Due to the importance of X^2 distribution, tables are available for values of the distribution function (1.6.17) for selected values of r and x.

(b) We leave as an exercise to show the following properties of X^2 random variable: mean $= r$, and variance $= 2r$.

The next distribution is widely used in many areas of research where statistical analysis is being used, particularly in statistics.

Definition 1.6.6

A continuous random variable X with pdf denoted by $f(x; \mu, \sigma^2)$ with two real parameters μ, $-\infty < \mu < \infty$, and σ^2, $\sigma > 0$, where:

$$f(x; \mu, \sigma^2) = \frac{1}{\sigma\sqrt{2\pi}} e^{-\frac{(x-\mu)^2}{2\sigma^2}}, \quad -\infty < x < \infty, \tag{1.6.18}$$

is called a *Gaussian* or *normal* random variable.

The notation ~ is used for distribution. The letter N and character Φ are used for normal cumulative distribution. Hence,

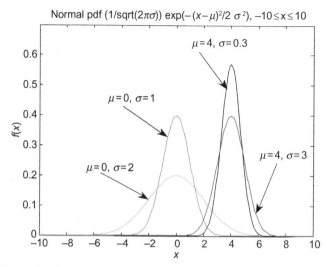

Figure 1.6.4. Normal pdf with different values for its parameters.

$$X \sim N\left(\mu, \sigma^2\right) \quad \text{or} \quad X \sim \Phi\left(\mu, \sigma^2\right) \tag{1.6.19}$$

is to show that the random variable X has a normal cumulative probability distribution with parameters μ and σ^2. The normal pdf has a *bell-shaped* curve and is *symmetric* about the line $f(x) = \mu$ (see Figure 1.6.4). The normal pdf is also asymptotic, that is, the tails of the curve from both sides get very close to the horizontal axis, but never touch it. Later, we will see that μ is the *mean* and σ^2 is the *variance* of the normal distribution function. The smaller the value of variance is, the narrower the shape of the "bell" would be. That is, the data points are clustered around the mean (i.e., the peak).

We leave it as an exercise to show that $f(x; \mu, \sigma^2)$, defined in (1.6.18), is indeed a pdf.

Definition 1.6.7

A continuous random variable Z with $\mu = 0$ and $\sigma^2 = 1$ is called a *standard normal* random variable. From (1.6.19), the cdf of Z, $P(Z \leq z)$, is denoted by $\Phi(z)$. The notation $N(0, 1)$ or $\Phi(0, 1)$ is used to show that a random variable has a standard normal distribution function, which means it has the parameters 0 and 1. The pdf of Z, denoted by $\phi(z)$, therefore, will be:

$$\phi(z) = \frac{1}{\sqrt{2\pi}} e^{-\frac{z^2}{2}}, \quad -\infty < z < \infty. \tag{1.6.20}$$

Note that any normally distributed random variable X with parameters μ and $\sigma > 0$ can be standardized using a substitution:

$$Z = \frac{X - \mu}{\sigma}. \tag{1.6.21}$$

The cdf of Z is:

$$\Phi(z) = P(Z \le z) = \frac{1}{\sqrt{2\pi}} \int_{-\infty}^{z} e^{-\frac{u^2}{2}} du, \qquad (1.6.22)$$

which is the integral of the pdf defined in (1.6.20). We leave it as an exercise to show that $\phi(x)$ defined in (1.6.20) is a pdf.

Note that the cumulative distribution function of a normal random variable with parameters μ and σ^2, $F(x; \mu, \sigma^2)$, whose pdf was given in (1.6.18), may be obtained by (1.6.22) as:

$$F(x; \mu, \sigma^2) = \Phi\left(\frac{x - \mu}{\sigma}\right) = P(Z \le z) = \frac{1}{\sqrt{2\pi}} \int_{-\infty}^{z} e^{-\frac{u^2}{2}} du. \qquad (1.6.23)$$

A practical way of finding normal probabilities is to find the value of z from (1.6.22), and then use available tables for values of the area under the curve of $\phi(x)$.

Figure 1.6.4 shows a graph of normal pdf for different values of its parameters. It shows the standard normal as well as the shifted mean and variety of values for standard deviation. Figure 1.6.5 shows a graph of normal cdf as area under pdf curve with different intervals.

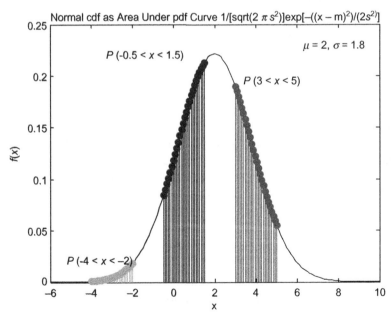

Figure 1.6.5. Normal cdf as area under pdf curve with different intervals.

Definition 1.6.8

A continuous random variable X, on the positive real line, is called a *Galton* or *lognormal* random variable if $\ln(X)$ is normally distributed.

Notes:

(a) If X is a normally distributed random variable, then $Y = e^X$ has a lognormal distribution.

(b) From Definition 1.6.8, X can be written as $X = e^{\mu + \sigma Z}$, where Z is a standard normal variable, and μ (location parameter) and σ (scale parameter) are the mean and standard deviation, respectively, of the natural logarithm of X.

The probability density function and cumulative distribution function of a lognormal random variable X, denoted by $f_X(x; \mu, \sigma)$ and $F_X(x; \mu, \sigma)$, respectively, are:

$$f_X(x; \mu, \sigma) = \frac{1}{x\sigma\sqrt{2\pi}} e^{-\frac{(\ln x - \mu)^2}{2\sigma^2}}, \quad x > 0, \tag{1.6.24}$$

and

$$F_X(x; \mu, \sigma) = \Phi\left(\frac{\ln x - \mu}{\sigma}\right), \tag{1.6.25}$$

where $\Phi(\cdot)$ is defined by (1.6.22).

We leave it as an exercise to show that mean and variance of a lognormal random variable X are, respectively, as

$$E(X) = e^{\mu + \sigma^2/2}, \quad \text{and} \quad Var(X) = \left(e^{\sigma^2} - 1\right)e^{2\mu + \sigma^2}. \tag{1.6.26}$$

Definition 1.6.9

A continuous random variable, X, with cumulative probability distribution function $F(x)$ defined as

$$F(x) = \frac{e^x}{1 + e^x}, \tag{1.6.27}$$

is called the *standard logistic probability distribution*. By adding location and scale parameters, we will have *general logistic cumulative probability distribution* and *probability density function*, respectively, defined as:

$$F(x) = \frac{e^{\frac{x-a}{b}}}{1 + e^{\frac{x-a}{b}}}, \quad x \geq 0, \tag{1.6.28}$$

and

$$f(x) = \frac{e^{\frac{x-a}{b}}}{b\left[1 + e^{\frac{x-a}{b}}\right]^2}, \quad x \geq 0, \tag{1.6.29}$$

where a and b are location (mean) and scale (a parameter proportion to the standard deviation), respectively.

In case there is no confusion, we refer to the general logistic as logistic distribution.

Notes:

(1) If we set $a = 0$ and $b = 1$ in (1.6.28), we obtain (1.6.29).

(2) For standard logistic, a and b are location (mean) and scale (a parameter proportion to the standard deviation), respectively.

(3) $f(x)$, defined in (1.6.29), is symmetric about $x = a$.

(4) $f(x)$ is increasing on $(-\infty, a)$ and decreasing on (a, ∞). This implies that the mode and median occur at $x = a$.

The graph of logistic distribution is very much like a normal distribution. Hence, we may approximate a logistic distribution by a normal distribution or vice versa. If we consider the standard logistic, one possibility is to set the mean of the normal distribution to zero so it is also symmetric about zero, as logistic is, then pick the variance of the normal distribution, $\sigma^2 = \pi^2/3$, so that both distributions have the same variance.

We leave it as an exercise to show that the mean and variance of standard logistic random variable X, respectively, are:

$$E(X) = a \quad \text{and} \quad Var(X) = \frac{1}{3}\pi^2 b^2. \tag{1.6.30}$$

Example 1.6.4
For values of x from -3 to 7 with increments of 0.1, $a = 2$, and $b = 2, 3$, and 4, the pdf of logistic is shown in Figure 1.6.6.

Example 1.6.5
Let there be 10 possible values available for x, in a logistic distribution, that is, the domain of the random variable X (the sample space) is $\Omega = \{1, 2, 3, 4, 5, 6, 7, 8, 9, 10\}$. If we choose $a = 5.5$ and $b = 1.5$ in (1.6.28), then we will have results as presented in Table 1.6.1.

With the chosen parameter values, we will have $F(5.50) = 0.50$ and $F(8.80) = 0.90$. In other words, we set the median of the logistic distribution at

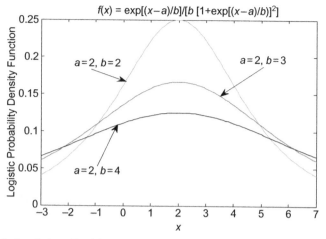

Figure 1.6.6. Logistic pdf with fixed location parameter and three values of scale parameter.

TABLE 1.6.1. Logistic cdf

Given x	1	2	3	4	5	6	7	8	9	10
Logistic cdf	0.047	0.088	0.159	0.269	0.417	0.583	0.731	0.841	0.912	0.953

5.5 and the 90th percentile at 8.80. Note that this choice of parameters centers the values of X.

We can find the inverse of the logistic distribution as follows. Let us write (1.6.28) as:

$$F(x) = \frac{1}{1+e^{-\frac{x-a}{b}}}.$$ (1.6.31)

By dropping x and cross-multiplying (1.6.28), we have $F + Fe^{-x-a/b} = 1$, or:

$$e^{-\frac{x-a}{b}} = \frac{1-F}{F},$$

or

$$-\frac{x-a}{b} = \ln\frac{1-F}{F},$$

or

$$\frac{x-a}{b} = \ln\frac{F}{1-F}.$$

TABLE 1.6.2. Inverse Logistic Cumulative Distribution Probability

Given values of logistic distribution probabilities, F	0.10	0.20	0.25	0.33	0.50	0.90
x values found	2.2042	3.4206	3.8521	4.4377	5.5000	8.7958

Hence, the inverse of logistic cdf (1.6.30) is:

$$x = a + b \ln\left(\frac{F}{1-F}\right). \tag{1.6.32}$$

If we choose values for F as 0.1, 0.2, 0.25, 0.33, 0.5, and 0.9, from (1.6.30) we will have results as presented in Table 1.6.2.

Definition 1.6.10
A continuous random variable X with cdf denoted by $F(x; \alpha, \beta, \gamma)$ as:

$$F(x; \alpha, \beta, \gamma) = 1 - e^{-\left(\frac{x-\gamma}{\alpha}\right)^{\beta}}, \quad \alpha, \beta > 0, x \geq \gamma \geq 0, \tag{1.6.33}$$

is called the *Weibull cumulative probability distribution function*, where α, β and γ are the *scale* (stretches/shrinks the graph), *shape* (such as skewness and kurtosis), and *location* (shifts the graph) parameters, respectively.

Without loss of generality, we let $\gamma = 0$. Thus, (1.6.33) will be reduced to the two-parameter Weibull cumulative distribution function as follows:

$$F(x; \alpha, \beta) = \begin{cases} 1 - e^{-\left(\frac{x}{\alpha}\right)^{\beta}}, & \alpha, \beta, > 0, x \geq 0, \\ 0, & \text{otherwise.} \end{cases} \tag{1.6.34}$$

From (1.6.34), the two-parameter Weibull pdf, denoted by $f(x; \alpha, \beta)$, is:

$$f(x; \alpha, \beta) = \begin{cases} \dfrac{\beta}{\alpha}\left(\dfrac{x}{\alpha}\right)^{\beta-1} e^{-\left(\frac{x}{\alpha}\right)^{\beta}}, & \alpha, \beta > 0, x \geq 0, \\ 0, & \text{otherwise.} \end{cases} \tag{1.6.35}$$

In this case, we write $X \sim \textit{Weibull}(\alpha, \beta)$. When $\alpha = 1$, (1.6.34) and (1.6.35) are referred to as *single-parameter* Weibull cdf and Weibull pdf, respectively. This case is referred to as *standard Weibull density* and *distribution*, respectively.

Note that when $\beta = 1$, we will have the exponential distribution, (1.6.4). In other words, exponential distribution function is a special case of Weibull distribution.

From (1.6.34), we find *the inverse*, that is, x, as follows:

$$e^{-\left(\frac{x}{\alpha}\right)^{\beta}} = 1 - F, \text{ or } -\left(\frac{x}{\alpha}\right)^{\beta} = \ln(1 - F), \text{ or }$$

$$\left(\frac{x}{\alpha}\right)^{\beta} = \ln\left(\frac{1}{1-F}\right), \text{ or } \frac{x}{\alpha} = \left[\ln\left(\frac{1}{1-F}\right)\right]^{\frac{1}{\beta}}, \text{ or } \qquad (1.6.36)$$

$$x = \alpha\left[\ln\left(\frac{1}{1-F}\right)\right]^{\frac{1}{\beta}}.$$

Example 1.6.6
For different pair values of (α, β), we have the results presented in Table 1.6.3.
  We leave it as an exercise to show that the mean and variance of the Weibull random variable, respectively, are

$$E(X) = \alpha\Gamma\left(1 + \frac{1}{\beta}\right), \text{ and } Var(X) = \alpha^2\left[\Gamma\left(1 + \frac{2}{\beta}\right) - \Gamma\left(1 + \frac{1}{\beta}\right)^2\right]. \qquad (1.6.37)$$

See also Figure 1.6.7 and Figure 1.6.8.

TABLE 1.6.3. Inverse Weibull Probability Distribution

Given Weibull cdf, $F(x, \alpha, \beta)$	0.10	0.20	0.25	0.33	0.50	0.90
Weibull parameters (α, β) (scale, shape)	(4.0,3.0)	(4.507, 3.25)	(5.014, 3.5)	(5.521, 3.75)	(8.028, 4.0)	(6.535, 4.25)
Inverse Weibull cdf, x	1.8892	2.8409	3.5123	4.3255	5.5002	7.9519

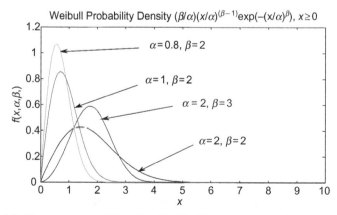

Figure 1.6.7. Two-parameter Weibull pdf with different scale and shape parameters.

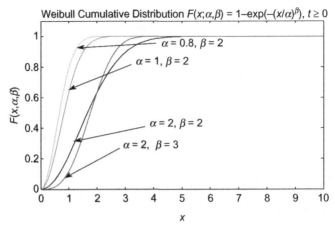

Figure 1.6.8. Two-parameter Weibull cdf with different scale and shape parameters.

Definition 1.6.11
A random variable X has an *extreme value distribution* if the distribution is of one the following three forms:

1. Type 1 (or *double exponential* or *Gumbel-type* distribution), with v as the location parameter and θ as the scale parameter:

$$F_X(x; v, \theta) \equiv P(X \le x) = e^{-e^{-(x-v)/\theta}}. \qquad (1.6.38)$$

2. Type 2 (or *Fréchet-type distribution*) with three parameters:

$$F_X(x; v, \theta, \alpha) \equiv P(X \le x) = e^{-\left(e^{-(x-v)/\theta}\right)^{\alpha}}, \quad x \ge v, \theta, \alpha > 0. \qquad (1.6.39)$$

3. Type 3 (or *Weibull-type distribution*) with two parameters:

$$F_X(x; v, \theta, \alpha) \equiv P(X \le x) = \begin{cases} e^{-\left(e^{-(x-v)/\theta}\right)^{\alpha}}, & x \le v, \theta, \alpha > 0, \\ 1, & x > v. \end{cases} \qquad (1.6.40)$$

Notes:

(a) If X is an extreme value random variable, so is $(-X)$.
(b) Type 2 and Type 3 can be obtained from each other by changing the sign of the random variable.
(c) Type 2 can be transformed to Type 1 by the following transformation: $Z = \ln(X - v)$.
(d) Type 3 can be transformed to Type 1 by the following transformation: $Z = -\ln(v - X)$.

(e) Of the three types, Type 1 is the most commonly used as *the extreme value distribution*. The pdf for Type 1 is:

$$f_X(x; v, \theta) = \frac{1}{\theta} e^{-e^{-(x-v)/\theta}} e^{-e^{-(x-v)/\theta}}.$$ (1.6.41)

(f) From (1.6.41), we have:

$$-\ln[-\ln P(X < x)] = \frac{x - v}{\theta}.$$ (1.6.42)

(g) All three types of extreme value distributions can be obtained from the following cdf:

$$F_X(x; v, \theta, \alpha) \equiv P(X \le x) = \left[1 + \alpha\left(\frac{x - v}{\theta}\right)\right]^{-1/\alpha},$$

$$1 + \alpha\left(\frac{x - v}{\theta}\right) > 0, -\infty < \alpha < \infty, \theta > 0.$$ (1.6.43)

Relation (1.6.43) is called a generalized extreme value or von Mises-type or von Mises–Jenkinson-type distribution.

1.7. RANDOM VECTOR

We have occasions where more than one random variable are considered at a time. Suppose that two distinct random experiments with sample spaces Ω_1 and Ω_2 can conceptually be combined into a single one with only one sample space, Ω. The new sample space, Ω, will be the Cartesian product of Ω_1 and Ω_2, that is, $\Omega_1 \times \Omega_2$, which is the set of all ordered pairs (ω_i, ω_j), where ω_i and ω_j are outcomes of the first and the second experiment, respectively, that is, $\omega_i \in \Omega_1$ and $\omega_j \in \Omega_2$. This idea may be extended to a finite number of sample spaces. In such cases, we are talking of a random vector. In other words, a *random vector* is defined as an n-tuple of random variables. More precisely, we have the following definition:

Definition 1.7.1
A *discrete random vector* $X(\omega) = [X_1(\omega), \ldots, X_r(\omega)]$, where $X_1, \ldots, X_r$ are r discrete random variables and $\omega \in \Omega$, is a function from the sample space Ω into the r-tuple (r-dimensional) real line, $\mathbb{R}^r$, such that for any r real numbers $x_1, x_2, \ldots, x_r$, the set $\{\omega \in \Omega: X_i(\omega) = x_i, i = 1, 2, \ldots, r\}$ is an event.

Example 1.7.1
Suppose a factory wants to conduct a quality control of its product by considering numerical factors $x_1, x_2, \ldots, x_n$ such as weight, height, volume, and color. Such a test can be done by the numerical value of the probability

$P(X_1 \leq x_1, \ldots, X_n \leq x_n)$, where the random vector $(X_1, X_2, \ldots, X_n)$ will describe the joint factors concerned by the factory.

When we have a random vector, the probability distribution of such a vector must give the probability for all the components at the same time. This is the joint probability distribution for the n component random variables which make up the random vector, such as the ones in previous example. Therefore, the necessary probabilistic information can be transferred to the range value from the original probability space.

Definition 1.7.2
Let X be a discrete bivariate random vector, that is, $X = (x_1, x_2)$. Suppose that x_1 and x_2 are two real numbers. Let $p_{x_1 x_2} = P(X_1 = x_1, X_2 = x_2)$. $p_{x_1 x_2}$ is then called the *joint probability mass function of x_1 and x_2*.

We have already discussed independence of two events. We now want to define it for random variables as follows:

Definition 1.7.3
Suppose (S, Ω, P) is an elementary probability space and the random vector $X = (X_1, X_2)$ is defined on the sample space S. The random variables X_1 and X_2 are *independent* if the partitions generated by X_1 and X_2 are independent.

The following theorem gives a better understanding of the independence concept.

Theorem 1.7.1
Two discrete random variables X and Y are independent if and only if the joint pmf of X and Y is the product of the marginal pmf of each X and Y. In other words, X and Y are independent if and only if:

$$P_{X,Y} = P_X P_Y, \tag{1.7.1}$$

where P_X and P_Y are the pmf of X and Y, respectively.

Proof:
See Haghighi et al. (2011a, p. 179).

The bivariate X can be extended to a random vector with discrete components $X_1, X_2, \ldots, X_r$. The joint probability mass function of $x_1, x_2, \ldots, x_r$ then, is similarly defined as:

$$P_{x_1, x_2, \ldots, x_r} = P(X_1 = x_1, X_2 = x_2, \ldots, X_r = x_r).$$

Notes:

(1) From the axioms of probability, a joint probability mass function has the following properties:

(a) $p_{x_1x_2,\cdots,x_r} \geq 0$, and

(b) $\Sigma_{x_1}\Sigma_{x_2}\cdots\Sigma_{x_r} p_{x_1x_2\cdots x_r} = 1$.

(2) The discrete bivariate, P_{x_i,y_j} means the probability that $X = x_i$, $Y = y_j$, and P is defined on the set of ordered pairs $\{(x_i, y_j), j \leq i \leq m, i \leq j \leq n\}$ by $p_{x_i,y_j} = P([X = x_i] \text{ and } [Y = y_j])$. We may obtain each individual distribution functions from the joint distribution. If $A_i = [X = x_i]$ and $B_j = [Y = y_j]$ are events, then $A_iB_j, i = 1, 2, \ldots, m$, are mutually exclusive events and $A_i = \cup A_iB_j$. Thus, we have:

$$p_{x_i} = P(A_i)\sum_{j=1}^{n} P(A_iB_j) = \sum_{j=1}^{n} p_{x_i,y_j}, \quad i = 1, 2, \ldots, m. \tag{1.7.2}$$

Similarly, we will have:

$$p_{y_j} = P(B_j)\sum_{i=1}^{m} P(A_iB_j) = \sum_{i=1}^{m} p_{x_i,y_j}, \quad i = 1, 2, \ldots, n. \tag{1.7.3}$$

Example 1.7.2

Let X and Y be two random variables with the following joint distribution:

$p_{X,Y}$		Y			
		-1	0	1	2
	-1	0	$\dfrac{1}{36}$	$\dfrac{1}{6}$	$\dfrac{1}{12}$
	0	$\dfrac{1}{18}$	0	$\dfrac{1}{18}$	0
X	1	0	$\dfrac{1}{36}$	$\dfrac{1}{6}$	$\dfrac{1}{12}$
	2	$\dfrac{1}{12}$	0	$\dfrac{1}{12}$	$\dfrac{1}{6}$

Questions:

(a) Find $P\{X \geq 1 \text{ and } Y \leq 0\}$.

(b) Find $P\{Y \leq 0 \mid X = 2\}$.

(c) Are X and Y independent? Why?

(d) Find the distribution of $Z = X \cdot Y$.

Answers

(a) For $X \geq 1$ and $Y \leq 0$, we have the following pairs $(1, 0)$, $(1, -1)$, $(2, 0)$, and $(2, -1)$, with probabilities $1/36, 0, 0$, and $1/12$, yielding:

$$P\{X \geq 1 \text{ and } Y \leq 0\} = \frac{1}{36} + 0 + 0 + \frac{1}{12} = \frac{1}{9}.$$

(b) For given X given as 2 and Y to be less than or equal to 0, we have the following pairs $(2, 0)$ and $(2, -1)$, with probabilities 0 and $1/12$, yielding:

$$P\{Y \leq 0 \,|\, X = 2\} = 0 + \frac{1}{12} = \frac{1}{12}.$$

(c) X and Y are dependent. Here is one reason why:

$$P\{X = -1 \text{ and } Y = -1\} = 0.$$

However,

$$P\{X = -1\} = 5/18 \quad \text{and} \quad P\{Y = -1\} = 5/36.$$

Hence,

$$P\{X = -1\} \cdot P\{Y = -1\} = \frac{5}{18} \cdot \frac{5}{36} \neq P\{X = -1, Y = -1\} = 0.$$

(d) Since the values for both X and Y from the table given are as $-1, 0, 1, 2$, we will have $Z = X \cdot Y = -2, -1, 0, 1, 2, 4$. Pairs comprising these values and corresponding probabilities are given in the following table:

XY	-2	-1	0	1	2	4
Possible pairs	$(-1,2), (2,-1)$	$(-1,1),$ $(1,-1)$	$(0,-1), (0,0), (0,1),$ $(0,2), (-1,0), (1,0),$ $(2,0)$	$(1,1),$ $(-1,-1)$	$(1,2), (2,1)$	$(2,2)$
$P(Z = X \cdot Y)$	$\frac{1}{12} + \frac{1}{12} = \frac{1}{6}$	$\frac{1}{6} + 0 = \frac{1}{6}$	$\frac{1}{18} + 0 + \frac{1}{18}$ $0 + \frac{1}{36} + \frac{1}{36} + 0 = \frac{1}{6}$	$\frac{1}{6} + 0 = \frac{1}{6}$	$\frac{1}{12} + \frac{1}{12} = \frac{1}{6}$	$\frac{1}{6}$

Definition 1.7.4

Each probability mass function p_X and p_Y, defined by (1.2.15) and (1.2.16), respectively, is called the *marginal probability mass function*. In other words, a marginal probability mass function, p_X or p_Y, can be found from the joint distribution function p_{XY} of X and Y by summing up the joint distribution over all values of the Y or X, respectively.

Now let $X_1, X_2, \ldots, X_r$ be r random variables representing the occurrence of r outcomes among $X_1, X_2, \ldots, X_n$ possible outcomes of a random experiment, which is being repeated independently n times. Suppose the corresponding probabilities of these outcomes are $p_1, p_2, \ldots, p_r$, respectively. The joint pmf of these random variables will then be:

$$P(X_1 = x_1, \ldots X_r = x_r) = \frac{n!}{n_1! \ldots n_r!} p_1^{n_1} \ldots p_r^{n_r}, \qquad (1.7.4)$$

where n_j ranges over all possible integral values subject to (1.7.4). The relation (1.7.4) does, indeed, represent a probability mass function (why?) and it is called the *multinomial probability mass function* for the random vector $(X_1, X_2, \ldots, X_r)$. We denote this distribution similar to the binomial distribution as: $m(n; r: n_1, \ldots, n_r)$, subject to $p_1 + p_2 + \ldots + p_r = 1$.

Marginal mass function for each one of the random variables $X_1, X_2, \ldots, X_r$ alone will be a binomial probability mass function and can be obtained from (1.7.4). For instance, for X_1 we have:

$$P(X_1 = x_1) = b(n; p, n) = \frac{n!}{n_1!(n-n_1)!} p^{n_1} (1-p)^{n-n_1}. \qquad (1.7.5)$$

As the reader recalls, we defined the conditional probability of the event and the law of total probability. Both these concepts may be extended to random variables. We leave the proof of the following theorem as an exercise.

Theorem 1.7.2. *The Law of Total Probability*

Let X be a random variable and let the event A be represented by a discrete random variable, then we have:

$$P(A) = \sum_x P(A \mid X = x) P(X = x). \qquad (1.7.6)$$

An important notion that measures the dependency of random variables is the notion of covariance, which will be defined in the following sections.

Definition 1.7.5

Let $E(X) = \mu_X$ and $E(Y) = \mu_Y$. The *covariance* of two random variables X and Y, denoted by $Cov(X, Y)$, is then defined by:

$$Cov(X, Y) \equiv E[(X - \mu_X)(Y - \mu_Y)]. \qquad (1.7.7)$$

The following properties of covariance can easily be proved and are left as exercises.

Properties of Covariance

1. $Cov(X, X) = Var(X)$.   (1.7.8)
2. $Cov(X, Y) = E[XY] - \mu_X \mu_Y$.   (1.7.9)
3. $Cov(X, Y) = Cov(Y, X)$.   (1.7.10)
4. If c is a real number, then:

$$Cov(cX, Y) = cCov(Y, X).$$ (1.7.11)

5. For two random variables X and Y, we have:

$$Var(X + Y) = Var(X) + Var(Y) + 2Cov(X, Y).$$ (1.7.12)

As an important application of covariant, we define the following:

Definition 1.7.6

The *coefficient of correlation* of two random variables X and Y, denoted by $\rho(X, Y)$, is given by:

$$\rho(X, Y) \equiv \frac{Cov(X, Y)}{\sigma(X)\sigma(Y)},$$ (1.7.13)

where $\sigma(X)$ and $\sigma(Y)$ are the standard deviations of X and Y, respectively, provided that the denominator is not zero.

It can be shown that:

$$-1 \le \rho(X, Y) \le 1.$$ (1.7.14)

We note that when ρ is negative, it means that the random variables are dependent oppositely, that is, if one increases, the other will decrease. When ρ is positive, it means that both random variables increase and decrease together. However, if $\rho = 0$, the random variables are called *uncorrelated*. Some other properties of the correlation coefficient are as follows:

$$\rho(X, Y) = \rho(Y, X),$$ (1.7.15)

$$\rho(X, X) = 1,$$ (1.7.16)

$$\rho(X, -X) = -1,$$ (1.7.17)

and

$$\rho(aX + b, cY + d) = \rho(X, Y),$$ (1.7.18)

where, a, b, c, and d are real numbers and $a, c \ne 0$.

Example 1.7.3

Suppose X and Y are two random variables representing the "on" or "off" situation of two automatic switches that work together with the following joint distribution: $P(X = 0, Y = 0) = 1/5$, $P(X = 0, Y = 0) = 1/4$, $P(X = 1, Y = 0) = 1/4$, and $P(X = 1, Y = 1) = 3/10$. We may tabulate these values as follows:

X \ Y	0	1	Sum
0	1/5	1/4	9/20
1	1/4	3/10	11/20
Sum	9/20	11/20	

The marginal distributions of X and Y, denoted by p_X and p_Y, respectively, are as follows:

X	0	1
p_X	9/20	11/20

Y	0	1
p_Y	9/20	11/20

Hence,

$$E(XY) = (0)(0)(1/5) + (0)(1)(1/4) + (1)(0)(1/4) + (1)(1)(3/10) = 3/10 = 0.3000,$$
$$E(X) = \mu_X = (0)(9/20) + (1)(11/20) = 11/20 = 0.5500,$$
$$E(Y) = \mu_Y = (0)(9/20) + (1)(11/20) = 11/20 = 0.5500,$$
$$E(X^2) = (0)(9/20) + (1)(11/20) = 11/20 = 0.5500,$$
$$E(Y^2) = (0)(9/20) + (1)(11/20) = 11/20 = 0.5500,$$
$$Var(X) = 0.5500 - 0.3025 = 0.2475,$$
$$Var(Y) = 0.5500 - 0.3025 = 0.2475,$$
$$\sigma(X) = 0.4975,$$
$$\sigma(Y) = 0.4975,$$
$$Cov(XY) = E(XY) - \mu_X \mu_Y = 0.3000 - (0.5500)(0.5500) = -0.0025, \text{ and}$$
$$r(X, Y) = (-0.0025)/(0.4975)(0.4975) = -0.0101.$$

The value of the correlation coefficient, in this case, is very close to 0. Hence, X and Y in this case are uncorrelated.

1.8. CONTINUOUS RANDOM VECTOR

Definition 1.8.1
The *joint bivariate pdf* of two continuous random variables X and Y with pdf $f_X(x)$ and $f_Y(y)$, respectively, is an integrable function, say $f_{X,Y}(x, y)$ or just $f(x, y)$, with the following properties:

(a) $P(X = x \text{ and } Y = y) \approx f_{X,Y}(x, y)dx\, dy.$ (1.8.1)

(b) $f_{X,Y}(x, y) \geq 0.$

(c) $\int_{-\infty}^{\infty}\int_{-\infty}^{\infty} f_{X,Y}(x, y)dx\, dy = 1.$

(d) $P\{(X, Y) \in A\} = \iint_A f_{X,Y}(x, y)dx\, dy,$ (1.8.2)

    where $P\{(X, Y) \in A\}$ is an event defined in the x–y plane.

We note that property (d) implies that properties of discrete joint pmf can be extended to a continuous case using the approximation (1.8.1).

Definition 1.8.2
The *marginal pdf* of X and Y can be obtained from (1.8.1), respectively, as:

$$f_X(x) = \int_{-\infty}^{\infty} f_{X,Y}(x, y)dy, \qquad (1.8.3)$$

and

$$f_Y(y) = \int_{-\infty}^{\infty} f_{X,Y}(x, y)dx. \qquad (1.8.4)$$

Definition 1.8.3
The *conditional probability density function of X given Y* is given by:

$$f_{X|Y}(x|y) = \frac{f_{X|Y}(x|y)}{f_Y(y)}. \qquad (1.8.5)$$

As in the discrete case, the joint pdf can be extended for finitely many random variables.

Definition 1.8.4
Let $\mathbf{X} = (X_1, X_2, \ldots, X_n)$ be a finite or denumerable random vector with joint pdf or pmf of $f_X(x_1, x_2, \ldots, x_n)$. We denote the marginal pdf or pmf of X_i,

$i = 1, 2, \ldots, n$, by $f_{X_i}(x_i)$. $X_1, X_2, \ldots, X_n$ are then *mutually independent random variables* if for every $(x_1, x_2, \ldots, x_n)$ we have:

$$f_{\mathbf{X}}(x_1, x_2, \ldots, x_n) = f_{X_1}(x_1) \times f_{X_2}(x_2) \times \cdots \times f_{X_n}(x_n). \qquad (1.8.6)$$

If $f_{X_i}(x_i)$ is parametric pdf or pmf with, say one parameter, θ, say, denoted by $f_{X_i}(x_i; \theta)$, then the joint parametric pdf or pmf is:

$$f_{\mathbf{X}}(x_1, x_2, \ldots, x_n; \theta) = f_{X_1}(x_1; \theta) \times f_{X_2}(x_2; \theta) \times \cdots \times f_{X_n}(x_n; \theta). \qquad (1.8.7)$$

Note that pairwise independence does not apply mutual independence.

1.9. FUNCTIONS OF A RANDOM VARIABLE

We now consider a random variable as a general term to include both discrete and continuous.

Definition 1.9.1

Suppose that $\phi(\cdot)$ is a function that associates real numbers onto real numbers. Next, the composite function $\phi[X(\cdot)]$ is defined and with each outcome ω, $\omega \in \Omega$, it associates the real number $\phi[X(\omega)]$. $Y(\omega) \equiv \phi[X(\omega)]$ is called the *function of the random variable X*. If $\mathbf{X} = (X_1, X_2, \ldots, X_n)$ is a random vector of n random variable that associates the sample space Ω to the space $\mathbb{R}^n$ of real n-tuples, then the function $\phi(\cdot, \ldots, \cdot)$ on n real variables associates with each point in $\mathbb{R}^n$ a real number. Hence, we define $\phi[\mathbf{X}] = \phi[X_1(\cdot), X_2(\cdot), \ldots, X_n(\cdot)]$ for each $\omega \in \Omega$ as the real number $\phi[X_1(\omega), X_2(\omega), \ldots, X_n(\omega)]$. $Y(\omega) \equiv \phi[X_1(\omega), X_2(\omega), \ldots, X_n(\omega)]$ is called the *function of n, n $\geq$ 1, random variables*.

It can be easily proved that:

Theorem 1.9.1

If Y is a function of a discrete random variable X, say $Y = \phi[X(\omega)]$ and $p(\omega) = P(X = \omega)$, then:

$$E(Y) = E[\phi(X)] = \sum_{\omega \in \Omega} \phi(\omega) p(\omega). \qquad (1.9.1)$$

Limiting distribution functions of certain functions of n random variables when n approaches infinity is an important class of problems in the theory of probability that serves mathematical statistics. In other words, let $\mathbf{X} = (X_1, X_2, \ldots, X_n)$ be a finite or denumerable random vector and $f_n(X_1, X_2, \ldots, X_n)$ is a function of $\mathbf{X} = (X_1, X_2, \ldots, X_n)$, which itself is a random variable. The question is to find the limit of the cumulative distribution function of $\mathbf{X} = f_n(X_1, X_2, \ldots, X_n)$ as n approaches infinity, and if this is not possible, maybe find some properties of cdf, if it exists. This idea leads to considering a *stochastic process* that is a sequence of random variables. We will discuss this process in detail in Chapter 5. But here we want to use it for a different purpose.

Definition 1.9.2
Let $(X_1, X_2, \ldots, X_n)$ be a stochastic process such that for an arbitrary small positive number ε,

$$\lim_{n \to \infty} P\{|X_n - X| > \varepsilon\} = 0. \tag{1.9.2}$$

We then say that $(X_1, X_2, \ldots, X_n)$ *converges stochastically* or *converges in probability* to the random variable X.

We note that if $P\{X = x_0\} = 1$, where x_0 is a constant, that is, X is a degenerate random variable, then $(X_1, X_2, \ldots, X_n)$ converges in probability to the constant x_0.

One of the most important examples of stochastic convergence is the weak law of large numbers. Before we state and prove this law, we present two important inequalities.

Theorem 1.9.2. *Markov's Inequality*
Let X be a nonnegative discrete random variable with a finite mean, $E(X)$. Let a be a fixed positive number. Hence,

$$P\{X \geq a\} \leq \frac{E(X)}{a}. \tag{1.9.3}$$

Proof:
By definition,

$$E(X) = \sum_x xf(x)$$

$$= \sum_{0 \leq x < a} xf(x) + \sum_{x=a}^{\infty} xf(x). \tag{1.9.4}$$

The first term on the right-hand side of (1.9.4) is positive. Hence,

$$E(X) \geq \sum_{x=a}^{\infty} xf(x). \tag{1.9.5}$$

Since a is the minimum value of x, from (1.9.5) we have:

$$E(X) \geq \sum_{x=a}^{\infty} af(x) = a \sum_{x=a}^{\infty} f(x). \tag{1.9.6}$$

Hence,

$$E(X) \geq aP\{X \geq a\}. \tag{1.9.7}$$

It is clear from (1.9.7) that (1.9.3) follows.

Theorem 1.9.3. *Chebyshev's Inequality*
Let X be a nonnegative random variable with finite mean μ and variance σ^2. Let k also be a fixed number. Hence:

$$P\{|X - \mu| > k\} \leq \frac{\sigma^2}{k^2}. \tag{1.9.8}$$

Proof:
Consider the random variable $(X - \mu)^2$, which is positive. From Marko's inequality (Theorem 1.9.2), we now have:

$$P\{(X - \mu)^2 \geq k^2\} \leq \frac{E\left[(X - \mu)^2\right]}{k^2}. \tag{1.9.9}$$

Now $(X - \mu)^2 \geq k^2$ implies that $|X - \mu| \geq k$. Hence,

$$P\{(X - \mu)^2 \geq k^2\} \leq \frac{E\left[(X - \mu)^2\right]}{k^2} = \frac{\sigma^2}{k^2}. \tag{1.9.10}$$

Example 1.9.1
Let us assume that the number of production of an item per week by a factory is a random variable with a mean of 600. We want to find (a) the probability that the number of production per week be at least 1200 and (b) the number of production to be between 500 and 600, if the variance of the number of production is 120.

Answer

(a) By Markov's inequality (Theorem 1.9.2) we have:

$$P\{X \geq 1200\} \leq \frac{E(X)}{1200} = \frac{600}{1200} = \frac{1}{2}.$$

(b) By Chebyshev's inequality (Theorem 1.9.3) we have:

$$P\{|X - 600| \geq 120\} \leq \frac{\sigma^2}{(120)^2} = \frac{1}{120}.$$

Therefore,

$$P\{|X - 600| < 120\} \geq 1 - \frac{1}{120} = \frac{119}{120}.$$

Theorem 1.9.4. *The Weak Law of Large Numbers*

Let $\{X_1, X_2, \ldots, X_n, \ldots\}$ be a sequence of independent and identically distributed (iid) random variables with mean μ. Next, for an arbitrary small positive number ε, we have:

$$\lim_{n \to \infty} P\left\{\left|\frac{X_1 + X_2 + \cdots + X_n}{n} - \mu\right| > \varepsilon\right\} = 0. \qquad (1.9.11)$$

Proof:

Since $\{X_1, X_2, \ldots, X_n, \ldots\}$ is iid, due to additivity property of expected value, we have:

$$E\left(\frac{X_1 + X_2 + \cdots + X_n}{n}\right) = \frac{n\mu}{n} = \mu.$$

Similarly,

$$Var\left(\frac{X_1 + X_2 + \cdots + X_n}{n}\right) = \frac{n\sigma^2}{n^2} = \sigma^2/n.$$

Next, by Chebyshev inequality, we have:

$$P\left\{\left|\frac{X_1 + X_2 + \cdots + X_n}{n} - \mu\right| \geq \varepsilon\right\} \leq \frac{Var(X_1 + X_2 + \cdots + X_n)}{\varepsilon^2} = \frac{\sigma^2}{n\varepsilon^2}. \qquad (1.9.12)$$

Thus, $\lim_{n \to \infty}(\sigma^2/n\varepsilon^2) = 0$, regardless how small ε is.

We note that (1.9.11) states that under the conditions of the theorem, the stochastic process $\{\bar{X}_1, \bar{X}_2, \ldots, \bar{X}_n, \ldots\}$ converges in probability to 0.

Theorem 1.9.4 may be stated differently and will be called differently, as follows:

Theorem 1.9.5. *The Strong Law of Large Numbers*

Let $\{X_1, X_2, \ldots, X_n, \ldots\}$ be a sequence of iid random variables with mean μ. We then have:

$$P\left\{\lim_{n \to \infty} \frac{X_1 + X_2 + \cdots + X_n}{n} = \mu\right\} = 1. \qquad (1.9.13)$$

Proof:

See Neuts (1973, p. 304).

We note that (1.9.13) states that under the conditions of the theorem, the sample mean, repeated infinitely many times, converges almost surely to the expected value. It is sometimes denoted as $\bar{X}_n \xrightarrow{a.s.} \mu$ when $n \to \infty$. Note also that in this convergence we are using "almost surely," while in the previous case, we used the expression "in probability." The Weal law essentially says that for a given small positive number, with a sufficient large number of sample, there will be a very high probability that the mean of observations will be within the given number of the expected value. This, of course, leaves open the possibility that $|\bar{X}_n - \mu| > \varepsilon$ happens many times, although at infrequent intervals. The strong law prevents this to happen, that is, $|\bar{X}_n - \mu| < \varepsilon$ will hold when n is large enough.

Example 1.9.2
Consider flipping a fair coin. Let A be an event. Let $X_i, i = 1, 2, \ldots,$ represent the ith trial such that:

$$X_i = \begin{cases} 1, & \text{if } A \text{ occurs on the } i^{th} \text{ trial,} \\ 0, & \text{otherwise.} \end{cases}$$

Based on the strong law of large numbers (Theorem 1.9.5), we then have:

$$\frac{X_1 + X_2 + \cdots + X_n}{n} \xrightarrow{n \to \infty} E(X) = P(A).$$

In other words, with probability 1, the limiting proportion of times the event A occurs is $P(A)$.

1.10. BASIC ELEMENTS OF STATISTICS

The purpose of most statistical studies is to obtain information about the population from a random sample by generalization. It is a common practice to identify a population and a sample with a distribution of their values: *parameters* and *statistic*, respectively. A *parameter* is a number that describes a character of the population. Numbers such as the mean and variance of a distribution are examples of a parameter. A characteristic of a sample is called a *statistic*. The smallest and largest values of a data set are referred to as the *minimum* and the *maximum*, respectively. The absolute value of difference between the maximum and minimum is called the *range*.

A set of data points may be summarized or grouped in a simple way or into a suitable number of classes (or categories). However, this grouping may cause the loss of some information in the data. This is because instead of knowledge of an individual data point, knowledge of its belonging to a group will be known.

A *simple frequency distribution* is a grouping of a data set according to the number of repetitions of a data point. The ratio of a frequency to the total number of observations is called *relative frequency*. This is the same terminology we used for outcomes earlier in this chapter. A relative frequency multiplied by 100 will yield a *percent relative frequency*.

Example 1.10.1
Consider the set of 30 observations of daily emission (in tons) of sulfur from an industrial plant that is given by:

20.2, 12.7, 18.3, 18.3, 23.0, 23.0, 12.7, 11.0, 21.5, 10.2, 17.1, 07.3, 11.0, 18.3, 17.1,

12.7, 18.3, 20.2, 20.2, 20.2, 18.3, 12.7, 21.5, 17.1, 11.0, 12.7, 11.0, 23.0, 07.9, 07.9

We rewrite the data in ordered points as follows:

7.3	7.9	7.9	10.2	11.0
11.0	11.0	11.0	12.7	12.7
12.7	12.7	12.7	17.1	17.1
17.1	18.3	18.3	18.3	18.3
18.3	20.2	20.2	20.2	20.2
21.5	21.5	23.0	23.0	23.0

Data Point	Frequency	Relative Frequency	Percent Frequency
07.3	1	0.033	03.3
0.79	2	0.067	06.7
10.2	1	0.033	03.3
11.0	4	0.133	13.3
12.7	5	0.167	16.7
17.1	3	0.100	10.0
18.3	5	0.167	16.7
20.2	4	0.133	13.3
21.5	2	0.067	06.7
23.0	3	0.100	10.0
Total	30	1.000	100

In presenting a set of data points (in the form of a table, in percents, or a graph, for instance), one must be careful for any misinterpretation. In fact, two very important words should be in mind for such presentation. They are *abuse* and *misuse*. The word "abuse" means the use of a wrong concept intentionally. But "misuse" is often referred to as the unintentional use of a wrong concept. However, the "misuse" of statistics is not limited to unintentionally using a wrong concept; it could be a misrepresentation of a set of data points for a regular audience who does not have much knowledge of statistics. Even scientists have been known to fool themselves with statistics due to lack of knowledge of *probability theory* and the lack of mathematical statistical concepts. Thus, it is important to make it clear for what purpose the statistical representation is for and avoid all ambiguity as much as possible.

Example 1.10.2

Just recently, the following table was posted on *Facebook* regarding the 2012 Summer Olympics in London, England. The table of medals received by different countries was to show the ranking of countries based on the percent of the number of medals versus the number of athletes that participated in that country. The following table was sourced from the *Los Angeles Times* (August 13, 2012):

London Olympics 2012 Medal Winners			
% of medals per athletes			
Source: LA Times Aug. 13, 2012			
Country	% of athletes with medals	Total Athletes	Total Medals
CHINA	22.83465	381	87
IRAN	22.64151	53	12
USA	19.29499	539	104
Russia	18.7643	437	82
Japan	12.45902	305	38
UK	11.883	547	65
Germany	11.11111	396	44
S. Korea	10.85271	258	28
France	10.36585	328	34
Italy	9.824561	285	28

Here is a simple frequency distribution and percent of ratios of two variables: "Total Athletes" and "Total Medals." The second column shows the percent ratios. The ranking caught the first author's attention since Iran is ranked second, receiving **12** medals with only **53** athletes, while the United States is ranked third, receiving **104** medals with **539** athletes participating in the games. To the author, this was an example of "misuse" of statistics. It gives a good feeling to a person from Iran, seeing his/her country ranked higher than the United States, but what does this ranking tell us? To get a sense of how other statisticians felt, the following question was posted on an online statistics site, *ResearchGate*, for statisticians to comment on (note that due to the limitation of the number of characters you are allowed in a post, the description of the question is brief):

Is it misuse of statistics?

Countries 1 and 2 send team to Olympic. C1 team has 1 athlete and C2 has 500. C1 gets a medal. C2 gets 100 medals. C1's percent of medals versus its athletes is 100% but C2 is 20%. Billboard ranks C1 # 1 and C2 # 2 for percents. Is it misuse of statistics? If so, how should it be corrected?

There were 13 responses by worldwide statisticians, ranging from academic statisticians to medical doctors. For instance, here are four excerpts:

Someone from the IFF of India said:

No, it is not misuse of statistics, this is the statistics.

Someone from the Danish methodological Institute said:

I like to worry about significance levels—if a big country gets 100 medals and a small one gets a single medal, then at the very least it is not certain that the small country with its single medal is a 'robust signal'—that small country might easily be replaced by any of several other small countries with a good chance of getting about one medal, while the big country with its 100 medals is almost certainly a 'stable' result.

A part of a reply from someone from the Université René Descartes of Paris reads as follows:

Then you must think if what you observed is really a random sample or not. I would personally say here that not in that context and that instead we have an exhaustive examination of the results, but let's assume than yes. In that case, you can ask "does a C1's athletes significantly has a higher probability to obtain a medal than a C2's athletes?". And here, you can apply "usual" statistical methods. But, with only n = 1 in the C1 sample, there cannot be any significant difference here at usual alpha levels. So, in that case, concluding that "C1 is better than C2" may be seen as misuse of statistics.

A medical doctor from Shiraz University of Medical Science said:

I believe that the interpretation of the data is more important. For instance, in Olympic Games, the total number of medals is important not the percentage of medal winners. But in another issue (e.g., incidence of a disease) the percentage is important not the number. So you should interpret the data based on your variable and background.

Hence, as it can be seen, statisticians are not unanimous in interpreting a statistical presentation. This, of course, is because they look much deeper in the presented than just a presentation itself. However, people not in the profession of statistics take it at face value, and that is where misleading and "misuse" enters the equation.

Another grouping of data is a *frequency distribution with classes* (or *intervals*). Here is how we construct it:

1. Find the range.
2. Decide how many classes you want to have. However, we don't want too many or too few classes.

3. Classes may be chosen with equal lengths. If that is the case, divide the range by the number of classes selected and choose the rounded up number. This number will be the *length* (or *size*) *of a class*.

4. Now to avoid overlaps, move 0.5 (if data are with integral digits and 0.05 if data are with no digits and start with the tenth decimal place, and 0.005 if the data start with the hundredth decimal place, and so on), down and up from each end of the interval (*class boundaries*) and choose the interval half-open on the right end.

5. If, however, the length of an interval is given, then divide the range by the length of an interval and choose the rounded up number to find the number of classes.

6. After the intervals have been determined, count the number of data points in each class to find the frequency for each class. We note that a sample point such that there is an unusually large gap between the largest and the smallest data point is called an *outlier* (or an *extreme data point*).

In addition to grouping data, there are several graphic ways that a set of data may be presented such as *histogram, dot plot, box plot, stem-and-leaves,* and *scatter plot*. The *histogram* is one of the most common graphic presentations of a data set. It displays data that have been summarized into class intervals. The histogram is a graph that indicates the "shape" of a sample. It can be used to assess the *symmetry* or *skewness* of the data. To construct a histogram: (1) the horizontal axis is divided into equal intervals, (2) a vertical bar (or a strip) is drawn at each interval to represent its frequency (the number of data points that fall within the interval), and (3) adjacent rectangles are constructed, with (4) the bases of the rectangles representing the end points of the class intervals and the heights representing the frequencies of the classes.

Using the concept of a random variable, we can redefine the random sample we defined in Section 1.1 as follows: a random sample $\mathbf{X} = (X_1, X_2, \ldots, X_n)$ is defined as a sample consisting of n independent random variables with the same distribution. Each component of a random sample is a random variable representing observations. The term "random sample" is also used for a set of observed values $x_1, x_2, \ldots, x_n$ of the random variables. We should, however, caution that it is not always easy to select a random sample, particularly when the population size is very large and the sample size is to be small. For instance, to select a sample of size 10 cartons of canned soup to inspect thousands of cartons in the storage, it is almost impossible to number all these cartons and then choose 10 at random. Hence, in cases like this, we do not have many choices; we have to do the best we can and hope that we are not seriously violating the randomness property of the sample.

A statistic is itself a random variable because the value of it is uncertain prior to gathering data. Denoted by, say λ, a statistic is usually calculated based on a random sample. In other words, a statistic is a function of the random vector $\mathbf{X} = (X_1, X_2, \ldots, X_n)$, say $\hat{\Lambda}(\mathbf{X}) = f(X_1, X_2, \ldots, X_n)$.

In Section 1.3, we discussed moments of a random variable that included mean and variance. We now define the concept of random sampling and its properties that includes sample mean and sample variance.

Definition 1.10.1
Let the random variables $X_1, X_2, \ldots, X_n$ be a sample of size n chosen from a population (of *infinite size*) in such a way that each sample has the same chance to be selected. This type of sampling is called *random sampling* (or *random sampling from an infinite population*) and the result is called a *random sample*. If the population is *finite of size N*, then a sample of size n from this population such that each sample of the combination:

$$\binom{N}{n},$$

would be referred to as a *random sample*.

The reader is encouraged to show that in sampling from a finite population, if selections are without replacement, then the random variables $X_1, X_2, \ldots, X_n$ are not mutually independent.

A population may be identified by its distribution $F(x)$. In that case, each X_i, $i = 1, 2, \ldots, n$ is an observation and has a marginal distribution $F(X)$. Additionally, each observation is taken such that its value has no effect or relationship with any other observation. In other words, $X_1, X_2, \ldots, X_n$ are mutually independent.

1.10.1. Measures of Central Tendency

Here in this section, we define the mean, median, and the mode for a distribution as examples of *measures of central tendency*. Each measure of central tendency defined loses some information of data points.

In Section 1.3, we defined *arithmetic average* or simply the *average*. In fact that is the mean of a sample as we define next.

Let $\mathbf{X} = (X_1, X_2, \ldots, X_n)$ be a random vector with $x_1, x_2, \ldots, x_n$ as values of observations. The statistic *sample mean* denoted by $\bar{x}$ is then defined as:

$$\bar{x} = \frac{1}{n} \sum_{i=1}^{n} x_i. \tag{1.10.1}$$

Now suppose we have a sample of size n, and p is a number between 0 and 1, exclusive. The $(100p)$th sample *percentile* is then a sample point at which there are approximately np sample points below it and $n(1 - p)$ sample point above it. The 25th *percentile* (also called the *first quartile*, denoted by Q_1) and the 75th *percentile* (also called the *third quartile*, denoted by Q_3) are the highest value for the lowest 25% of the data points and the lowest value for the highest 25% of the data points, respectively. Similarly, we could have *deciles* that are the percentiles in the 10th increments, such as the *first deciles as the 10th percentile, the fifth deciles as the 50th percentile* or *median*, and so on. The *interquartile range*, denoted by IQR, is the range of the middle 50% of the data points and is calculated as the difference between Q_3 and Q_1, that is, $Q_3 - Q_1$. The *median* (also called the *second quartile*) of n (nonmissing) observations $x_1, x_2, \ldots, x_n$ is loosely defined as the *middlemost* of the observed values.

To find the median, arrange $x_1, x_2, \ldots, x_n$ according to their values in ascending or descending order, and then pick the middle value if n is odd, that is, $(n + 1)/2$, and the average (mean) of the middle two if n is even, that is, $[(n/2 + (n + 2)/2]/2$. In general, to find a percentile, we consider three cases for the $(n + 1)p$, namely, (1) $(n + 1)p$ as an integer, (2) $(n + 1)p$ as an integer plus a proper fraction, and (3) $(n + 1)p < 1$. Then:

1. If $(n + 1)p$ is an integer, the $(100p)$th sample percentile is the $(n + 1)p$th ordered data point display.
2. If $(n + 1)p$ is not an integer, but is equal to $r + a$, where r is the whole part and a is the proper fraction part of $(n + 1)p$, then take the weighted average of the rth and $(r + 1)$st ordered data points. In other words, let $D_r =$ the rth ordered data point and the $D_{r+1} = (r + 1)$st ordered data point. Next, denoting the weighted average by $\overline{\pi_p}$, we will have $\pi_p = D_r + a(D_{r+1} - D_r) = (1 - a)D_r + aD_{r+1}$.
3. If $(n + 1)p < 1$, then the sample percentile is not defined.

The median is less sensitive to extreme values than the mean. Thus, when data contain outliers, or are *skewed* (lack of symmetry), the median is used instead of the mean. If one tail extends farther than the other, we say the distribution is *skewed*. Outliers cause *skewness*. An outlier on the far left will cause a *negative or left skew*, while on the far right will cause a *positive* or *right skew*. To avoid outliers, it is customary to *trim* the data. The *trimmed mean* is the mean of the *trimmed data*, that is, of the *remaining data set* by cutting the data about 5–10% (rounded to the nearest integer) on each end (after being sorted in ascending or descending order). The measure of the sharpness of the peak of a distribution is referred to as *kurtosis*. Similar to skewness, positive or negative values of kurtosis will cause a peak flatter than or sharper than the peak of the normal curve.

The most frequent of the observed values is called *mode*. When a distribution has more than one mode, the data set is said to have *multimodes*. If this number is two, it is called *bimodal*.

1.10.2. Measure of Dispersion

Now that we have discussed measures of central tendency, we will discuss measures of dispersion. Range, as defined before, is a statistic that is often used to describe dispersion in data sets. As another measure of dispersion, let

$\mathbf{x} = (x_1, x_2, \ldots, x_n)$ be a random vector of observations with sample mean $\bar{x}$. The statistic *sample variance* (sometimes called *mean square*) denoted by S^2 with its values as s^2, is then defined as:

$$s^2 = \frac{1}{n-1} \sum_{i=1}^{n} (x_i - \bar{x})^2. \tag{1.10.2}$$

Note that in the denominator of (1.10.2), the term $n - 1$ instead of n has been used. Although logically we should use n (and some authors in their publications a couple of decades ago used n), it tends to underestimate the population variance σ^2. Since the primary use of the sample variance s^2 is to estimate the population variance σ^2, by replacing $n - 1$ and, thus, enlarging s^2, the tendency is corrected. The same correction is made in the use of sample *standard deviation*, which is the positive square root of the sample variance.

In practice, the following formula (Formula 1.10.3) for sample variance is used instead of (1.10.2). It is easy to see that (1.10.2) is equivalent to (1.10.3):

$$s^2 = \frac{1}{n-1} \left[\sum_{i=1}^{n} x_i^2 - \frac{1}{n} \left(\sum_{i=1}^{n} x_i \right)^2 \right]. \tag{1.10.3}$$

In case that data are grouped in a simple frequency format, the sample variance would be calculated using:

$$s^2 = \frac{1}{n-1} \left[\sum_{i=1}^{k} f_i x_i^2 - \frac{1}{n} \left(\sum_{i=1}^{n} f_i x_i \right)^2 \right], \quad 1 \leq k \leq n, \tag{1.10.4}$$

where $f_i, 1 \leq k \leq n$ denote frequencies of each group of data points.

We note that it is usually desired to have the variance as small as possible so that data points are gathered around the mean. However, there might be cases that we have to deal with highly scattered data such that the mean is smaller than the variance or the standard deviation. There is no theory avoiding such cases. Particularly, since mean and variance are with different units of measurement, it would be difficult to compare the two.

In the following example, we will construct a histogram using the statistical software MINITAB. Using MINITAB has the advantage of providing information such as the measure of central tendencies. We note that this example is taken from Haghighi et al. (2011a, p. 265).

Example 1.10.3
Consider Example 1.7.1. The median for this data set is Q_2, that is, $p = 0.50$. Hence, $(n + 1)p = (31)(0.5) = 15.5$. Thus, $r = 15$ and $a = 0.5$. Therefore, Q_2 is the average of the 15th and the 16th data points, which is 17.1. In contrast, for the 53rd percentile, $(31)(0.53) = 16.43$, yielding $r = 16$ and $a = 0.43$. Hence:

$$\overline{\pi_{0.53}} = (1 - 0.43)D_{16} + 0.43D_{17} = (0.57)(17.1) + (0.43)(18.3) = 17.616.$$

In addition, the 95th percentile is:

$$\overline{\pi}_{0.95} = (1-0.45)D_{29} + 0.45D_{30} = (0.55)(23) + (0.45)(23) = 23.$$

Using MINITAB, we obtain the following information as well as a histogram. Note that the "confidence interval" will be discussed later in this chapter.

Descriptive Statistics: Observations

Variable	Mean	SE Mean	TrMean	StDev	Variance	CoefVar	Sum of Squares
Observations	15.880	0.899	15.969	4.92	24.231	31.00	8267.940
Variable	Minimum	Q1	Median	Q3	Maximum	Range	IQR Skewness
Observations	7.300	11.000	17.100	20.200	23.000	15.700	9.200 −0.19

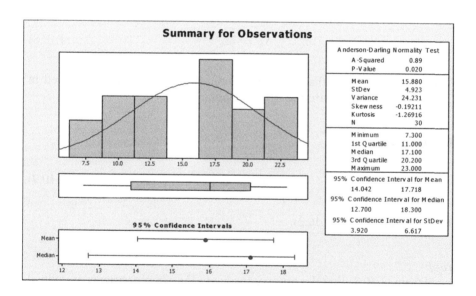

1.10.3. Properties of Sample Statistics

Suppose $X_1, X_2, \ldots, X_n$ is a random sample from a population with mean and variance μ and $\sigma^2 < \infty$, respectively. Denoting the sample mean and variance by $\overline{X}$ and S^2, respectively, we then leave it as an exercise to prove the following properties of sample mean:

$$E(\overline{X}) = \mu, \tag{1.10.5}$$

and

$$Var(\bar{X}) = \frac{\sigma^2}{n}. \tag{1.10.6}$$

A measure of variability of the sampling distribution of the sample mean is called the *standard error of the sample mean*. If the population standard deviation, σ, is *known*, then the standard error of the sample mean, denoted by $\sigma_{\bar{x}}$, is defined by:

$$\sigma_{\bar{x}} = \frac{\sigma}{\sqrt{n}}. \tag{1.10.7}$$

If the population standard deviation is *unknown*, and denoting S as the *sample standard deviation*, then the *standard error of the sample mean*, denoted by $S_{\bar{x}}$, is:

$$S_{\bar{x}} = \frac{S}{\sqrt{n}}. \tag{1.10.8}$$

We now want to show that the mean of sample variance, as we defined in (1.10.2), is, indeed, the variance of the population.

Theorem 1.10.1

$$E(S^2) = \sigma^2, \tag{1.10.9}$$

where S^2 is defined in (1.10.2).

Proof:
Recall from (1.10.2) that the observed value of the sample variance is defined as:

$$s^2 = \frac{1}{n-1}\sum_{i=1}^{n}(x_i - \bar{x})^2,$$

where

$$\bar{x} = \frac{1}{n}\sum_{i=1}^{n}x_i,$$

as defined in (1.10.1). Now, we add and subtract μ in $\sum_{i=1}^{n}(x_i - \bar{x})^2$ and do some algebra as follows:

$$\sum_{i=1}^{n}(x_i - \bar{x})^2 = \sum_{i=1}^{n}(x_i - \mu + \mu - \bar{x})^2 = \sum_{i=1}^{n}[(x_i - \mu) - (\bar{x} - \mu)]^2$$

$$= \sum_{i=1}^{n}[(x_i - \mu)^2 - 2(x_i - \mu)(\bar{x} - \mu) + n(\bar{x} - \mu)^2]$$

$$= \sum_{i=1}^{n}(x_i - \mu)^2 - 2(\bar{x} - \mu)\sum_{i=1}^{n}(x_i - \mu) + n(\bar{x} - \mu)^2$$

$$= \sum_{i=1}^{n}(x_i - \mu)^2 - 2(\bar{x} - \mu)\left[n\frac{\sum_{i=1}^{n}x_i}{n} - n\mu\right] + n(\bar{x} - \mu)^2$$

$$= \sum_{i=1}^{n}(x_i - \mu)^2 - 2(\bar{x} - \mu)(nx_i - n\mu) + n(\bar{x} - \mu)^2$$

$$= \sum_{i=1}^{n}(x_i - \mu)^2 - n(\bar{x} - \mu)^2.$$

Thus,

$$E(s^2) = E\left[\frac{\sum_{i=1}^{n}(x_i - \bar{x})^2}{n-1}\right] = \frac{1}{n-1}\left(E\left[\sum_{i=1}^{n}(x_i - \bar{x})^2\right] - nE[(\bar{x} - \mu)^2]\right)$$

$$= \frac{1}{n-1}[n\sigma^2 - nVar(\bar{x})] = \frac{n}{n-1}[\sigma^2 - Var(\bar{x})]$$

$$= \frac{n}{n-1}\left(\sigma^2 - \frac{\sigma^2}{n}\right) = \sigma^2.$$

Now consider a population with a mean and variance of μ and σ^2, respectively. As n increases without bound, we see from (1.10.6) that the variance of $\bar{X}$ will decrease. Hence, the distribution of $\bar{X}$ depends on the sample size n. In other words, for different sizes of n, we will have a sequence of distributions to deal with. To see the limit of such a sequence as n increases without bound, let us consider the random variable W defined by:

$$W = \frac{X - \mu}{\sigma/\sqrt{n}}, \quad n = 1, 2, \ldots. \tag{1.10.10}$$

We then leave it as an exercise to show that W is a standard normal random variable for each positive integer n. Therefore, the limiting distribution of W

is also a standard normal, which we will state as the following theorem known as the *central limit theorem*, and we refer the reader to Hogg and Tanis (1993) for the proof.

Theorem 1.10.2. *The Central Limit Theorem*
Suppose $X_1, X_2, \ldots, X_n$ is a random sample of size n from a distribution with finite mean μ and a finite positive variance σ^2. Then the limiting distribution (i.e., as $n \rightarrow \infty$) of

$$W = \frac{\bar{X} - \mu}{\sigma/\sqrt{n}} = \frac{\sum_{i=1}^{n} X_i - n\mu}{\sigma\sqrt{n}}$$

is the standard normal.

As we noted before, we are using $n - 1$ instead of n in the sample variance. We stated the reason for such a choice. That choice actually causes the obtaining of (1.10.10), which makes the sample variance the so-called *unbiased* estimator for the population variance.

To find the variance of the sample variance, we remind the reader of equations (1.10.2) and (1.10.9). We leave it as an exercise to show that the second moment of S^2 is:

$$E\left[\left(S^2\right)^2\right] = \frac{\mu_4}{n} + \frac{(n-1)^2 + 2}{n(n-1)}\sigma^4, \tag{1.10.11}$$

where μ_4 is the fourth central moment of X. Hence, from (1.10.2) we will have:

$$\begin{aligned} Var(\sigma^2) &= \frac{\mu_4}{n} + \frac{(n-1)^2 + 2}{n(n-1)}\sigma^4 - \left(\sigma^2\right)^2 \\ &= \frac{\mu_4}{n} + \sigma^4\left[\frac{(n-1)^2 + 2}{n(n-1)} - 1\right] \\ &= \frac{1}{n}\left(\mu_4 - \frac{n-3}{n-1}\sigma^4\right). \end{aligned} \tag{1.10.12}$$

Example 1.10.4
As an example, let us find the mean and variance of the sampling distribution of the standard deviation for Gaussian (normal) distributions.

Answer
What the question says is that we have a standard normal population. Take a sequence of samples from this population, calculate the standard deviation for each sample taken, and consider the values of these standard deviations as a set of data or new sample points. The question is, what are the mean and standard deviation of this set of data?

Let $X_1, X_2, \ldots, X_n$ be a simple random sample (i.e., a set of n independent random variables) from a normal population with mean μ and variance σ^2. Now, multiplying both sides of (1.10.2) by $(n-1)$ and dividing by σ^2, we will have:

$$\frac{(n-1)S^2}{\sigma^2} = \frac{\sum_{i=1}^{n}(X_i - \bar{X})^2}{\sigma^2}.$$ (1.10.13)

We leave it as an exercise to show that:

$$\frac{(n-1)S^2}{\sigma^2} \sim \chi^2(n-1).$$ (1.10.14)

Therefore, let:

$$W \equiv \frac{(n-1)S^2}{\sigma^2}.$$ (1.10.15)

From (1.10.13) and (1.10.15), W is $\chi^2(n-1)$. Thus, the pdf of W, denoted by $g_W(w)$ is:

$$g_W(w) = \frac{w^{\frac{n-1}{2}-1} e^{\frac{-w}{2}}}{2^{\frac{n-1}{2}} \Gamma\left(\frac{n-1}{2}\right)}, \quad w \geq 0$$

or

$$g_W(w) = \frac{w^{\frac{n-3}{2}} e^{\frac{-w}{2}}}{2^{\frac{n-1}{2}} \Gamma\left(\frac{n-1}{2}\right)}, \quad w \geq 0.$$ (1.10.16)

From (1.10.15), we have:

$$S^2 = \frac{\sigma^2 W}{n-1}.$$ (1.10.17)

Letting S denote the sample standard deviation for the normal population, from (1.10.17) we have:

$$S = \frac{\sigma}{\sqrt{n-1}}\sqrt{W}.$$ (1.10.18)

We define the random variable Y as:

$$Y \equiv \sqrt{W}$$ (1.10.19)

To find the pdf of Y, we note that since W is a continuous-type random variable with pdf of $g_W(w)$ as defined in (1.10.15), for $w \in [0, \infty)$, Y is defined as a function of W, as in (1.10.19), say $Y = r(W)$, and Y and its inverse $W = f(Y)$ are increasing continuous functions, then the pdf of Y is:

$$h_Y(y) = g[f(y)]f'(y). \tag{1.10.20}$$

Hence, since from (1.10.19), $W = f(Y) = Y^2$ and thus $f'(Y) = 2Y$. From (1.10.20), we have:

$$h_Y(y) \equiv g_W(y^2)2y = \frac{y^{n-2}e^{\frac{-y^2}{2}}}{2^{\frac{n-3}{2}}\Gamma\left(\frac{n-1}{2}\right)}, \quad y \geq 0. \tag{1.10.21}$$

Then:

$$E(Y) = \int_0^\infty yh(y)dy = \int_0^\infty \frac{y^{n-1}e^{\frac{-y^2}{2}}}{2^{\frac{n-3}{2}}\Gamma\left(\frac{n-1}{2}\right)}dy$$

$$= \frac{1}{2^{\frac{n-3}{2}}\Gamma\left(\frac{n-1}{2}\right)}\int_0^\infty y^{n-1}e^{\frac{-y^2}{2}}dy. \tag{1.10.22}$$

Now, let $u = y^2/2$, from which we have $du = ydy$ and $y^{n-1} = 2^{n-1}u^{(n-1)/2}$. Then, from (1.10.22), we have:

$$E(Y) = \frac{1}{2^{\frac{n-3}{2}}\Gamma\left(\frac{n-1}{2}\right)}\int_0^\infty 2^{\frac{n-1}{2}}u^{\frac{n-1}{2}}e^{-u}\frac{du}{2^{1/2}u^{1/2}}$$

$$= \frac{2^{\frac{n-2}{2}}}{2^{\frac{n-3}{2}}\Gamma\left(\frac{n-1}{2}\right)}\int_0^\infty u^{\frac{n-1}{2}}e^{-u}u^{-\frac{1}{2}}du$$

$$= \frac{2^{\frac{n-2}{2}}}{2^{\frac{n-3}{2}}\Gamma\left(\frac{n-1}{2}\right)}\int_0^\infty u^{\frac{n}{2}-1}e^{-u}du \tag{1.10.23}$$

$$= \frac{2^{\frac{n-2}{2}}\Gamma\left(\frac{n}{2}\right)}{2^{\frac{n-3}{2}}\Gamma\left(\frac{n-1}{2}\right)}$$

$$= \frac{2^{\frac{1}{2}}\Gamma\left(\frac{n}{2}\right)}{\Gamma\left(\frac{n-1}{2}\right)}.$$

Now, from (1.10.18) and (1.10.19) we have:

$$E(S) = \frac{\sigma}{\sqrt{n-1}} E(Y). \tag{1.10.24}$$

Substituting (1.10.23) in (1.10.24), the mean of standard deviation of the normal population is:

$$E(S) = \sqrt{\frac{2}{n-1}} \frac{\sigma\Gamma\left(\frac{n}{2}\right)}{\Gamma\left(\frac{n-1}{2}\right)}. \tag{1.10.25}$$

It is known that:

$$Var(S) = E(S^2) - (E(S))^2. \tag{1.10.26}$$

Since the mean of $\chi^2(n-1)$ is $n-1$, from (1.10.14), we will have:

$$E(S^2) = \frac{\sigma^2}{n-1} E(W) = \sigma^2. \tag{1.10.27}$$

Thus, from (1.10.25) and (1.10.27), the variance of the standard deviation of the normal population is:

$$Var(S) = \sigma^2 \left\{ 1 - \frac{2}{n-1} \left[\frac{\Gamma\left(\frac{n}{2}\right)}{\Gamma\left(\frac{n-1}{2}\right)} \right]^2 \right\}. \tag{1.10.28}$$

For the standard normal population, $\mu = 0$ and $\sigma = 1$. Therefore, the mean of standard normal standard deviation remains as in (1.10.25), and its variance from (1.10.26) will be:

$$Var(S) = 1 - \frac{2}{n-1} \left[\frac{\Gamma\left(\frac{n}{2}\right)}{\Gamma\left(\frac{n-1}{2}\right)} \right]^2. \tag{1.10.29}$$

1.11. INFERENTIAL STATISTICS

The purpose of most statistical investigations is to generalize information contained in samples to the populations from which the samples were drawn. This is the essence of statistical inference. The term *inference* means a conclusion or a deduction. Two major areas are main components of the methods of *statistical inference* in the classical approach: (1) *hypotheses testing* and (2) *estimation* (*point* and *interval*). In summary, an *estimate* of a distribution based on a sample and drawing a conclusion about a parameter based on a sample is called a *statistical inference*. Of the different methods of estimation that are available, we discuss *point* and *interval estimations*.

1.11.1. Point Estimation

Definition 1.11.1
The most plausible value of a parameter μ (population value) is called the *point estimate* of μ. The statistic (sample value) that estimates this parameter is called the *point estimator* of μ and is denoted by $\hat{\mu}$. In other words, we want to find a number $\hat{\mu}$ as a function of observations $(x_1, x_2, \ldots, x_n)$, that is, $\hat{\mu} = Y(x_1, x_2, \ldots, x_n)$. The function Y is a statistic that estimates μ, that is, a *point estimator* for μ. We want the computed value $\hat{\mu} = Y(x_1, x_2, \ldots, x_n)$ to be closed to the actual value of the parameter μ.

Note that Y is a random variable and thus has a pdf by its own.

It is known that a good estimator is the one that, on the average, is equal to the parameter (*unbiased*) and its *variance is as small as possible* (*minimum variance*). In other words, a point estimator, denoted by $\hat{\mu}$, of a parameter μ is *unbiased* if $E(\hat{\mu}) = \mu$, otherwise it is said to be *biased*. The amount of *bias* of $\hat{\mu}$, denoted by $B(\hat{\mu})$, is defined by $B(\hat{\mu}) = E(\hat{\mu}) - \mu$.

Let μ be a parameter and $\hat{\mu}$ an estimator of it. Then the *mean square error* of $\hat{\mu}$, denoted by *MSE*, (and if there is confusion, by $MSE(\hat{\mu})$) is defined by:

$$MSE(\hat{\mu}) \equiv MSE = [E(\hat{\mu}) - \mu]^2 + Var(\hat{\mu}). \qquad (1.11.1)$$

There are different methods of point estimation. One common one is the *method of moments*. However, the maximum likelihood estimation method (MLE), which is defined later, is the most widely used method of estimation, especially when the sample size is large. It is a method that allows one to choose a value for the unknown parameter that most likely is the closest to the observed data.

We note that in the maximum likelihood function described later, for discrete distribution function we use the probability mass function, and for continuous distribution function we use the probability density function. However, we will use the pdf notation for both cases. Hence, we will use $f(x; \theta)$ for a density function with one parameter and $f(x; \theta_1, \theta_2)$ for a density function with two parameters.

Definition 1.11.2
Suppose $(X_1, X_2, \ldots, X_n)$ is a random sample of size n, with observed values $(x_1, x_2, \ldots, x_n)$ from a cumulative distribution function (cdf) $F(x; \theta)$ with pdf $f(x; \theta)$, where θ is a vector of k parameters $\theta_1, \theta_2, \ldots, \theta_k$. The cdf of this random sample denoted by $F_n(x_1, \ldots, x_n; \theta_1, \theta_2, \ldots, \theta_k)$ is:

$$F_n(x_1, \ldots, x_n; \theta_1, \theta_2, \ldots, \theta_k) = \prod_{i=1}^{n} F(x_i; \theta_1, \theta_2, \ldots, \theta_k). \quad (1.11.2)$$

For the given sample $(x_1, x_2, \ldots, x_n)$, the quantity $dF_n(x_i; \theta) = \prod_{i=1}^{n} dF(x_i; \theta)$ is called the *likelihood function* of $\theta_1, \theta_2, \ldots, \theta_k$ for $(x_1, x_2, \ldots, x_n)$. We denote the likelihood function of θ by $L(\theta)$,

$$L(\theta) = L(x_1, \ldots, x_n; \theta_1, \theta_2, \ldots, \theta_k) = \prod_{i=1}^{n} f(x_i; \theta_1, \theta_2, \ldots, \theta_k), \quad (1.11.3)$$

that is, the joint density function of n random variables and k parameters. For each sample point x, let $\hat{\theta}(x)$ be a value of the parameter that maximizes $L(\theta)$. The *maximum likelihood estimator (MLE)* of the parameter θ based on a sample $(X_1, X_2, \ldots, X_n)$ is denoted by $\hat{\theta}(X)$, where $X = (X_1, X_2, \ldots, X_n)$.

We note that in order find the value of θ that maximizes $L(\theta)$, we take the derivative of $L(\theta)$, set it equal to zero, and find θ. If θ is a vector of, say, size k, then we need to take partial derivatives with respect to each element of θ, set it equal to zero, and solve the system of k equations with k unknowns. That is, if the likelihood function is differentiable with respect to θ, then letting x represent the random sample, potential candidates for the MLE are the values of $(\theta_1, \theta_2, \ldots, \theta_k)$ that solve:

$$\frac{\partial L(x; \theta_i)}{\partial \theta_i} = 0, \quad i = 1, 2, \ldots, k.$$

In most cases, it is easier to use derivatives of the natural logarithm of $L(\theta)$ rather than $L(\theta)$, which is called the *log likelihood function*. This is because the logarithmic function is strictly increasing on $(0, \infty)$ and that implies that the extrema of $\ln L(\theta)$ and $L(\theta)$ coincide (we leave the proof of this statement as an exercise).

Example 1.11.1. *Estimating Poisson Parameter by MLE*
Suppose $x_1, x_2, \ldots, x_n$ are observed values of a Poisson random variable with parameter λ representing the random sample $X_1, X_2, \ldots, X_n$ of size n. We want to estimate λ using MLE.

Answer
From the Poisson assumption with parameter λ, the probability of observing x_i events in the ith trial is:

$$p_X(x_i; \lambda) \equiv P(X_i = x_i; \lambda) = \frac{\lambda^{x_i} e^{-\lambda}}{x_i!}, \quad i = 1, 2, \ldots, n; \; x_i = 0, 1, 2, \ldots. \quad (1.11.4)$$

We now take a random sample of size n, say $X_1, X_2, \ldots, X_n$. Let $(x_1, x_2, \ldots, x_n)$ denote the set of observations of $(X_1, X_2, \ldots, X_n)$. Then, from (1.11.3) and (1.11.4), we have the likelihood function as:

$$L(\lambda) = L(x_1, \ldots, x_n; \lambda) = \prod_{i=1}^{n} \left(\frac{\lambda^{x_i} e^{-\lambda}}{x_i!} \right), \tag{1.11.5}$$

and the log likelihood function as:

$$\ln L(\lambda) = \sum_{i=1}^{n} (x_i \ln \lambda) - n\lambda - \sum_{i=1}^{n} \ln(x_i!). \tag{1.11.6}$$

We regard (1.11.6) as a function of λ and will find the value of λ that maximizes this likelihood function. Taking derivative with respect to λ of (1.11.6) and set it equal to zero, we obtain:

$$\frac{d \ln L(\lambda)}{d\lambda} = \frac{1}{\lambda} \sum_{i=1}^{n} x_i - n = 0,$$

from which

$$\lambda = \frac{1}{n} \sum_{i=1}^{n} x_i = \bar{x}. \tag{1.11.7}$$

Thus, MLE for λ is $\bar{x}$, which is the sample mean.

Example 1.11.2. *Estimating Weibull Parameters by MLE*

In Example 1.11.1, we estimated the parameter of a discrete distribution with one parameter. In this example, we want to show how to apply MLE to estimate parameters of a continuous-type distribution with two parameters.

We note that the contents of this example are straightforward materials, and perhaps this is why similar formulae and discussions appear in different textbooks in the literature without any reference to each other. It should also be noted that unfortunately, some properties of Weibull distribution have appeared in some published papers that are incorrect. Hence, readers must be careful when referring to such papers without making necessary corrections.

We consider the two-parameter Weibull distribution function defined in (1.6.34) and its pdf by (1.6.35). Thus, from (1.11.3), the likelihood function is:

$$L(\alpha, \beta) = L(x_1, \cdots, x_n; \alpha, \beta) = \prod_{i=1}^{n} f(x_i; \alpha, \beta)$$

$$= \prod_{i=1}^{n} \frac{\beta}{\alpha} \left(\frac{x_i}{\alpha} \right)^{\beta-1} e^{-\left(\frac{x_i}{\alpha} \right)^{\beta}}. \tag{1.11.8}$$

Hence, for $n \geq 1$, using log likelihood functions, the *likelihood equations* for α and β are as follows:

$$\begin{cases} \dfrac{\partial \ln L}{\partial \beta} = \dfrac{n}{\beta} - n \ln \alpha + \sum_{i=1}^{n} \ln x_i - \sum_{i=1}^{n} \left[\left(\dfrac{x_i}{\alpha} \right)^{\beta} \ln \left(\dfrac{x_i}{\alpha} \right) \right] = 0, \\[4mm] \dfrac{\partial \ln L}{\partial \alpha} = -\dfrac{n\beta}{\alpha} + \dfrac{\beta}{\alpha} \sum_{i=1}^{n} \left(\dfrac{x_i}{\alpha} \right)^{\beta} = 0. \end{cases} \qquad (1.11.9)$$

To solve the system (1.11.9), we eliminate α from the first equation of (1.11.9). From the second equation of (1.11.9), we have:

$$\sum_{i=1}^{n} \left(\frac{x_i}{\alpha} \right)^{\beta} = n \quad \text{and} \quad \frac{1}{\alpha^{\beta}} = \frac{1}{n} \sum_{i=1}^{n} x_i^{\beta}. \qquad (1.11.10)$$

Now from the first equation of (1.11.9), we have:

$$\frac{n}{\beta} - n \ln \alpha + \sum_{i=1}^{n} \ln x_i - \sum_{i=1}^{n} \left(\frac{x_i}{\alpha} \right)^{\beta} \ln x_i + \ln \alpha \sum_{i=1}^{n} \left(\frac{x_i}{\alpha} \right)^{\beta} = 0. \qquad (1.11.11)$$

Using (1.11.10) and cancelling $n \ln \alpha$ terms, (1.11.11) will yield:

$$\frac{n}{\beta} + \sum_{i=1}^{n} \ln x_i - \sum_{i=1}^{n} \left(\frac{x_i}{\alpha} \right)^{\beta} \ln x_i = \frac{n}{\beta} + \sum_{i=1}^{n} \ln x_i - \frac{1}{\alpha^{\beta}} \sum_{i=1}^{n} x_i^{\beta} \ln x_i = 0,$$

$$\frac{n}{\beta} + \sum_{i=1}^{n} \ln x_i - \frac{1}{\dfrac{\sum_{i=1}^{n} \ln x_i^{\beta}}{n}} \sum_{i=1}^{n} x_i^{\beta} \ln x_i = 0.$$

Hence:

$$\frac{\displaystyle\sum_{i=1}^{n} x_i^{\hat{\beta}} \ln x_i}{\displaystyle\sum_{i=1}^{n} x_i^{\hat{\beta}}} - \frac{1}{\hat{\beta}} - \frac{1}{n} \sum_{i=1}^{n} \ln x_i = 0, \qquad (1.11.12)$$

from which $\hat{\beta}$ should be found. Substituting $\hat{\beta}$ in the second equation of (1.11.9), will give $\hat{\alpha}$ as:

$$\hat{\alpha} = \left[\frac{1}{n} \sum_{i=1}^{n} x_i^{\hat{\beta}} \right]^{1/\hat{\beta}}. \qquad (1.11.13)$$

Since (1.11.12) cannot be solved analytically, we apply Newton's iterative method:

$$x_{n+1} = x_n - \frac{f(x_n)}{f'(x_n)}, \tag{1.11.14}$$

where

$$f(\hat{\beta}) = \frac{1}{\hat{\beta}} + \frac{1}{n}\sum_{i=1}^{n}\ln x_i - \frac{\sum_{i=1}^{n} x_i^{\hat{\beta}}(\ln x_i)}{\sum_{i=1}^{n} x_i^{\hat{\beta}}}, \tag{1.11.15}$$

and

$$f'(\hat{\beta}) = -\frac{1}{\hat{\beta}^2} - \frac{\sum_{i=1}^{n} x_i^{\hat{\beta}}(\ln x_i)^2}{\sum_{i=1}^{n} x_i^{\hat{\beta}}} + \left(\frac{\sum_{i=1}^{n} x_i^{\hat{\beta}}(\ln x_i)}{\sum_{i=1}^{n} x_i^{\hat{\beta}}}\right)^2, \tag{1.11.16}$$

to find $\hat{\beta}$ numerically. To do that, we would need to choose a sample of observations. That can be done by simulation.

We should note that we went through the lengthy process to show the MLE
method. However, we could use the MATLAB computer program, for instance,
to find the MLE estimate of parameters numerically.

1.11.2. Interval Estimation

Although point estimation has its usage, it also has its limitations. For instance,
point estimation needs to be accompanied by an MSE, as we mentioned
earlier. In reality, a point estimate cannot be expected to coincide with the
quantity we are to estimate. Hence, to avoid limitations, that is, to determine
the precision of the estimate, the *interval estimator* is used. That is, we can
assert with some level of certainty that the interval contains the parameter
under consideration. Such an interval is called a *confidence interval* for the
parameter. The lower and upper limits for this interval are constructed in such
a way that the true value of the estimate falls within the interval. The upper
end point is usually obtained by adding a multiple value of the standard deviation to the estimated value and the lower end point is obtained by subtracting
the same multiple values from the estimated mean. The likeliness of the true
value falling within the interval is referred to as the *level* (or *degree*) *of
confidence*.

Example 1.11.1
We use this example to show how to construct a confidence interval. Suppose
we have a large random sample of size n (i.e., $n \geq 30$), say $X_1, X_2, \ldots, X_n$, from
a normal population with a *known variance* σ^2 and we want to estimate the
unknown mean, μ. We want to estimate μ and find a 95% confidence for the
estimate.

Answer

Let $\bar{X}$ be the sample mean. We have seen that it is an unbiased point estimator of the population mean μ. We also have seen that:

$$Z = \frac{\bar{X} - \mu}{\sigma/\sqrt{n}} \tag{1.11.17}$$

is a random variable that has an approximately standard normal distribution. The end points of the interval for which the value of the estimator falls within, with, say, 95% probability, can be looked up from the standard normal tables. Hence, we will see that $\bar{X}$ falls within the 1.96 standard deviation of mean 0.95. That is,

$$P(-1.96 < z < 1.96) = 0.95. \tag{1.11.18}$$

This is because 2.5% of the area under the standard normal density curve is to the right of the point $z = 1.96$ ($z_{\alpha/2} = z_{.025} = 1.96$) and 2.5% of the area is to the left of -1.96. Hence, 95% of the area is between -1.96 and $+1.96$. Thus, from (1.11.17) and (1.11.18), we have:

$$P\left(-1.96 < \frac{\bar{X} - \mu}{\sigma/\sqrt{n}} < 1.96\right) = 0.95, \tag{1.11.19}$$

or

$$P\left(\bar{X} - 1.96\frac{\sigma}{\sqrt{n}} < \mu < \bar{X} + 1.96\frac{\sigma}{\sqrt{n}}\right) = 0.95. \tag{1.11.20}$$

Therefore, since 1.96 is 97.5% of the standard normal distribution, a 95% confidence interval estimator of the population mean μ is:

$$\left(\bar{X} - 1.96\frac{\sigma}{\sqrt{n}}, \bar{X} + 1.96\frac{\sigma}{\sqrt{n}}\right). \tag{1.11.21}$$

In other words, 95 times out of every 100 times we sample, μ will fall within the interval $\bar{X} \pm 1.96(\sigma/\sqrt{n})$. For instance, suppose X represent the lifetime of a light bulb. Suppose also that X is normally distributed with mean μ and standard deviation of 42. The manufacturer chose 36 light bulbs and lighted them to observe their lifetime until each burned out. The mean of this sample was 1500 hours. To estimate the mean of all such light bulbs, a 95% confidence interval is decided. Thus, from (1.11.21) we will have:

$$\left(1500 - 1.96\frac{42}{\sqrt{36}}, 1500 + 1.96\frac{42}{\sqrt{36}}\right) = (1486.28, 1513.72).$$

We note that we can use the central limit theorem to approximate the confidence interval. This is because when the sample size is large enough, the ratio

$$Z = \frac{\bar{X} - \mu}{\sigma/\sqrt{n}}$$

is $N(0,1)$. Hence, a $100(1 - \alpha)\%$ confidence interval for μ is:

$$P\left(-z_{\alpha/2} < \frac{\bar{X} - \mu}{\sigma/\sqrt{n}} < z_{\alpha/2}\right) = 1 - \alpha, \tag{1.11.22}$$

and that implies that the *standard normal confidence interval* is:

$$\left(\bar{x} - z_{\alpha/2}\frac{\sigma}{\sqrt{n}}, \bar{x} + z_{\alpha/2}\frac{\sigma}{\sqrt{n}}\right), \tag{1.11.23}$$

where $z_{\alpha/2}$ is the value of the area under the normal curve to its right. We note that the amount of error of the estimate is $|\bar{X} - \mu|$ and:

the maximum error of estimate $< z_{\alpha/2}\sigma/\sqrt{n}$, with probability of $1 - \alpha$.
$$\tag{1.11.24}$$

Other percent confidence intervals can be found similarly.

In case the *variance is unknown*, the sample variance, S^2, can be used. In this case, however, we can again use the quantity $(\bar{X} - \mu)/(\sigma/\sqrt{n})$ with the population standard deviation σ replaced by the sample standard deviation, S, that is,

$$T \equiv \frac{\bar{X} - \mu}{S/\sqrt{n}} \tag{1.11.25}$$

The T defined in (1.11.25) is a random variable since it depends upon S and $\bar{X}$; however, it is not normally distributed; rather, it has *Student's t-distribution* (or just simply, *t-distribution*). T may be treated as a standard normal as long as the sample size is large (30 or more), since in that case there is a high probability that S is very close to σ.

So far, we have discussed the confidence interval for the population mean based on large sample sizes ($n \geq 30$). Those discussions should ensue regardless of the distribution because of the central limit theorem. When the sample size is small ($n < 30$), S may not be close to σ and so $\bar{X}$ may not be normally distributed, unless the sample is selected from a normal population. For a small sample size, there is no good general method for finding the confidence interval. In such a case, the Student's t-distribution will be used along with the t-distribution tables. The *Student's t-distribution confidence interval*, then, would be:

$$\left(\bar{x}-t_{n-1,\alpha/2}S\sqrt{1+\frac{1}{n}},\ \bar{x}+t_{n-1,\alpha/2}S\sqrt{1+\frac{1}{n}}\right). \tag{1.11.26}$$

Example 1.11.3

Suppose that to determine a value of interest (the melting point of tin, for instance), an experiment is conducted six times (sample size). The mean and standard deviation are found to be 0.14 and 232.26, respectively. If the researcher is to use the mean to estimate the actual value of interest (meting point) with high probability, say 98%, what would be the expected maximum error?

Answer

Since $1-\alpha = 98\%$, $\alpha/2 = 0.01$, thus, from the t-table, for $n-1 = 5$ (degrees of freedom) is $i_{0.01} = 3.365$. Hence, from (1.8.13), with probability of 98%, the maximum error will be less than $3.365\times\left(0.14/\sqrt{6}\right)=0.19$.

We note that one must be careful about rounding off numbers; the least significant digit should reflect the precision of the estimate. The precision depends on the variable, on the technique of measurement, and also on the sample size. It is not sufficient to round off just by taking into consideration the sample size alone.

1.12. HYPOTHESIS TESTING

A question such as if the average lifetime of a light bulb is more than 2000 hours is answered by first hypothesizing it and then testing it. This is the objective of this section.

By a *statistical hypothesis*, we mean a statement about the nature of a parameter of a population. By *testing of a hypothesis*, we mean to determine the truth or falsity of the statement. The way the testing is set up is that a default position is set, called the *null hypothesis*, denoted by H_0. The negation of the null hypothesis, H_0, is called the *alternative hypothesis*, denoted by H_1. A statistic whose value is determined from the sample observations is called a *test statistic*. The test statistic is used to decide to reject or accept the null hypothesis. *Critical region* (or *rejection region*) is the set of values of the test statistics for which H_0 is rejected. The *critical value* is the dividing point of the region between values for which the null hypothesis is rejected and not rejected. Testing a null hypothesis may cause two types of errors, called *Type I error* and *Type II error*.

If H_0 is erroneously rejected, it is said that the Type I error has occurred. In contrast, if H_0 is erroneously accepted, it is said that the Type II error has occurred. To assure that the decision of rejecting H_0 is the correct one, one requires that the probability of rejection be less than or equal to a small number, say α. The preassigned small number α, which is the *probability of a*

Type I error (i.e., the probability that H_0 is true and H_1 was not rejected), and is often chosen as 0.01, 0.05, or 0.10, is called the *level of significance of the test* (or *significance level of the test*). The smallest significance level at which the data lead to the rejection of H_0 is called the *p-value*. A small *p*-value (0.05 or less) strongly suggests that H_0 is not true. In contrast, if the *p*-value is large, it would mean that there is strong evidence not to reject H_0.

Let the *probability of Type II error* be denoted by β (i.e., the probability that H_1 is true and H_0 was not rejected). The probability of erroneously rejecting H_0, $1 - \beta$, is then called the *power of the test*. The ideal power function is $\beta = 0$. However, that is only possible in trivial cases. Hence, the practical hope is that the power function is near 1.

We note that increasing the sample size will reduce Type II error and increase power. However, it will not affect Type I error.

The comparison of means of several populations is called the *analysis of variance (ANOVA)*. By that, it is meant estimating the means of more than one population simultaneously. ANOVA is one of the most widely used statistical methods. It is not about variance; rather, it is about analyzing variations in means. In statistical design, ANOVA is used to determine how one can get the most information on the most populations with a fewer observations. Of course, this desire is extremely important in the industry due to minimizing the cost.

In addition to ANOVA, the *t-test* or *error bar* (a graphical representation of the variability of data, which gives a general idea of how accurate a measurement is), when one uses multipliers based on the Student distribution, may be used for the same purpose. The error bar may also be used with normal, Poisson, and other distributions as well. For most straightforward situations, the multiplier lies between 1.95 and 2.00 for *p*-value less than the 0.05 test (two sided against a known chance/null baseline or one sided against an alternative result expected to beat). Usually, the multiplier is applied to the standard error of the cases one regards as the baseline or null hypothesis, but even when comparing two or more arbitrary conditions/systems in an ad hoc way with no strong a priori idea that a particular one should be better, perhaps it is preferred to make the error bars 1 standard error around points from each source, and regard differences as tentatively significant if the error bars do not cross, corresponding to an averaging of the conditions where each is considered the baseline for the other. By "tentative," is it meant that this acts as a filter and formal tests can be done for those that are interesting and potentially significant.

For a two-sided test, for a *p*-value to be considered significant, one needs to judge that 1.5 standard errors would not meet. This problem may be taken care of by including additional whiskers, so at ±1 and ±1.5 standard error.

A word of caution: One should not compare means in cases like repeated measures or pre- or posttest designs. In such cases, the individual differences across the system should be used. Comparing means is often too conservative and is not sensitive to correlation between the individual measurements.

Often there is interest in knowing the relationship between two variables. Finding the relationship and degree of dependence are what *regression analysis* does. Regression (in different forms such as linear, nonlinear, parametric, nonparametric, simple, and multiple) is widely used in different disciplines such as engineering, biological sciences, social and behavioral sciences, and business. For instance, we might be interested in analyzing the relation between two variables such as the strength of and stress on a beam in building a bridge; or a student's grade point average and his/her grade in a statistics course. The analysis of regression is not only used to show the relationship between variables, but it is also for prediction.

The simplest relation between two variables is a straight line, say $Y = \beta_0 + \beta_1 x$. Hence, for each value of x, the value of Y will be predicted, but not exactly when estimations are involved. In those cases, the values are subject to *random error*. Thus, in general, the linear equation considered is of the type $Y = \beta_0 + \beta_1 x + \varepsilon$, where β_0 and β_1 are *parameters*, called *regression coefficients*, and ε is what is called the *random error* with mean 0.

We note that the error terms ε for different trials are assumed to be uncorrelated. Thus, the outcome in any trial has no effect on the error term for any other trial. This implies that Y_i and Y_j, $i \neq j$, are uncorrelated. It should also be noted that, essentially, regression and correlation (that we discussed earlier) are the same. However, we cannot conclude causation from either one. Causes can only be established by an experiment. Causality cannot be determined by statistics; it needs to be assumed. In fact R.A. Fisher, one of the pioneers of statistics, has tried and failed. Correlation and regression are analyses of relationships among mathematical constructs. Causal models are hypothetical statements which guide both the choice of which variables should be measured as potentially relevant and the interpretation of associations that are found in the data.

We also make a general note on rounding numbers that usually causes error of some type. There are no standards to how to round a number to the nearest tenth or others. Essentially, the precision depends on the variable, on the technique of measurement, and also on the sample size. One cannot rule just by sample size alone. Some have set as their basic underlying principle the effect of rounding should always be less than 1% of the standard error of the observation or estimate. In other words, set the rounding error to be less than 1% of the calculated standard error.

One should make sure that rounding numbers is done only after all computations have been completed.

Some use the following rule, called the *rounding interval*, denoted, say, by r. It is defined as the smallest possible positive difference between two rounded values for the same statistic as reported in a study report. If results are reported in two decimal places, the rounding interval $r = 0.01$. Admissible values for r are powers of 10 only, that is, they all belong to the set $\{\ldots; 0.001; 0.01; 0.1; 1; 10; 100; 1000; \ldots\}$. For r, the maximum value should be selected from this set that does not exceed half the standard error of the observation or of the statistic. So $r \leq 1/2$ (standard error) $< 10r$. For instance, if some statistic has a standard error of 0.042, then r should be selected as 0.01, since $0.01 < 0.021 < 0.1$.

Numbers are rounded to the nearest rounded value. If the choice is not confusing, the last digit after rounding should be even. For instance, for $r = 0.01$, we should have the following: 5.233 rounded to 5.23; 4.268 rounded to 4.27; 6.445 rounded to 6.43; and 6.435 rounded to 6.44.

1.13. RELIABILITY

It is now well known to the industry that for manufacturing goods, along with all factors considered such as ease of manufacturing, cost, size, weight, and maintenance, they must pay great attention to the reliability of their products. If a system is multicomponent, then reliability of each component should be considered.

In simple words, the *reliability* of a product is the probability that the product will function within specified limits for at least a specified period of time under specified environmental conditions. These components may be installed in parallel, series, or a mixture of both. A system and each of its components may be in *functioning*, or *partially functioning*, or *failed* state.

If we assume a system with n independent components that are connected in series and only with two states, namely *function* and *failed*, then the probability of the system in series, denoted by R_S, is the product of reliabilities of its components, denoted by R_i, $i = 1, 2, \ldots, n$. To increase the reliability of a system, the components may be connected in parallel. In such a case, the system fails only if all components fail. Hence, if we denote the probability of failure of each component by $F_i = 1 - R_i$, then the probability of the system failure, denoted by F_P, will be the product of F_i's, that is, the reliability of the system in parallel, denoted by R_P, is $R_P = 1 - F_P$.

Suppose that each component x_i, $i = 1, 2, \ldots, n$, of the system has only two states, namely *functioning*, denoted by 1, and *nonfunctioning (failed)*, denoted by 0. If we assume that the state of the system is determined completely by the states of its components and denote the state of the system by Ψ, then:

$$\Psi = \Psi(\mathbf{x}), \quad \mathbf{x} = (x_1, x_2, \ldots, x_n). \tag{1.13.1}$$

The function $\Psi(\mathbf{x})$ defined by (1.13.1) is called the *structure function* of the system and the number of components, n, is called the *order of the system*. A *series structure* (*n out of n*) functions if and only if each component functions, while a *parallel structure* (**1** *out of n*) functions if and only if at least one component functions. Thus, for these two cases, structure functions are given by:

$$\Psi(\mathbf{x}) = \prod_{i=1}^{n} x_i = \min(x_1, x_2, \ldots, x_n), \text{ series structure.} \tag{1.13.2}$$

and

$$\Psi(\mathbf{x}) = \coprod_{i=1}^{n} x_i \equiv 1 - \prod_{i=1}^{n}(1 - x_i) = \max(x_1, x_2, \ldots, x_n), \text{ parallel structure.}$$
$$(1.13.3)$$

A *k-out-of-n structure* functions if and only if k out of n components function. In this case, the structure function is:

$$\Psi(\mathbf{x}) = \begin{cases} 1, & \text{if } \displaystyle\sum_{i=1}^{n} x_i \geq k, \\ 0, & \text{if } \displaystyle\sum_{i=1}^{n} x_i < k. \end{cases} \qquad (1.13.4)$$

Relation (1.13.4) is equivalent to:

$$\Psi(\mathbf{x}) = \prod_{i=1}^{k} x_i, \; k\text{-out-of-}n \text{ structure.} \qquad (1.13.5)$$

As mentioned earlier, the probability R that the system functions is called the *reliability of the system*, that is,

$$R \equiv P\{\Psi(\mathbf{X}) = 1\}, \mathbf{X} = (X_1, X_2, \ldots, X_n). \qquad (1.13.6)$$

Thus,

$$E[\Psi(\mathbf{X})] = R. \qquad (1.13.7)$$

We define $\mathbf{r}$ as the random vector, $\mathbf{r} = (r_1, r_2, \ldots, r_n)$. Based on the independence assumption, the system reliability R will be a function of $\mathbf{r}$, that is,

$$R = R(\mathbf{r}). \qquad (1.13.8)$$

Let a random variable, say T, represent the lifetime of a component or a system with $f_T(t)$ and $F_T(t)$ as the pdf and cdf, respectively. The *reliability function*, at a time t, denoted by $R(t)$, is the probability that the component or the system is still functioning at time t, that is,

$$R(t) = P(T > t). \qquad (1.13.9)$$

From (1.13.9), it is clear that:

$$R(t) = P(T > t) = 1 - P(T \leq) = 1 - F_T(t). \qquad (1.13.10)$$

Hence,

$$R'(t) = -f_T(t). \qquad (1.13.11)$$

Also, the *mean functioning* time or *survival* is:

$$E(T) = \int_0^\infty f_T(t)\,dt = \int_0^\infty R_T(t)\,dt. \tag{1.13.12}$$

The *failure rate function* (or *force of mortality*), denoted by $r(t)$, is defined by:

$$r(t) = f_T(x \mid T > t)_{x=t}. \tag{1.13.13}$$

The function $r(t)$ is an increasing function of t, that is, the older the unit is, the better the chance of failure within a short interval of length h, namely $r(t)h$. Thus, (1.13.13) is equivalent to:

$$r(t) = f_T(t \mid T > t) = \frac{-R'(t)}{R(t)}. \tag{1.13.14}$$

In general, the failure rate function and the reliability are related by:

$$R(t) = e^{-\int_0^t r(\tau)\,d\tau}, \tag{1.13.15}$$

and

$$f_T(t) = r(t)e^{-\int_0^t r(\tau)\,d\tau}. \tag{1.13.16}$$

As an example, suppose that the time to failure of two units 1 and 2 of a system is exponentially distributed with parameters $\lambda_1 = 2$ and $\lambda_2 = 2.5$, respectively. From (1.13.15), the reliability functions of the units are $R_1(t) = e^{-0.5t}$ and $R_2(t) = e^{-0.4t}$, respectively. Assume that the units fail independently. Thus, the reliability functions for the series and parallel systems, respectively, are:

$$R_S = \prod_{i=1}^n R_i(t) = \left(e^{-0.5t}\right)\left(e^{-0.4t}\right) = e^{-0.9t},$$

and

$$R_P = \prod_{i=1}^n R_i(t) = 1 - \left(e^{-0.5t}\right)\left(e^{-0.4t}\right) = 1 - \left(e^{-0.4t} - e^{-0.5t} + e^{-0.9t}\right)$$

$$= e^{-0.4t} + e^{-0.5t} - e^{-0.9t}.$$

For instance, if a unit of time is 2000 hours, then the probability that the series and parallel systems function for more than a unit of time each is $R_S(1) = e^{-0.9} = 0.4066$ and $R_P = e^{-0.4t} + e^{-0.5t} - e^{-0.9t} = 0.8703$, respectively.

The estimation of the reliability has become a concern for many quality control professionals and statisticians. When Y represents the random value of a stress (or supply) that a device (or a component) will be subjected to in service, X represents the strength (or demand) that varies from product to product in the population of devices. The device fails at the instant that the stress applied to it exceeds the strength and functions successfully whenever $X > Y$. The reliability (or the measure of reliability) R is then defined as $P(Y < X)$, that is, the probability that a randomly selected device functions successfully.

Consider a system with only one component. Suppose the random variable X that represents the *strength* takes place in the lifetime, T, of a device. Suppose also that Y that represents the random value of a *stress* takes place in the time of failure. The reliability, then, that a randomly selected device survives under the stress exerted is the probability that the device functions successfully, that is,

$$R = P(Y < X). \tag{1.13.17}$$

The Weibull distribution is commonly used as a life-lengths model and in the study of breaking strengths of materials. The value of R has been calculated in the literature under the assumed distributions and using different methods of estimations such as MLE, shrinkage estimation procedures, and method of moments.

As an example, for the random variables X and Y as strength and stress, respectively, we choose a random sample for each, say, $X_1, X_2, \ldots, X_m$ and Y_1, $Y_2, \ldots, Y_n$. To conduct a simulation, due to the involved and tedious calculations of the MLE of α, we choose $\alpha = 2, 3$, and 4 when the ratio of β_1/β_2 is 1, $1/2$, and 2, with β_1 and β_2 as parameters for X and Y, respectively. We choose the ratio of the scale parameters to compare the effect measures of the reliability. We also choose $m = n$, and $m = 5(1)10$, (5–10 with increment 1). For instance, based on 1000 runs, estimated values of the different estimators can be calculated.

EXERCISES

1.1. Suppose we want to seat two persons on two chairs.
   a. What are the possible outcomes?
   b. List all possible simple events.

1.2. Three pairs of "before" and "after" pictures of three different people are given. Holding the "before" pictures and randomly matching the "after" pictures, what is the probability of:
   a. all matching?
   b. none matching?

1.3. Suppose an election is to choose a set of 4 persons from a group of 10 people without replacement. What is the number of ways to choose the set?

1.4. Suppose a repairman has a diagnostic instrument which can identify problems with a machine three out of four times. Suppose he runs the diagnostic eight times. What is the probability that he identifies the problem

   a. less than four times?

   b. exactly four times?

1.5. A "fair" coin is tossed three times. What is the probability that two heads turn up?

1.6. A random number generator generates numbers at random from the unit interval $[0, 1]$. Find the following:

   a. $P(A)$, where $A = (1/2, 2/3]$.

   b. $P(B)$, where $B = [0, 1/2)$.

   c. the partition generated by A and B.

   d. the event algebra.

1.7. For mutually exclusive events $A_1, A_2, \ldots, A_n$ from a probability space $(\Omega, \mathcal{B}, P)$, show that: $P(\bigcup_i A_i) = \sum_i P(A_i)$.

1.8. If A and B are events and A is a subset of B, prove that $P(B - A) = P(B) - P(A)$ and $P(A) \leq P(B)$.

1.9. If A and B are two events, prove that $P(A \cup B) = P(A) + P(B) - P(A \cap B)$.

1.10. Manufactured articles in a manufacturing company are required to pass two inspections by two inspectors. Experience shows that one inspector will miss 5% of the defective articles, whereas the second inspector will miss 4% of them. If good articles always pass inspection and if 10% of the articles turned out in manufacturing process are defective, what percentage of the articles that passes both inspections will be defective?

1.11. Suppose an institution of higher learning has equal numbers of male and female students, The chance of male students majoring in science, technology, engineering and mathematics (STEM) is 1/5 and for female students this probability is 1/25. A student is chosen at random; what is the probability that:

   a. the chosen student will be a male majoring in STEM?

   b. the chosen student will be a STEM major?

   c. a science student selected at random will be a male student?

1.12. Suppose two technicians T_1 and T_2 independently inspect a unit for a problem. The probability that technician T_1 finds the problem is 1/3 and technician T_2 is 1/6. What is the probability that:

a. both T_1 and T_2 find the problem?

b. at least one will find the problem?

c. neither will find the problem?

d. only T_2 finds the problem?

1.13. A gambler has three coins in his pocket, two of which are fair and one is two headed. He selects a coin. If he tosses the selected coin

a. and it turns up a head, what is the probability that the coin is a fair?

b. five times, what is the probability that he gets five heads?

c. six times, what is the probability of getting five heads followed by a tail?

1.14. Suppose a polling institution finds the approval rating of the handling of foreign policy by the president of the United States during the last 4 years is 23%. Ten Americans are chosen randomly and each is asked about the president's handling of foreign policy. What is the probability that there are at least 2 among the 10 who approve of the handling of foreign policy by the President?

1.15. Suppose that telephone calls enter the department of mathematics' switchboard, on the average, 2 every 10 minutes. Arriving calls are known to have a Poisson pmf with parameter 3. What is the probability of less than 7 calls in a 20-minute period?

1.16. A material is known for 20% of breakage under stress. Five different samples of such materials are tested. What is the probability that:

a. the second sample resists the stress?

b. the second and third samples resist the stress?

1.17. Show that if a random sample of size n, $X_1, X_2, \ldots, X_n$, is drawn from a finite population without replacement, then the random variables $X_1, X_2, \ldots, X_n$ are not mutually independent.

1.18. In a game, a fair die is rolled. To win or lose dollar amounts equal to the number that shows up depends on whether an even or odd number turns up, respectively.

a. What is the expected win or loss in the game?

b. What is the probability if the order of winning and losing is changed?

1.19. Suppose that X is a discrete random variable whose values are 0, 1, 2, 3, 4, and 5, with probabilities 0.2, 0.3, 0.1, 0.1, 0.2, and 0.1, respectively. Find the mean and standard deviation of X.

1.20. Let X be the number of modules with programming errors in a piece of computer software. Let Y be the number of days it takes to debug the software. Suppose X and Y have the following joint probability mass function:

	p_X	X				
		0	1	2	3	4
	0	0.20	0.08	0.03	0.02	0.01
	1	0	0.06	0.09	0.04	0.01
Y	2	0	0.04	0.09	0.06	0.02
	3	0	0.02	0.06	0.04	0.03
	4	0	0	0.03	0.02	0.02
	5	0	0	0	0.02	0.01

Find the following:

a. $E(XY)$

b. The marginal probability mass function for each X and Y

c. The $Cov(X, Y)$

1.21. Let X denote the number of hotdogs and Y the number of sodas consumed by an individual at a game. Suppose X and Y have the following joint probability mass function:

	$p_{X.Y}$	X			
		0	1	2	3
	0	0.06	0.15	0.06	0.03
Y	1	0.04	0.20	0.12	0.04
	2	0.02	0.08	0.06	0.04
	3	0.01	0.03	0.04	0.02

a. Find $E(X)$, $STD(X)$, $E(Y)$, and $STD(Y)$.

b. Find $Cov(X, Y)$ and $Corr(X, Y)$.

c. If hotdogs cost $4.00 each and soda cost $2.50 a can, find an individual's expected value and standard deviation of total costs for sodas and hotdogs at a game.

1.22. A number is to be chosen at random from the interval $[0, 1]$. What is the probability that the number is less than 2/5?

1.23. Let X is a standard logistic random variable. Show that

$$Y = \frac{1}{1+e^{-X}}, \quad -\infty < X < \infty,$$

has a uniform distribution with mean 0 and variance 1.

1.24. Show that for n events $A_1, A_2, \ldots, A_n$, we have:

$$P(A_1 A_2 \cdots A_n) = P(A_1) \times P(A_2|A_1) \times P(A_3|A_1 A_2) \times \cdots \times P(A_n|A_1 A_2 \cdots A_{n-1}).$$

1.25. Show that if events A and B are independent and $P(B) > 0$, then $P(A \mid B) = P(A)$.

1.26. Show the following properties of random variables:

    a. if X and Y are two discrete random variables, then $X \pm Y$ and XY are also random variables, and

    b. if $\{Y = 0\}$ is empty, then X/Y is also a random variable.

1.27. Let X be a binomial random variable with distribution function B_k. Prove that:

$$B_k = \binom{n}{k} p^k q^{n-k}, \quad k = 0, 1, 2, \ldots, n,$$

where $q = 1 - p$.

1.28. Show that (1.2.11) is equivalent to (1.2.9).

1.29. Let X be a binomial random variable with distribution function B_k and $\lambda = np$ be fixed. Show that:

$$B_k = \lim_{\substack{n \to \infty, \\ p \to 0}} = \frac{\lambda^k e^{-k}}{k!}, \quad k = 0, 1, 2, \ldots.$$

1.30. Prove:

    a. The expected value of the indicator function $I_A(\omega)$ defined in (1.2.3) is $P(A)$, that is, $E(I_A) = P(A)$.

    b. If c is a constant, then $E(c) = c$.

    c. If c, c_1, and c_2 are constants and X and Y are two random variables, then $E(cX) = cE(X)$ and $E(c_1X + c_2Y) = c_1E(X) + c_2E(Y)$.

    d. If $X_1, X_2, \ldots, X_n$ are n random variables, then $E(X_1 + X_2 + \ldots + X_n) = E(X_1) + E(X_1) + \ldots + E(X_n)$.

    e. Let X_1 and X_2 be two independent random variables with marginal mass (density) functions p_{x1} and p_{x2}, respectively. Then, if $E(X_1)$ and $E(X_2)$ exist, we will have $E(X_1X_2) = E(X_1)E(X_2)$.

    f. For a finite number of random variables, that is, if $X_1, X_2, \ldots, X_n$ are n independent random variables, then:

$$E(X_1 X_2 \cdots X_n) = E(X_1) E(X_2) \cdots E(X_n).$$

1.31. Show that the variance of a binomial random variable X with parameters n and p is $Var(X) = np(1 - p)$.

1.32. Verify that $F_X(-\infty) = 0$ and $F_X(\infty) = 0$.

1.33. Show that $f_X(x; \mu, \alpha)$ given by:

$$f_X(x) = \begin{cases} \mu e^{-\mu x}, & x \geq 0, \\ 0, & \text{elsewhere,} \end{cases}$$

indeed, defines a probability density function.

1.34. If n is a natural number, show that $\Gamma(n) = (n-1)!$, $n = 1, 2, \ldots$, where $n!$ is defined by $n! = n(n-1)(n-2)\ldots(2)(1)$.

1.35. Show that $0! = 1$.

1.36. Show that $\int_0^\infty e^{-x^2}\,dx = \sqrt{2\pi}$.

1.37. Show that $\Gamma(1/2) = \sqrt{\pi}\,\Gamma(1/2) = \sqrt{\pi}$.

1.38. Show the following properties of X^2 (chi-square) random variable: mean $= r$, and variance $= 2r$.

1.39. Show that $f(x; \mu, \sigma^2)$ given by:

$$f(x; \mu, \sigma^2) = \frac{1}{\sigma\sqrt{2\pi}} e^{-\frac{(x-\mu)^2}{2\sigma^2}}, \quad -\infty < x < \infty$$

is indeed a pdf.

1.40. Show that $\phi(z)$ given by:

$$\phi(z) = \frac{1}{\sqrt{2\pi}} e^{-\frac{z^2}{2}}, \quad -\infty < z < \infty$$

is indeed a pdf.

1.41. Show that the mean and variance of a lognormal random variable X are, respectively,

$$E(X) = e^{\mu + \sigma^2/2} \quad \text{and} \quad Var(X) = \left(e^{\sigma^2} - 1\right)e^{2\mu + \sigma^2}.$$

1.42. Show that the mean and variance of standard logistic random variable X, respectively, are $E(X) = a$ and $Var(X) = 1/3(\pi^2 b^2)$.

1.43. Show that the mean and variance of the Weibull random variable, respectively, are:

$$E(X) = \alpha\Gamma\left(1 + \frac{1}{\beta}\right) \quad \text{and} \quad Var(X) = \alpha^2\left[\Gamma\left(1 + \frac{2}{\beta}\right) - \Gamma\left(1 + \frac{1}{\beta}\right)^2\right].$$

1.44. Show that the relation (1.7.4) represents a probability mass function.

1.45. Prove Theorem 1.7.2, the law of total probability.

1.46. Prove the following properties of covariance:

a. $Cov(X, X) = Var(X)$.
b. $Cov(X, Y) = E[XY] - \mu_X \mu_Y$.
c. $Cov(X, Y) = Cov(Y, X)$.
d. If c is a real number, then $Cov(cX, Y) = cCov(Y, X)$.
e. For two random variables X and Y we have:

$$Var(X + Y) = Var(X) + Var(Y) + 2Cov(X, Y).$$

1.47. Prove the following properties of sample mean $E(\bar{X}) = \mu$ and $Var(\bar{X}) = \sigma^2/n$.

1.48. Show that the random variable

$$W = \frac{X - \mu}{\sigma/\sqrt{n}}, \quad n = 1, 2, \dots.$$

is a standard normal random variable for each positive integer n.

1.49. Show that the second moment of S^2 is:

$$E\left[(S^2)^2\right] = \frac{\mu_4}{n} + \frac{(n-1)^2 + 2}{n(n-1)}\sigma^4,$$

where μ_4 is the fourth central moment of X.

1.50. Show that:

$$\frac{(n-1)S^2}{\sigma^2} \sim \chi^2(n-1).$$

1.51. Use derivatives of the natural logarithm of $L(\theta)$ rather than ln $L(\theta)$ ln $L(\theta)$ to show that the extrema of ln $L(\theta)$ and ln $L(\theta)$ coincide.

1.52. Is it possible for the standard deviation to be 0? If so, give an example; if not, explain why not.

1.53. Let X be a continuous random variable with pdf $f(x) = 1 - |x - 1|, 0 \leq x \leq 2$.

a. Find the 20th and 95th percentiles and show them on the graph of cdf of X.
b. Find the IQR.

1.54. Suppose a random variable X represents a large population consisting of three measurements 0, 3, and 12 with the following distribution:

X	0	3	12
P_X	1/3	1/3	1/3

a. Write every possible sample of size 3.

b. Assuming equiprobable property for every possible sample of size 3, what is the probability of each?

c. Find the sample mean, $\bar{X}$, for each possible sample.

d. Find the sample median, M_m, for each possible sample.

e. Find the sampling distribution of the sample mean, $\bar{X}$.

1.55. For Exercise 1.54, show that $\bar{X}$ is an unbiased estimator of the population parameter μ.

1.56. Let $X_1, X_2, \ldots, X_n$ represent a random sample with its n independent observed values $x_1, x_2, \ldots, x_n$ from a normal random variable with mean μ and variance σ^2. Find the MLE of $\theta(\mu, \sigma^2)$.

1.57. Suppose $X_1, X_2, \ldots, X_n$ is a random sample of size n with pdf $f(x; \lambda) = (1/\lambda)e^{-x/\lambda}, x \geq 0, \lambda > 0$. Find the MLE for the parameter λ.

1.58. Consider the pdf of a one-parameter *Cauchy distribution* as:

$$f(x; \lambda) = \frac{\lambda}{\pi(x^2 + \lambda^2)}, \lambda > 0.$$

Find the maximum likelihood estimate of the parameter λ.

1.59. Find $z_{\alpha/2}$ for each of the following values of α:

a. 0.10

b. 0.05

c. 0.01

1.60. A random sample of 80 shoppers at an automotive part store showed that they spent an average of $20.5 with variance of $39.2. Find a 95% confidence interval for the average amount spent by a shopper at the store.

1.61. The time it takes for a manufacturer to assemble an electronic instrument is a normal random variable with mean of 1.2 hours and a variance of 0.04 hour. To reduce the assembly time, the manufacturer implements a new procedure. A random sample of size 35 is then taken and it shows the mean assembly time as 0.9 hour. Assuming the variance remains unchanged, form a 95% confidence interval for the mean assembly time under the new procedure.

1.62. Find the value of $t_{n-1,\alpha}$ to construct a two-sided confidence interval of the given level with the indicated sample size:

a. 95% level, $n = 5$.

b. 99% level, $n = 29$.

1.63. Suppose that at a company it is known that over the past few years, employees' sick days averaged 5.4 days per year. To reduce this number, the company introduces telecommuting (allowing employees to work at home on their computers). After implementing the new policy, the Human Resource Department chooses a random sample of 50 employees at the end of the year and found an average of 4.5 sick days with a standard deviation of 2.7 days. Let μ be the mean sick days of all employees of the company. Find the p-value for testing hypothesis $H_0: \mu \geq 5.4$ versus $H_1: \mu < 5.4$.

1.64. In a comparison of the effectiveness of online leaning with the traditional in-classroom instruction, 12 students enrolled in a business course online and 14 enrolled in a course with traditional in-classroom instruction. The final exam scores were as follows:

Classroom	80	77	74	64	71	80	68	85	83	59	55	75	81	81
Online	64	66	74	69	75	72	77	83	77	91	85	88		

Does the mean score differ between the two type of course?

CHAPTER TWO

Transforms

Quite often solving a differential equation (ordinary or partial), differential–difference equation, or difference equation is much easier to deal with if it is transformed to an algebraic equation or a simpler form through some type of transform such as Fourier, Laplace, generating function, and/or Z-transform. Fourier transform applies to partial differential equations. In engineering fields, from a practical point of view, Laplace transform is preferred over classical methods for fields such as the propagation of currents and voltages along transmission lines. The Laplace transform now has applications in solving problems that include topics such as improper integrals and asymptotic series. It works similar to a logarithm operator on multiplication and division that transforms them to addition and subtraction and then inverts the results. In case of Laplace transform, which is an example of the integral transform family, the process acts the same, that is, it transforms, manipulates, and then inverts. As the Laplace transform is a continuous transform operator, for the discrete case, the tool we will use, among many, is the generating function, which has many applications in queueing models (when the exponents are nonnegative), and the Z-transform, which essentially is a Laurent series. This transform may be used in areas such as signal processing. We will see applications of these transforms in the next chapters. Thus, this chapter is divided into four parts: (1) Fourier transform, (2) the Laplace transform, (3) the Z-transform, and (4) the probability generating function.

2.1. FOURIER TRANSFORM

The Fourier series is the sum of trigonometric or periodic functions, as will be formally defined later. The Fourier series are used to solve various ordinary and partial differential equations. The foundation of the periodic function goes back to the Babylonians, who used it to predict astronomical events. However, the introduction of the subject to modern mathematics goes back to Euler in

Difference and Differential Equations with Applications in Queueing Theory, First Edition.
Aliakbar Montazer Haghighi and Dimitar P. Mishev.
© 2013 John Wiley & Sons, Inc. Published 2013 by John Wiley & Sons, Inc.

1748 (see Dym and McKean, 1972). The use of the sum of trigonometric functions in describing signals was disputed in the eighteenth century even by Euler and Lagrange, two prominent mathematicians. Fifty years later, Fourier introduced his finding, which was published after a long fight among mathematicians. The reader is referred to publications on the history of mathematics and the development of Fourier series, some of which are cited in the References section at the end of this book. We will provide examples that apply the Fourier series after we have discussed differential equations in the coming chapters.

Definition 2.1.1
The trigonometric series:

$$\frac{a_0}{2} + \sum_{n=1}^{\infty} [a_n \cos nt + b_n \sin nt], \qquad (2.1.1)$$

with

$$a_n = \frac{1}{\pi} \int_{-\pi}^{\pi} f(t) \cos nt \, dt, \quad n = 0, 1, \ldots \qquad (2.1.2)$$

and

$$b_n = \frac{1}{\pi} \int_{-\pi}^{\pi} f(t) \sin nt \, dt, \quad n = 1, 2, \ldots, \qquad (2.1.3)$$

is called a *Fourier series*, where f is a function defined on interval $[-\pi, \pi]$. Coefficients a_n and b_n in (2.1.1) are called *Fourier coefficients*.

Note that a Fourier series being trigonometric is a periodic function.

Example 2.1.1
Suppose that function $f(t)$ is defined on the interval $(-\pi, \pi)$ as:

$$f(t) = \begin{cases} t, & \text{if } t \in [0, \pi), \\ 0, & \text{otherwise.} \end{cases} \qquad (2.1.4)$$

We want to write the Fourier transform of $f(t)$ defined in (2.1.4).

Answer
From (2.1.2) and (2.1.3) we have:

$$a_0 = \frac{1}{\pi} \left[\int_{-\pi}^{0} 0 \cos nt \, dt + \int_{0}^{\pi} t \cos 0 \, dt \right] = \frac{1}{\pi} \left[0 + \frac{t^2}{2} \right]_0^{\pi} = \frac{\pi}{2}, \qquad (2.1.5)$$

$$a_n = \frac{1}{\pi} \int_0^\pi t \cos nt \, dt, = \frac{\cos nt + nt \sin nt}{\pi n^2} \bigg|_0^\pi$$

$$= \frac{\cos n\pi - 1}{\pi n^2} = \frac{(-1)^n - 1}{\pi n^2}, \quad n = 1, 2, \ldots \qquad (2.1.6)$$

$$= \begin{cases} -\dfrac{2}{\pi n^2}, & \text{if } n \text{ is odd} \\[2mm] 0, & \text{if } n \text{ is even,} \end{cases}$$

$$b_n = \frac{1}{\pi} \int_0^\pi t \sin nt \, dt, = \frac{\sin nt - nt \cos nt}{\pi n^2} \bigg|_0^\pi = -\frac{(-1)^n}{n}, \quad n \geq 1. \qquad (2.1.7)$$

Hence, substituting (2.1.5), (2.1.6), and (2.1.7) into (2.1.1) yields the answer as:

$$f(t) \sim \frac{\pi}{4} + \sum_{n=1}^{\infty} \left[\frac{(-1)^n - 1}{\pi n^2} \cos nt - \frac{(-1)^n}{n} \sin nt \right], \quad t \in (-\pi, \pi). \qquad (2.1.8)$$

We leave it as an exercise to show that the Fourier series found in (2.1.8) converges to $f(t)$ given in (2.1.4). Thus, it follows that (2.1.8) represents the periodic extension of $f(t)$. The function $f(t)$ is discontinuous at $t = \pm\pi, \pm 3\pi, \ldots$ (Why?) We leave it as an exercise to show that sum of the series at each one of these points of discontinuity is $\pi/2$.

Example 2.1.2

Suppose that function $f(t)$ is defined on the interval $(-\pi, \pi)$ as:

$$f(t) = \begin{cases} e^t, & \text{if } t \in (-\pi, \pi), \\ 0, & \text{otherwise.} \end{cases} \qquad (2.1.9)$$

We want to write the Fourier transform of $f(t)$ defined in (2.1.9).

Answer

We remind the reader that a function $f(t)$ is called even if $f(-t) = f(t)$ over the domain of $f(t)$. If $f(t)$ is integrable over $(0, b)$, then $\int_{-b}^{b} f(t)\,dt = 2\int_0^b f(t)\,dt$. A function $f(t)$ is odd if $f(-t) = -f(t)$ over the domain of $f(t)$. If $f(t)$ is integrable over $(0, b)$, then $\int_{-b}^{b} f(t)\,dt = 0$. For instance, the functions t^2 and $\cos t$ are even, while t and $\sin t$ are odd functions. Any function $f(t)$ over its domain, say $(-b, b)$, can be expressed as sum of an even and odd function such as:

$$f(t) = \frac{1}{2}[f(t) + f(-t)] + \frac{1}{2}[f(t) - f(-t)]. \qquad (2.1.10)$$

The Fourier coefficients a_n and b_n defined in (2.1.2) and (2.1.3) in case that $f(t)$ *is an even function* in $(-\pi, \pi)$ are as follows:

$$a_n = \frac{2}{\pi} \int_0^\pi f(t)\cos nt\, dt, \quad n = 0, 1, \ldots \tag{2.1.11}$$

and

$$b_n = 0, \quad n = 1, 2, \ldots. \tag{2.1.12}$$

Hence, for an even real-valued function $f(t)$, the Fourier series defined by (2.1.1) will reduce to:

$$\frac{1}{\pi} \int_0^\pi f(\tau)\, d\tau + \frac{2}{\pi} + \sum_{n=1}^\infty \left[\cos n\tau + \int_0^\pi f(\tau)\cos n\tau\, d\tau \right], \tag{2.1.13}$$

which is called the *Fourier cosine series* corresponding to $f(t)$ on $(0, \pi)$.

The Fourier coefficients a_n and b_n defined in (2.1.2) and (2.1.3) in case that $f(t)$ *is an odd function* in $(-\pi, \pi)$ are as follows:

$$a_n = 0, \quad n = 0, 1, 2, \ldots \tag{2.1.14}$$

and

$$b_n = \frac{2}{\pi} \int_0^\pi f(t)\sin nt\, dt, \quad n = 1, 2, \ldots. \tag{2.1.15}$$

Hence, for an odd real-valued function $f(t)$, the Fourier series defined by (2.1.1) will reduce to:

$$f(t) \sim \frac{2}{\pi} + \sum_{n=1}^\infty \left[\sin n\tau + \int_0^\pi f(\tau)\sin n\tau\, d\tau \right], \tag{2.1.16}$$

which is called the *Fourier sine series* corresponding to $f(t)$ on $(0, \pi)$.

Definition 2.1.2
By an *integral transform*, it is meant an operator (or transform), say T, of a function to a new function such that:

$$T[f(v)] = \int_a^b K(\mu, v) f(u)\, du, \quad v \in \mathbb{R}, \tag{2.1.17}$$

where $K(\mu, v)$ is a *kernel function*, or simply a *kernel*.

Definition 2.1.3
The *Fourier transform* of an integrable function $f : \mathbb{R} \to \mathbb{C}$, denoted by *FT*, is defined by:

$$\hat{f}(\omega) = \int_{-\infty}^{\infty} f(t) e^{-i\omega t} dt. \tag{2.1.18}$$

The *inverse Fourier transform* is defined by:

$$f(t) = \frac{1}{2\pi} \int_{-\infty}^{\infty} \hat{f}(\omega) e^{it\omega} d\omega. \tag{2.1.19}$$

We note that in (2.1.18) and (2.1.19), the kernels are $K(t, \omega) = e^{-i\omega t}$ and $K(\omega, t) = e^{i\omega t}$, respectively.

2.2. LAPLACE TRANSFORM

Laplace transform is used for analysis of continuous-time systems. It is a powerful tool to solve differential equations and differential–difference equations. It may be thought of as a generalization of the continuous-time Fourier transform. This is because in signal processing analysis, the Laplace transform can be applied to a broader class of signals since there are signals for which the Fourier transform does not converge, but Laplace transform does.

As we mentioned earlier, transforms are used to convert a nonalgebraic equation or a system of equations to algebraic. Another advantage of transforms is that they change convolutions to products. One further important advantage of transforms is that both the transient and steady state characterizations of a discrete system may be obtained by analyzing the poles (roots of denominator) and zeros (roots of numerator) of the system. To obtain functions from their Laplace transforms, the inverse Laplace transform has to be introduced.

As a historic note, Pierre-Simon, marquis de Laplace (March 23, 1749– March 5, 1827), was a French mathematician and astronomer.

Definition 2.2.1
Let $f(t)$ be a function defined on $[0, \infty)$. The *Laplace transform* of $f(t)$, denoted by $F(s)$ or $\mathcal{L}\{f(t)\}$, is then defined by:

$$F(s) = \mathcal{L}\{f(t)\} = \int_{0}^{\infty} e^{-st} f(t) dt, \tag{2.2.1}$$

provided the integral in (2.2.1) exists. That is, $f(t)$ is a complex-valued function of a real variable such that its value is zero on the negative part of the real line. The function $f(t)$ is sometimes referred to as the *original function*, or just the *original*, and $F(s)$ is referred to as the *image function*, or just the *image*. The *inverse Laplace transform* of $F(s)$, denoted by $f(t)$ or $\mathcal{L}^{-1}\{F(s)\}$, is defined by:

$$f(t) = \mathcal{L}^{-1}\{F(s)\}, \tag{2.2.2}$$

if $F(s)$ is determined from $f(t)$ in (2.2.1).

Note that the Laplace transform that we defined in Definition 2.2.1 is for a continuous function $f(t)$ as defined by (2.2.1). The discrete version of (2.2.1) is the Laplace transform of a sequence $f_n, n = 0, 1, 2, \ldots$, defined as:

$$\mathcal{L}f_n \equiv F^*(s) = \sum_{n=0}^{\infty} f_n e^{-ns}, \tag{2.2.3}$$

provided that the series converges for some value of s. The sequence f_n may be extended for $f_{nt}, t > 0$, denoted by $F_t^*(s)$, as:

$$\mathcal{L}f_{nt} \equiv F_t^*(s) = \sum_{n=0}^{\infty} f_{nt} e^{-nts}, \quad t > 0, \tag{2.2.4}$$

provided that the series converges for some value of s. Of course, if $t = 1$, then (2.2.4) will reduce to (2.2.3).

Properties of Laplace Transform

Property 2.2.1. Linearity

Laplace transform is a linear operator. In other words:

$$\mathcal{L}\{a_1 f_1(t) + a_2 f_2(t)\} = a_1 \mathcal{L}\{f_1(t)\} + a_2 \mathcal{L}\{f_2(t)\} \tag{2.2.5}$$

for arbitrary constants a_1 and a_2, provided that both $\mathcal{L}\{f_1(t)\}$ and $\mathcal{L}\{f_2(t)\}$ exist. The linearity property can be extended to more than two functions. That is,

$$\mathcal{L}\left[\sum_{k=1}^{n} a_k f_k(t)\right] = \left[\sum_{k=1}^{n} a_k \mathcal{L}\{f_k(t)\}\right]. \tag{2.2.6}$$

Example 2.2.1

Let $f(t) = e^{\lambda t}, t \geq 0$, where λ is a constant. We then have:

$$\mathcal{L}(e^{\lambda t}) = \int_0^{\infty} e^{-st} e^{\lambda t} dt = \int_0^{\infty} e^{-(s-\lambda)t} dt = \left.\frac{e^{-(s-\lambda)t}}{-(s-\lambda)}\right|_{t=0}^{\infty} = \frac{1}{s-\lambda}, \quad s > \lambda.$$

Example 2.2.2

Let $f(t) = t^n, n = 0, 1, 2, \ldots$. From (2.2.1), we then have:

$$\mathcal{L}(1) = \int_0^{\infty} e^{-st} dt = \left.\frac{1}{-s} e^{-st}\right|_{t=0}^{\infty} = \frac{1}{s}, \quad n = 0. \tag{2.2.7}$$

Hence, from (2.2.7):

$$\mathcal{L}^{-1}\left(\frac{1}{s}\right) = 1.$$

For $n = 1, 2, \ldots$, using integration by parts, we will have:

$$\mathcal{L}(t^n) = \int_0^\infty e^{-st} t^n dt = -\frac{1}{s} t^n e^{-st}\Big|_{t=0}^\infty + \frac{n}{s}\int_0^\infty e^{-st} t^{n-1} dt$$

$$= \frac{n}{s}\mathcal{L}(t^{n-1}), \quad n = 1, 2, \ldots.$$

(2.2.8)

Thus, for $n = 1$, from (2.2.7) and (2.2.8), we will have:

$$\mathcal{L}(t) = \frac{1}{s^2},$$

that is,

$$\mathcal{L}^{-1}\left(\frac{1}{s^2}\right) = t.$$

Continuing recursively and by mathematical induction, we will have:

$$\mathcal{L}^{-1}\left(\frac{n!}{s^{n+1}}\right) = t^n, \quad n = 0, 1, \ldots. \tag{2.2.9}$$

Tables for Laplace transforms of commonly used elementary functions are readily available in many textbooks and handbooks. Table 2.2.1 contains a few functions and their Laplace transforms that we might use in this book.

Example 2.2.3

$$\mathcal{L}(t^2 - 4e^{-t} + 2\cos 2t) = \mathcal{L}(t^2) - 4\mathcal{L}(e^{-t}) + 2\mathcal{L}(\cos 2t)$$

$$= \frac{2!}{s^3} - 4\frac{1}{s+1} + 2\frac{s}{s^2+4}$$

$$= \frac{2}{s^3} - \frac{4}{s+1} + \frac{2s}{s^2+4}.$$

Example 2.2.4
Let us find the Laplace transform of both $\cos \omega t$ and $\sin \omega t$.

Answer
From Euler's formula:

$$e^{i\omega t} = \cos \omega t + i\sin \omega t, \tag{2.2.10}$$

TABLE 2.2.1. Laplace Transforms of Some Elementary Functions

Formula #	$f(t)$	$F(s) \equiv \mathcal{L}\{f(t)\}$
1	$f(at)$	$(1/s)F(s/a), s > a$
2	$f'(t)$	$s\mathcal{L}\{f(t)\} - f(0)$
3	$f''(t)$	$s^2\mathcal{L}\{f(t)\} - sf(0) - f'(0)$
4	1	$\dfrac{1}{s}$
5	$\dfrac{t^n}{n!}, n = 0, 1, 2, \ldots$	$\dfrac{1}{s^n}, n = 0, 1, 2, \ldots$
6	$\dfrac{t^{a-1}}{\Gamma(a)}, a \geq 0$	$\dfrac{1}{s^a}, a > 0$
7	$u(t - a)$, Heaviside function	$\dfrac{1}{s}e^{-as}$
8	$u(t - a)f(t - a)$	$e^{-as}F(s)$
9	$\delta(t)$, unit impulse at $t = 0$	1
10	$\delta(t - a)$, Dirac delta function	e^{-as}
11	$e^{at}, a \geq 0$	$\dfrac{1}{s-a}, s > a$
12	$e^{at}f(t), a \geq 0$	$F(s - a), s > a,$
13	$\dfrac{(ac+d)e^{at} - (bc+d)e^{bt}}{a-b}$	$\dfrac{cs+d}{(s-a)(s-b)}, a \neq b$
14	$\dfrac{1}{t}(e^{at} - e^{bt})$	$\ln\left(\dfrac{s-b}{s-a}\right)$
15	$\sin(at)$	$\dfrac{a}{s^2 + a^2}$
16	$\cos(at)$	$\dfrac{s}{s^2 + a^2}$

we see that $\cos \omega t$ is the real part and $\sin \omega t$ is the imaginary part of $e^{i\omega t}$, respectively. Hence,

$$\mathcal{L}(e^{i\omega t}) = \frac{1}{s - i\omega} = \frac{s + i\omega}{(s - i\omega)(s + i\omega)} = \frac{s + i\omega}{s^2 + \omega^2} = \frac{s}{s^2 + \omega^2} + i\frac{\omega}{s^2 + \omega^2}.$$

Thus,

$$\mathcal{L}(\cos \omega t) = \frac{s}{s^2 + \omega^2},$$

and

$$\mathcal{L}(\sin \omega t) = \frac{\omega}{s^2 + \omega^2}.$$

Example 2.2.5
We want to find the inverse Laplace transform of $(s - 7)/(s^2 - s - 20)$.

Answer
To find the inverse Laplace transform, in this case, we use the partial fraction method. Hence,

$$\frac{s-7}{s^2-s-20} = \frac{s-7}{(s+4)(s-5)}. \tag{2.2.11}$$

Now,

$$\frac{s-7}{s^2-s-20} \equiv \frac{A}{s+4} + \frac{B}{s-5}$$
$$= \frac{(A+B)S - 5A + 4B}{(S+4)(s-5)}. \tag{2.2.12}$$

Thus, from (2.2.11) and (2.2.12), we have:

$$\begin{cases} A+B=1 \\ -5A+4B=-7. \end{cases} \tag{2.2.13}$$

Therefore, from (2.2.13), we have $A = -1/3$ and $B = 4/3$. Thus, (2.2.11) can be rewritten as:

$$\frac{s-7}{s^2-s-20} = -\frac{1}{3(s+4)} + \frac{4}{3(s-5)}. \tag{2.2.14}$$

Hence, using Formula #5 from Table 2.2.1, the inverse transform of (2.2.14) is:

$$\mathcal{L}^{-1}\left\{\frac{s-7}{s^2-s-20}\right\} = -\frac{1}{3}\mathcal{L}^{-1}\left\{\frac{1}{s+4}\right\} + \frac{4}{3}\mathcal{L}^{-1}\left\{\frac{1}{s-5}\right\}$$
$$= -\frac{1}{3}e^{-3t} + \frac{4}{3}e^{5t}. \tag{2.2.15}$$

We note that we could use Formula #6 and directly obtain the answer (2.2.15).

Property 2.2.2. *Existence of Laplace Transform*
The following theorem states the sufficient condition for existence of the Laplace transform for a given function $f(t)$.

Theorem 2.2.1. *Existence of Laplace Transform*
Let $f(t)$ be a function that:

1. is defined on $[0, \infty)$;
2. is piecewise continuous on $[0, \infty)$; and
3. for some constants M and k, with $M > 0$, it satisfies:

$$|f(t)| \le Me^{kt}, \quad t \in [0, \infty). \tag{2.2.16}$$

Then, the Laplace transform of $f(t)$, $F(s)$ exists for $s > k$.

Proof:
The function $f(t)$ being piecewise continuous implies that $e^{-st}f(t)$ is integrable on $[0, \infty)$. Thus,

$$|F(s)| = |\mathcal{L}\{f(t)\}| = \left| \int_0^\infty e^{-st} f(t)\,dt \right| \le \int_0^\infty e^{-st} |f(t)|\,dt$$

$$\le \int_0^\infty e^{-st} \cdot Me^{kt}\,dt = \frac{M}{s-k}, \quad s > k.$$

Property 2.2.3. *The First Shifting or s-Shifting*

Theorem 2.2.2. *The First Shifting or s-Shifting Theorem*
Let $F(s), s > k$, be the Laplace transform of $f(t)$. Then:

$$\mathcal{L}\{e^{at} f(t)\} = F(s-a), \quad s-a > k, \tag{2.2.17}$$

or equivalently:

$$\mathcal{L}^{-1}\{F(s-a)\} = e^{at} f(t). \tag{2.2.18}$$

Proof:
From (2.2.1), the definition of Laplace transform, we have"

$$\mathcal{L}\{e^{at} f(t)\} = \int_0^\infty e^{-st} e^{at} f(t)\,dt = \int_0^\infty e^{-(s-a)t} f(t)\,dt = F(s-a).$$

Property 2.2.4. *Laplace Transform of Derivatives*
Using integration by parts, the following theorems give the Laplace transforms of derivatives of a function.

Theorem 2.2.3. *Laplace Transform of the First Order Derivative*
If $f(t)$ satisfies all three conditions in Theorem 2.2.1, and $f'(t)$ is piecewise continuous on $[0, \infty)$, then:

$$\mathcal{L}\{f'(t)\} = s\mathcal{L}\{f(t)\} - f(0). \tag{2.2.19}$$

Proof:

$$\mathcal{L}\{f'(t)\} = \int_0^\infty e^{-st} f'(t) dt = e^{-st} f(t)\Big|_{t=0}^{t=\infty} + s\int_0^\infty e^{-st} f(t) dt$$
$$= s\mathcal{L}\{f(t)\} - f(0).$$

Theorem 2.2.4. *Laplace Transform of High Order Derivative*

If $f(t)$ and all its derivatives $f'(t), \ldots, f^{(n-1)}(t)$ satisfy all three conditions in Theorem 2.2.1, and $f^{(n)}(t)$ is piecewise continuous on $[0, \infty)$, then:

$$\mathcal{L}\{f^{(n)}(t)\} = s^n \mathcal{L}\{f(t)\} - s^{n-1} f(0) - s^{n-2} f'(0)$$
$$- \cdots - s f^{(n-2)}(0) - f^{(n-1)}(0). \tag{2.2.20}$$

Proof:

We leave it as an exercise to show that the proof can follow the proof of Theorem 2.2.3 and be generalized by mathematical induction.

Property 2.2.5. *Laplace Transform of Integral*

Let $g(t) = \int_0^t f(x) dx$. Denote the Laplace transform of $g(t)$ by $G(s)$. Then, by (2.2.1) and by interchanging the order of integration in the double integral, we have:

$$G(s) = \int_0^\infty e^{-st} g(t) dt$$
$$= \int_0^\infty e^{-st} \int_0^t f(x) dx \, dt = \int_0^\infty f(x) \int_x^\infty e^{-st} dt \, dx.$$

Now by evaluating the inner integral (assuming that $\Re(s) > 0$), we have:

$$\int_0^\infty f(x) \int_x^\infty e^{-st} dt \, dx = \int_0^\infty f(x) \frac{e^{-sx}}{s} dx = \frac{1}{s} \int_0^\infty f(x) e^{-sx} dx$$
$$= \frac{F(s)}{s}. \tag{2.2.21}$$

Property 2.2.6. *The Second Shifting or t-Shifting Theorem*

Before stating the second shifting theorem, we will first define the Heaviside step function. The *Heaviside step function* or the *unit step function* is one of the important functions in engineering. It was named after the British mathematician Oliver Heaviside (May 18, 1850–February 3, 1925).

The Heaviside function is used in control theory and signal processing to represent a signal for the "on and off" switch that stays "on" or "off" for a specific time. Another application of the Heaviside function is in structural mechanics together with the Dirac delta function to express different types of structural loads. The Dirac delta function is the one that is zero everywhere on the real line except at zero, where the area under its curve is one.

Mathematically speaking, as it can be seen from this definition, the Dirac delta function is not really a function since the integral at a point should be 0 and not 1. It is a discrete analog to the Kronecker delta function.

We first define the Dirac delta function so that we can define the Heaviside function in terms of the Dirac delta function.

Definition 2.2.2

The *Dirac delta function*, denoted by $\delta(x)$, is defined as a function on the real line, which is zero everywhere except at the origin, that is,

$$\delta(x) = \begin{cases} +\infty, & \text{if } x = 0, \\ 0, & \text{otherwise,} \end{cases} \qquad (2.2.22)$$

and satisfies the following relation:

$$\int_{-\infty}^{+\infty} \delta(x)\,dx = 1. \qquad (2.2.23)$$

Definition 2.2.3

The *Heaviside function* or *unit step function*, denoted by $u(t-a)$, is a discontinuous function defined by:

$$u(t-a) = \begin{cases} 0, & \text{if } t < a, \\ 1, & \text{if } t > a. \end{cases} \qquad (2.2.24)$$

The value of $u(0)$ is not important. Since $u(t-a)$ is usually used as a distribution function, $u(0)$ is rarely needed. If we denote the Heaviside function by $H(x)$, then $u(t-a)$ can be rewritten as:

$$H(x) = \int_{-\infty}^{x} \delta(t)\,dt, \qquad (2.2.25)$$

where $\delta(x)$ is the Dirac delta function defined in Definition 2.2.2.

Theorem 2.2.5. *The Second Shifting or t-Shifting Theorem*

Let $F(s)$ be the Laplace transform of a function $f(t)$. Then,

$$\mathcal{L}\{f(t-a)u(t-a)\} = e^{-as}F(s), \qquad (2.2.26)$$

or equivalently,

$$\mathcal{L}^{-1}\{e^{-as}F(s)\} = f(t-a)u(t-a)\},$$

where $u(t-a)$ is the Heaviside function defined in (2.2.24).

Proof:

By definition, we have:

$$L\{f(t-a)u(t-a)\} = \int_0^\infty e^{-st} f(t-a)u(t-a)dt. \qquad (2.2.27)$$

From (2.2.27) and (2.2.24), we will have:

$$L\{f(t-a)u(t-a)\} = \int_0^a e^{-st} \cdot (0)dt + \int_a^\infty e^{-st} f(t-a)dt$$
$$= \int_a^\infty e^{-st} f(t-a)dt. \qquad (2.2.28)$$

To evaluate the last integral in (2.2.28), we let $x = t - a$, which leads to:

$$L\{f(t-a)u(t-a)\} = \int_a^\infty e^{-s(x+a)} f(x)dx = e^{-as} \int_a^\infty e^{-sx} f(x)dx$$
$$= e^{-sx} F(s).$$

Example 2.2.6

We want to find the Laplace transform of $f(t) = 1 - e^{-2t}$ $(0 < t < 2)$.

Answer

We rewrite $f(t)$ as:

$$f(t) = 1 - e^{-2t}, 0 < t < 2$$
$$= (1 - e^{-2t})[1 - u(t-2)]$$
$$= 1 - e^{-2t} - u(t-2) + e^{-4} e^{-2(t-2)} u(t-2).$$

Then:

$$L\{1 - e^{-2t}, 0 < t < 2\} = \frac{1}{s} - \frac{1}{s+2} - \frac{e^{-2s}}{s} + \frac{e^{-2s}}{s+2}$$
$$= \frac{2 - 2e^{-2s}}{s(s+2)}.$$

Example 2.2.7

Let us find the inverse Laplace transform of:

$$F(s) = \frac{2(1 - e^{-\pi s})}{s^2 + 16}.$$

Answer

We rewrite $F(s)$ as:

$$F(s) = \frac{2(1-e^{-\pi s})}{s^2 + 16} = 2 \times \frac{1}{4} \times \frac{4}{s^2 + 16} - 2 \times \frac{1}{4} e^{-\pi s} \times \frac{4}{s^2 + 16}.$$

Then:

$$\mathcal{L}^{-1}\left[\frac{2(1-e^{-\pi s})}{s^2 + 16}\right] = \frac{1}{2}\sin 4t - \frac{1}{2}u(t-\pi)\sin 4(t-\pi)$$

$$= \frac{1}{2}\sin 4t + \frac{1}{2}u(t-\pi)\sin 4t.$$

Property 2.2.7. *Laplace Transform of Convolution of Two Functions*

Definition 2.2.4

By convolution of two functions $f(t)$ and $g(t)$, denoted by $(f*g)(t)$, with $t > 0$, it is meant that:

$$(f * g)(t) = \int_0^t f(t-x)g(x)dx. \tag{2.2.29}$$

We note that convolution operator is *associative*, *commutative*, and *distributive*. That is, for three functions $f(t)$, $g(t)$, and $h(t)$, we have:

$$(f * g * h)(t) = [(f * g) * h](t) = f * (g * h)(t),$$

$$(f * g)(t) = (g * f)(t),$$

and

$$f * (g + h)(t) = (f * g)(t) + (f * h)(t).$$

Theorem 2.2.6. *Convolution Theorem*

Let $F(s)$ and $G(s)$ be the Laplace transforms of $f(t)$ and $g(t)$, respectively. Then:

$$\mathcal{L}[(f * g)(t)] = F(s) \cdot G(s). \tag{2.2.30}$$

Proof:

We start from the right-hand side of (2.2.30) to reach the left-hand side. Hence:

$$F(s) \cdot G(s) = \int_0^\infty e^{-st} f(t)dt \int_0^\infty e^{-st} g(t)dt$$

$$= \iint_S e^{-s(u+v)} f(u)g(v)du\,dv, \tag{2.2.31}$$

where $S = \{(u, v) | 0 \leq u < \infty, 0 \leq v < \infty\}$ is the domain of the double integral. To evaluate (2.2.31), we make the change of variables $t = u + v$ and $x = v$, that is, we change (u, v) to (t, x) with the transformation differential element

$$du\,dv = \left| \frac{\partial(u, v)}{\partial(t, x)} \right| dt\,dx$$

and domain of the transformed double integral as $S_1 = \{(t, x) | 0 \leq t < \infty, 0 \leq x < t\}$. Thus:

$$\iint_S e^{-s(u+v)} f(u) g(v)\,du\,dv = \iint_{S_1} e^{-st} f(t-x) g(x)\,dt\,dx,$$

$$= \int_0^\infty e^{-st} \int_0^t f(t-x) g(x)\,dx\,dt = \mathcal{L}[(f * g)(t)].$$

2.3. $\mathcal{Z}$-Transform

Historically, it seems Laplace knew the generating function that was reintroduced by Hurewicz (1947). He used it as a mean to solve linear constant-coefficient difference equations. Ragazzini and Zadeh (1952) used the idea and dubbed it as $\mathcal{Z}$-*transform* in the sampled-data control group. The modified or advanced $\mathcal{Z}$-transform was later developed by E.I. Jury.

The $\mathcal{Z}$-transform has applications not only in mathematics but also in signal processing. It can also be considered as a discrete-time Laplace transform. As an example of a discrete-time process, we can consider a digital thermometer that is a device that gives a digital output for an analog input (a signal that varies continuously with time). A conversion from an analog to digital signal requires the signal to be sampled at intervals of time, that is, samples of input waveform are taken every t units of time. Hence, for intervals of small length, only the value at the end of the interval is reported and the rest of the interval is ignored.

In fact, the $\mathcal{Z}$-transform maps a discrete sequence to a function of complex variable $\mathcal{Z}$. Thus, if $f(t)$ describes a sampling process for a continuous-time signal, the output, denoted by $f^*(t)$, is referred to as f_n, $n = 0, 1, 2, \ldots$, where f_n is a series of pulses at epochs $0, t, 2t, \ldots, nt, \ldots, t > 0$. These pulses may be represented as the sequence $f_0, f_t, f_{2t}, \ldots, f_{nt}, \ldots, n = 0, 1, 2, \ldots$. In fact, the main difference between Laplace and Fourier transforms on one hand and $\mathcal{Z}$-transform on the other hand is that the $\mathcal{Z}$-transform operates on a sequence $\{f_n\}$ of integer-valued argument n rather than on a piecewise continuous function, as in Laplace and Fourier transforms.

The $\mathcal{Z}$-transform could be viewed as the Fourier transform of an exponentially weighted sequence. There is a close relationship between the $\mathcal{Z}$-transform and the discrete-time Fourier transform. In fact, if the magnitude of $\mathcal{Z}$ is 1, the $\mathcal{Z}$-transform reduces to the Fourier transform. The $\mathcal{Z}$-transform may be thought of as the discrete-time counterpart of the Laplace transform.

Hence, properties of the Z-transform closely parallel those of the Laplace transform.

The complex plane associated with the Z-transform is referred to as the Z-plane. Of particular interest in the Z-plane is the circle of radius 1, concentric with the origin that is referred to as the unit circle.

There are different definitions for the Z-transform, one of which is often referred to the generating function. Therefore, it seems the basic idea of a Z-transform is old and it is what is known as the *generating function*. It can be traced back as early as 1730 when it was introduced by de Moivre in conjunction with probability theory.

We now define the version we will use in this book.

Definition 2.3.1

Let $\{f_n\}$ be an integer-valued sequence. Then the Z-*transform* (or *unilateral Z-transform*) of the sequence $\{f_n\}$, denoted by $Z\{f_n\} \equiv F(z)$ is a function of the complex variable z such that:

$$F(z) \equiv \sum_{n=0}^{\infty} f_n z^{-n}. \tag{2.3.1}$$

If the limit of the series in (2.3.1) extends from negative infinity, that is,

$$F(z) \equiv \sum_{n=-\infty}^{\infty} f_n z^{-n}, \tag{2.3.2}$$

where n is an integer and z is a complex variable, then (2.3.2) is called a *bilateral Z-transform*. The *region of convergence*, often denoted as *ROC*, of the Z-transform is the region where the transform exits.

Notes:

1. Since $z = re^{i\omega}$, where i is the complex unit and r is the magnitude of z, the substitution of z, if $r = 1$, in (2.3.2) leads to the Fourier transform of the sequence $\{f_n\}$ on the unit circle, $|z| = 1$, as:

$$F(e^{i\omega}) = \sum_{n=-\infty}^{\infty} f_n (e^{i\omega})^{-n}$$

$$= \sum_{n=-\infty}^{\infty} f_n (e^{-i\omega n}). \tag{2.3.3}$$

2. The inverse transform of Z-transform , denoted by $Z^{-1}\{F(z)\}$, is

$$Z^{-1}\{F(z)\} = f_n. \tag{2.3.4}$$

3. Since the $\mathcal{Z}$-transform defined by (2.3.1) is a power series, it converges when $f_n z^{-n}$ satisfies:

$$\sum_{n=0}^{\infty} |f_n z^{-n}| < \infty. \tag{2.3.5}$$

Example 2.3.1

For $n = 0, 1, 2, 3, 4, 5$, let the sequence $\{f_n\}$ be given by $\{4, 5, 0, 6, 7, 8\}$. We want to find the $\mathcal{Z}$-transform of this sequence.

Answer

From (2.3.1), we have:

$$F(z) = \sum_{n=0}^{5} f_n z^{-n} = 4z^{-0} + 5z^{-1} + 0z^{-2} + 6z^{-3} + 7z^{-4} + 8z^{-5}$$
$$= 4 + 5z^{-1} + 6z^{-3} + 7z^{-4} + 8z^{-5}. \tag{2.3.6}$$

Note that if (2.3.6) was given as the $\mathcal{Z}$-transform of sequence $\{f_n\}$ and the sequence $\{f_n\}$ was being determined, the coefficient of the term missing in the sequence with ascending order of powers should be taken as 0 in terms of the sequence.

Example 2.3.2

We want to find the bilateral $\mathcal{Z}$-transform of a signal given by:

$$f(n) = 8\left(\frac{1}{3}\right)^n u_n - 6\left(\frac{1}{2}\right)^n u_n, \tag{2.3.7}$$

where u_n represents the Heaviside step function given in Definition 2.2.3.

Answer

Applying (2.3.2) on (2.3.7), we will have the following:

$$F(z) = \sum_{n=-\infty}^{\infty} \left\{ 8\left(\frac{1}{3}\right)^n u_n - 6\left(\frac{1}{2}\right)^n u_n \right\} z^{-n}$$
$$= 8\sum_{n=-\infty}^{\infty} \left(\frac{1}{3}\right)^n z^{-n} u_n - 6\sum_{n=-\infty}^{\infty} \left(\frac{1}{2}\right)^n z^{-n} u_n \tag{2.3.8}$$
$$= 8\sum_{n=-\infty}^{\infty} \left(\frac{1}{3}z^{-1}\right)^n - 6\sum_{n=-\infty}^{\infty} \left(\frac{1}{2}z^{-1}\right)^n.$$

The series on the left-hand side of (2.3.8) are geometric with ratios $1/3(z^{-1})$ and $1/2(z^{-1})$, respectively. Hence, we will have:

$$F(z) = \frac{8}{1 - \frac{1}{3}z^{-1}} - \frac{6}{1 - \frac{1}{2}z^{-1}} = \frac{2(1 - z^{-1})}{\left(1 - \frac{1}{3}z^{-1}\right)\left(1 - \frac{1}{2}z^{-1}\right)}. \qquad (2.3.9)$$

Multiplying the numerator and denominator of (2.3.9) by z^2, the bilateral Z-transform in for this example will be:

$$F(z) = \frac{2z(z-1)}{\left(z - \frac{1}{3}\right)\left(z - \frac{1}{2}\right)}. \qquad (2.3.10)$$

Regarding the convergence of (2.3.10), we note from (2.3.8) that both infinite series on the right-hand side must converge. Hence, we must have

$$\left|\frac{1}{3}z^{-1}\right| < 1 \quad \text{and} \quad \left|\frac{1}{2}z^{-1}\right| < 1.$$

That is, $|z| > 1/3$ and $|z| > 1/2$. Hence, the ROC is common ROC of both, that is, $|z| > 1/2$.

From (2.3.10), we note that the bilateral Z-transform for this example has two poles and one zero. They are $1/3$, $1/2$, and 1, respectively.

Example 2.3.3
We want to find the bilateral Z-transform of a signal given by:

$$f_n = \left(\frac{1}{2}\right)^n u_n, \qquad (2.3.11)$$

where u_n represents the Heaviside step function.

Answer
From (2.3.11), we have:

$$f_n = \left\{\ldots, 0, 0, 0, 1, \left(\frac{1}{2}\right), \left(\frac{1}{2}\right)^2, \left(\frac{1}{2}\right)^3, \ldots\right\}. \qquad (2.3.12)$$

Thus:

$$F(z) = \sum_{n=0}^{\infty} \left(\frac{1}{2}\right)^n z^{-n} = \sum_{n=0}^{\infty} \left(\frac{1}{2z}\right)^n. \qquad (2.3.13)$$

The series on the right-hand side of (2.3.13) is a geometric series with ratio $1/(2z)$. Hence, we have:

$$F(z) = \frac{1}{1 - \frac{1}{2z}}, \quad \left| \frac{1}{2z} \right| < 1 \quad \text{or} \quad |z| > \frac{1}{2}.$$

(2.3.14)

Therefore, the ROC is $|z| > 1/2$.

Example 2.3.4
This example is to show that the condition for the transform is not satisfied. We consider a revised version of Example 2.3.3 as follows. Let the bilateral $\mathcal{Z}$-transform of a signal be given by:

$$f_n = \left(\frac{1}{2} \right)^n, \quad \text{for all integers } n.$$

(2.3.15)

Then from (2.3.15) we have:

$$f_n = \left\{ \ldots, \left(\frac{1}{2} \right)^{-3}, \left(\frac{1}{2} \right)^{-2}, \left(\frac{1}{2} \right)^{-1}, 1, \left(\frac{1}{2} \right), \left(\frac{1}{2} \right)^2, \left(\frac{1}{2} \right)^3, \ldots \right\}$$

$$= \left\{ \ldots, 2^3, 2^2, 2, 1, \left(\frac{1}{2} \right), \left(\frac{1}{2} \right)^2, \left(\frac{1}{2} \right)^3, \ldots \right\}.$$

(2.3.16)

Thus:

$$F(z) = \sum_{n=-\infty}^{\infty} \left(\frac{1}{2} \right)^n z^{-n}$$

(2.3.17)

diverges, that is, there is no ROC. The condition cannot be satisfied for any values of z.

Example 2.3.5
We want to find the bilateral $\mathcal{Z}$-transform of a signal given by:

$$f_n = \left(\frac{1}{5} \right)^n \sin\left(\frac{\pi}{3} n \right) u_n,$$

(2.3.18)

where $u(n)$ represents the time unit.

Answer
We use Euler's formula $e^{i\theta} = \cos\theta + i\sin\theta$. It is easy to see that:

$$\sin\theta = \frac{1}{2i} e^{i\theta} - \frac{1}{2i} e^{-i\theta}.$$

Thus, we rewrite (2.3.18) as:

$$f_n = \frac{1}{2i}\left(\frac{1}{5}e^{i\frac{\pi}{3}}\right)^n u_n - \frac{1}{2i}\left(\frac{1}{5}e^{-i\frac{\pi}{3}}\right)^n u_n. \tag{2.3.19}$$

Applying (2.3.2) to (2.3.19), we will have:

$$F(z) = \sum_{n=-\infty}^{\infty} \left\{ \frac{1}{2i}\left(\frac{1}{5}e^{i\frac{\pi}{3}}\right)^n u_n - \frac{1}{2i}\left(\frac{1}{5}e^{-i\frac{\pi}{3}}\right)^n u_n \right\} z^{-n}$$

$$= \frac{1}{2i}\sum_{n=-\infty}^{\infty}\left(\frac{1}{5}e^{i\frac{\pi}{3}}z^{-1}\right)^n - \frac{1}{2i}\sum_{n=-\infty}^{\infty}\left(\frac{1}{5}e^{-i\frac{\pi}{3}}z^{-1}\right)^n \tag{2.3.20}$$

$$= \frac{1}{2i}\frac{1}{1-\frac{1}{5}e^{i\frac{\pi}{3}}z^{-1}} - \frac{1}{2i}\frac{1}{1-\frac{1}{5}e^{-i\frac{\pi}{3}}z^{-1}}$$

$$= \frac{\dfrac{\sqrt{3}}{10}z^{-1}}{\left(1-\dfrac{1}{5}e^{i\frac{\pi}{3}}z^{-1}\right)\left(1-\dfrac{1}{5}e^{-i\frac{\pi}{3}}z^{-1}\right)} \tag{2.3.21}$$

$$= \frac{\dfrac{\sqrt{3}}{10}z}{\left(z-\dfrac{1}{5}e^{i\frac{\pi}{3}}\right)\left(z-\dfrac{1}{5}e^{-i\frac{\pi}{3}}\right)}.$$

For convergence of (2.3.21), we note from (2.3.20) that both infinite series on the right-hand side must converge. Hence, we must have:

$$\left|\frac{1}{5}e^{i\frac{\pi}{3}}z^{-1}\right| < 1 \quad \text{and} \quad \left|\frac{1}{5}e^{-i\frac{\pi}{3}}z^{-1}\right| < 1.$$

Since $\left|e^{-i\frac{\pi}{3}}\right| = \left|e^{i\frac{\pi}{3}}\right| = 1$, the ROC is $|z| > 1/5$.

From (2.3.21), we see that bilateral $\mathcal{Z}$-transform for this example has one pole and one zero. They are 1/5 and 1, respectively.

Example 2.3.6

Suppose we are given the following bilateral $\mathcal{Z}$-transform:

$$F(z) = 5z^4 + 3z^3 - z^2 - 6 + 6z^{-2}, \quad 0 < |z| < \infty. \tag{2.3.22}$$

We want to find the inverse transform of (2.3.22).

Answer
Based on (2.3.2) and (2.3.4), the inverse transform of (2.3.22), that is, the sequence $\{f_n\}$, is as follows:

$$\{f_n\} = \begin{cases} 5, & n = -4, \\ 3, & n = -3, \\ -1, & n = -2, \\ -6, & n = 0, \\ 6, & n = 2 \\ 0, & \text{otherwise.} \end{cases}$$

Formally, the inverse $\mathcal{Z}$-transform as defined in (2.3.4) can be found as follows:

$$f_n = \mathcal{Z}^{-1}\{F(z)\} = \frac{1}{2\pi i}\oint_C F(z)z^{n-1}dz, \tag{2.3.23}$$

where C in the contour integral is a counterclockwise closed path encircling the origin and in the ROC. In particular, if C is the unit circle. That is because if $F(z)$ is stable, the ROC includes the unit circle. In that case, all poles are within the unit circle. In this case, the inverse $\mathcal{Z}$-transform leads to the following inverse discrete-time Fourier transform:

$$f_n = \frac{1}{2\pi}\int_{-\pi}^{\pi} F\left(e^{i\omega}\right)e^{i\omega n}d\omega. \tag{2.3.24}$$

We note that discrete-time Fourier transform is a special case of a $\mathcal{Z}$-transform with z within the unit circle.

Let $F(z)$ be the $\mathcal{Z}$-transform of sequence $\{f_n\}$. The following are properties of transform (and we leave the proofs as exercises):

Property 2.3.1. Let ζ be a nonzero complex number. The $\mathcal{Z}$-transform of $\{\zeta^n f_n\}$ is then $F(\zeta z)$.

Property 2.3.2. Let k be a nonnegative integer. The $\mathcal{Z}$-transform of the sequence $\{n^k f_n\}$, denoted by $F_k(z)$, is then:

$$\begin{cases} F_0(z) = \text{the } z \text{ transform of the sequence } \{1\}, \\ F_k(z) = (-z)\dfrac{dF_{k-1}(z)}{dz}, \quad k = 1, 2, \dots. \end{cases} \tag{2.3.25}$$

Example 2.3.7
Let $\{f_n\} \equiv \{1\} = 1, 1, \dots$. Let us denote by $F_0(z)$ and $F_1(z)$ the $\mathcal{Z}$-transforms of the sequences $\{n^0 f_n\} = \{1\} = 1, 1, \dots$ and $\{n^1 f_n\} = \{n\} = n, n, \dots$, respectively. We want to find the last two sequences.

Solution

For $k = 0$, the $\mathcal{Z}$-transform of $\{n^0 f_n\} = \{1\} = 1, 1, \ldots$ is:

$$F_0(z) = \sum_{n=0}^{\infty} z^{-n} = 1 + \frac{1}{z} + \frac{1}{z^2} + \cdots = \frac{1}{1 - \frac{1}{z}},$$

$$= \frac{z}{z-1}, \quad |z| > 1. \tag{2.3.26}$$

Now, for $k = 1$, the $\mathcal{Z}$-transform of $\{n^1 f_n\} = \{n\} = n, n, \ldots$ is:

$$F_1(z) = -z \frac{d}{dz} F(z) = -z \frac{d}{dz} \frac{z}{z-1} = (-z) \frac{z - 1 - z}{(z-1)^2}$$

$$= \frac{z}{(z-1)^2}. \tag{2.3.27}$$

Property 2.3.3. Let k be a positive integer. Then the $\mathcal{Z}$-transform of f_{n+k}, denoted by $F_{n_k}(z)$, that is, $\{f_n\}$ is shifted k units to the left, is:

$$F_{n_k}(z) = z^k F(f) - \sum_{n=0}^{k-1} f_n z^{k-n}. \tag{2.3.28}$$

2.4. PROBABILITY GENERATING FUNCTION

Probability generating functions play an extremely useful role in solving difference equations. We will see its applications later in this book. For instance, we will see in Chapter 4 and Chapter 5 that applying a probability generating function on a set of difference equations will result in either an algebraic or a functional equation.

Definition 2.4.1

Let X be a discrete random variable with probability mass function (pmf) p_n, where n is a nonnegative integer. Then the *probability generating function* (or if there is no confusion, a *generating function*) (pgf) of X, denoted by $G(z)$, is the power series representation of the pmf of X, that is,

$$G(z) = \sum_{n=0}^{\infty} p_n z^n, \tag{2.4.1}$$

which converges for $|z| < 1$, where z is a complex variable.

Note that if n is a value of X, then z^n will be a value of z^X. In this case, z^X could be considered to be a random variable, whose expected value is given by (2.4.1), that is,

$$G(z) \equiv E(z^X). \tag{2.4.2}$$

2.4.1. Some Properties of a Probability Generating Function

Property 2.4.1. If the z-variable in (2.4.1) is replaced with 1/z, then we obtain (2.3.1), the unilateral $\mathcal{Z}$-transform.

Property 2.4.2. Since sum of probabilities must be 1, we have:

$$G(1) = \sum_{n=0}^{\infty} p_n = 1. \tag{2.4.3}$$

Notes:

1. If (2.4.3) does not hold, (2.4.1) will still be a generating function, but not a probability generating function.
2. The probability generating function is intended to encapsulate all the information about the random variable, one of which was stated in (2.4.3).

Example 2.4.1
Consider the sequence of all 1's, that is, $\{a_n\} = 1, 1, \ldots \equiv \{1\}, \forall n = 1, 2, \ldots$. The generating function of this sequence will then be:

$$G(z) = \sum_{n=0}^{\infty} 1 \cdot z^n = 1 + z + z^2 = \cdots = \frac{1}{1-z}.$$

This is because by dividing 1 by $(1 - z)$, we obtain the infinite series $1 + z + z^2 + z^3 + \ldots + z^n + \ldots$, for $|z| < 1$, where all coefficients equal to 1. However, this series is not a probability generating function since $G(1) \neq 1$.

Property 2.4.3. The pmf can be recovered by derivatives of the probability generating function (2.4.1), evaluated at zero, that is,

$$p_n = P(X = n) = \frac{G^{(n)}(0)}{n!}, \tag{2.4.4}$$

where $G^{(n)}(0)$ denotes the nth derivative of $G(z)$ evaluated at $z = 0$.

It can be observed from (2.4.4) that if two random variables X and Y have the same pgf, then their pmfs are the same, that is, they are identically distributed.

Property 2.4.4. The nth *factorial moment* of X, that is, $E[X(X-1)\ldots(X-n+1)]$, can be obtained by the nth derivative of the generating function evaluated at 1, that is, since $X! = X(X-1)\ldots 2\cdot 1$, we can complete the expression $X(X-1)\ldots(X-n+1)$ and divide by the terms added. Hence:

$$E[X(X-1)\ldots(X-n+1)] = E\left[\frac{X!}{(X-n)!}\right]$$

$$= \frac{d^n G(z)}{dz^n}\bigg|_{z=1} = \sum_{n=k}^{\infty} n!\, p_n \frac{z^{n-k}}{(n-k)!}\bigg|_{z=1} \qquad (2.4.5)$$

$$= \sum_{n=k}^{\infty} \frac{n!\, p_n}{(n-k)!}.$$

Thus, the first factorial moment, that is, the expected value of X, is obtained when $n = 1$. That is,

$$E(X) = G'(1). \qquad (2.4.6)$$

Moreover, since $Var(x) = E(X^2) - (E(X))^2$, from (2.4.5) and (2.4.6) we will have:

$$Var(X) = E[X(X-1)] + E(X) - [E(X)]^2$$
$$= G''(1) + G'(1) - [G'(1)]^2. \qquad (2.4.7)$$

Example 2.4.2
Let X be a Poisson random variable with parameter λ. The probability generating function of X will then be:

$$G(z) = \sum_{k=0}^{\infty} \frac{\lambda^k e^{-\lambda}}{k!} z^k = e^{-\lambda}\sum_{k=0}^{\infty} \frac{(\lambda z)^k}{k!} = e^{-\lambda} e^{\lambda z} = e^{-\lambda(1-z)}. \qquad (2.4.8)$$

Hence, from (2.4.6) we have:

$$E(X) = G'(1) = e^{-\lambda}\lambda e^{\lambda z}\big|_{z=1} = \lambda. \qquad (2.4.9)$$

As it can be seen from (2.4.9), the parameter of the Poisson distribution is, in fact, the expectation of X. To find the variance, from (2.4.7) we have:

$$Var(X) = G''(1) + G' - [G']^2$$
$$= e^{-\lambda}\lambda^2 e^{\lambda z}\big|_{z=1} + e^{-\lambda}\lambda e^{\lambda z}\big|_{z=1} + \left[e^{-\lambda}\lambda e^{\lambda z}\big|_{z=1}\right]^2 = \lambda. \qquad (2.4.10)$$

Again, we see from (2.4.10) that the parameter of the Poisson distribution is also the variance of X.

We state other properties of the generating function in the theorems and notes that follow.

Theorem 2.4.1

The *generating function of the sum* of n independent random variables is the product of the generating function of each.

Proof:

Let $X_1, X_2, \ldots, X_n$ be n independent random variables with corresponding generating functions, respectively, as $G_1(z) = E(z^{X_1}), G_2(z) = E(z^{X_2}) \ldots,$ $G_n(z) = E(z^{X_n})$. Let also $X = X_1 + X_2 + \ldots + X_n$. Then:

$$E(z^X) = E(z^{X_1 + X_2 + \cdots + X_n}) = E(z^{X_1} z^{X_2} \cdots z^{X_n})$$
$$= E(z^{X_1}) E(z^{X_2}) \ldots E(z^{X_n})$$
$$= G_1(z) G_2(z) \ldots G_n(z).$$

Theorem 2.4.2

The *product of two generating functions* is another generating function.

Proof:

Let $G(z)$ and $H(z)$ be two generating functions, respectively, defined as $G(z) = \sum_{i=0}^{\infty} a_i z^i$ and $H(z) = \sum_{j=0}^{\infty} b_j z^j$, with $|z| < 1$. Then, their product is:

$$G(z) \cdot H(z) = \left(\sum_{i=0}^{\infty} a_i z^i \right) \left(\sum_{j=0}^{\infty} b_j z^j \right) = \sum_{i=0}^{\infty} \sum_{j=0}^{\infty} a_i b_j z^{i+j}. \qquad (2.4.11)$$

Now let $k = i + j$. Then, (2.4.11) can be rewritten as a single power series as follows:

$$G(z) \cdot H(z) = \sum_{k=0}^{\infty} c_k z^k, \qquad (2.4.12)$$

where

$$c_k = \sum_{i=0}^{k} a_i b_{k-i}, \quad k = 1, 2, \ldots. \qquad (2.4.13)$$

Definition 2.4.2

The sequence $\{c_k\}$ given in (2.4.13) is called the *convolution* of the two sequences $\{a_i\}$ and $\{b_j\}$.

Definition 2.4.3

Let X be a discrete random variable with *pmf* p_X. Let also t be a nonnegative real number or a complex number with real part nonnegative. Assume that the expected value $E(e^{tX})$ exists. We then define the *moment generating function* of X, denoted by $M(t)$, as:

$$M(t) = E(e^{tX}) = \sum_x e^{tX} p_X.$$
(2.4.14)

Note that the moment generating function of a discrete random variable is a probability generating function when z is replaced by e^t.

The following are consequences of (2.4.14):

1. If $t = 0$, then $M(0) = \sum_{n=1}^{\infty} p_n = 1.$
(2.4.15)

2. $\lim_{t \to 0^-} (d^n M(t)/dt^n) = \sum_x x^n p_n = E(X^n).$
(2.4.16)

Thus, the moment generating function generates all the moments of X.

Example 2.4.3

Let X be a random variable over $\{1, 2, \ldots, n\}$ with $p_x = 1/n, n = 1, 2, \ldots$. That is, X is uniformly distributed. Thus, the moment generating function of X is:

$$M(t) = \sum_{k=1}^{n} e^{tk} \frac{1}{n} = \frac{e^t + e^{2t} + \cdots + e^{nt}}{n}$$

$$= \frac{e^t}{n}(1 + e^t + \cdots + e^{(n-1)t}).$$
(2.4.17)

The terms in the parentheses on the right-hand side of (2.4.17) is a geometric progression with ratio e^t. Hence:

$$M(t) = \frac{e^t}{n}\frac{e^{nt} - 1}{e^t - 1}.$$
(2.4.18)

We note that although equation (2.4.18) looks simpler than (2.4.17) for the moment generating function, for the purpose of generating moments of X using (2.4.16), an indeterminate form will appear that would require l'Hôpital's rule to find values of the moments. Thus, in practice, (2.4.17) is preferable. Thus, the mean and variance of X are as follows:

$$E(X) = M'(0) = \left. \frac{e^t + 2e^{2t} + \cdots + ne^{nt}}{n}\right|_{t=0} = \frac{1 + 2 + \cdots + n}{n} = \frac{n(n+1)}{2n}$$

$$= \frac{n+1}{2},$$
(2.4.19)

$$E(X^2) = M''(0) = \frac{e^t + 2^2 e^{2t} + \cdots + n^2 e^{nt}}{n}\bigg|_{t=0}$$

$$= \frac{1 + 2^2 + \cdots + n^2}{n} = \frac{n(n+1)(2n+1)}{6n}$$

$$= \frac{(n+1)(2n+1)}{6},$$

$$Var(X) = M''(0) - [M'(0)]^2 = \frac{(n+1)(2n+1)}{6} - \left(\frac{n+1}{2}\right)^2$$

$$= \frac{(n^2-1)}{12}. \tag{2.4.20}$$

We made a remark about (2.4.18) and why we used (2.4.17). In general, for computational purposes, it is more convenient to use the logarithm of $M(t)$. Thus, we will use the following for the mean and variance of a random variable using the moment generating function:

$$E(X) = \frac{d[\log M(t)]}{dt}\bigg|_{t=0} \tag{2.4.21}$$

and

$$Var(X) = \frac{d^2[\log M(t)]}{dt^2}\bigg|_{t=0}. \tag{2.4.22}$$

EXERCISES

2.1. Show that the Fourier series:

$$\frac{\pi}{4} + \sum_{n=1}^{\infty}\left[\frac{(-1)^n - 1}{\pi n^2}\cos nt - \frac{(-1)^n}{n}\sin nt\right], \quad t \in (-\pi, \pi)$$

converges to:

$$f(t) = \begin{cases} t, & \text{if } t \in [0, \pi), \\ 0, & \text{otherwise.} \end{cases}$$

2.2. Show that $f(t)$ defined in Exercise 2.1 is discontinuous at $t = \pm\pi, \pm 3\pi, \ldots$.

2.3. Show that sum of the series in Exercise 2.1 at each one of the discontinuity in Exercise 2.2 is $\pi/2$.

2.4. Prove Theorem 2.2.4.

2.5. Show that:

$$\sin\theta = \frac{1}{2i}e^{i\theta} - \frac{1}{2i}e^{-i\theta}.$$

2.6. Expand $f(t) = x, -\pi < t < \pi$ in a Fourier series.

2.7. Find the Fourier coefficients corresponding to the function $g(x)$, where $g(x)$ is given by:

$$g(x) = \begin{cases} -1 & \text{if } -\pi < x < 0, \\ 1 & \text{if } 0 < x < \pi. \end{cases}$$

2.8. Write the corresponding Fourier series of:

$$f(x) = \begin{cases} 0 & \text{if } -3 < x < 0, \\ 5 & \text{if } 0 < x < 3, \end{cases} \quad L = 3.$$

2.9. Expand $f(x) = 3x^2, -\pi < x < \pi,$ in a Fourier series.

2.10. Expand $g(x) = |\sin t|, -\pi < t < \pi$ in a Fourier series.

2.11. Find the Laplace transform of:

a. $f(t) = \begin{cases} 2, & t \in [0, 2], \\ 0, & \text{otherwise.} \end{cases}$

b. $f(t) = t^2 e^t + 3te^{-t} + 3\cos(4t)$.

c. $f(t) = \begin{cases} t, & 0 \le t < 3, \\ 6-t, & 3 \le t < 6, \\ 0, & 6 \le t. \end{cases}$

d. $f(t) = \begin{cases} e^{-2t}, & 0 \le t < 3, \\ 0, & 3 \le t. \end{cases}$

e. $f(t) = \begin{cases} 3, & 0 \le t < 3, \\ 5, & 3 \le t < 6, \\ 1, & 6 \le t. \end{cases}$

f. $f(t) = \begin{cases} \cos t, & 0 \le t < 3\pi \\ 0, & 3\pi \le t. \end{cases}$

g. $f(t) = e^{-3s}/(s^2 + 9)$.

h. $f(t) = e^{-s}/(s^2 + 6s + 9)$.

2.12. Find the inverse Laplace transform of:

a. $F(s) = 2/(1-s)$.

b. $F(s) = (s+1)/(s^2 - 2s)$.

c. $F(s) = 2/(1-s)^3$.

d. $F(s) = 3/(s^2 + 3s + 2)$.

e. $F(s) = (s+1)/s^2(s+2)^3$.

2.13. Find $L\{f'(t)\}$ of the following functions:

   a. $f(t) = e^{3t}$

   b. $f(t) = e^{-3t} + 3$

   c. $f(t) = t^2 + 2t + 5$

2.14. Find $L\{f''(t)\}$ of the following functions:

   a. $f(t) = e^{3t}$

   b. $f(t) = t^2 + \sin(2t) + 5$

2.15. Using the first shifting theorem, find the Laplace transform of the following functions:

   a. $f(t) = t^3 e^{5t}$

   b. $f(t) = e^{5t} \cos 2t$

   c. $f(t) = e^{5t} \sin 2t$

2.16. Using the first shifting theorem, find the inverse Laplace transform of the following functions:

   a. $F(s) = (2s+1)/(s^2 + 2s + 4)$

   b. $F(s) = 1/(s^2 - 2s + 5)$

2.17. Let $F(z)$ be the Z-transform of the sequence $\{f_n\}$.

   a. Let ζ be a nonzero complex number. Prove that the Z-transform of $\{\zeta^n f_n\}$ is $F(\zeta z)$.

   b. Prove that the Z-transform of the sequence $\{nf_n\}$, denoted by $F_1(z)$, is:

$$F_1(z) = -z \frac{dF(z)}{dz}.$$

   c. Let k be a positive integer. Prove that the Z-transform of the sequence $\{n^k f_n\}$, denoted by $F_k(z)$, is:

$$\begin{cases} F_0(z) = \text{the } z \text{ transform of the sequence } \{1\}, \\ F_k(z) = (-z)\dfrac{dF_{k-1}(z)}{dz}, \quad k = 1, 2, \ldots. \end{cases}$$

   d. Let k be a positive integer. Prove, then, that $F^{(k)}(z)$ is the Z-transform of:

$$(-1)^k u(n-1-k)(n-1)(n-2)\cdots(n-k)f_{n-k},$$

where $u(t - a)$ is defined in (2.2.24).

e. Let k be a positive integer. Prove that the $\mathcal{Z}$-transform of f_{n+k}, denoted by $F_{n_k}(z)$, is:

$$F_{n_k}(z) = z^k F(f) - \sum_{n=0}^{k-1} f_n z^{k-n}.$$

2.18. Let ζ be a nonzero complex number. Show that the $\mathcal{Z}$-transform of $f_n = \zeta^n$ is:

$$F(z) = \frac{z}{z - \zeta}, |z| > \frac{1}{|\zeta|}.$$

2.19. In Example 2.3.7, show that the $\mathcal{Z}$-transform of n^2 is:

$$F(z) = \frac{z(z+1)}{(z-1)^3}.$$

2.20. Use Example 2.3.7 to show that the $\mathcal{Z}$-transform of n^3 is:

$$F(z) = \frac{z(z^2 + 4z + 1)}{(z-1)^4}.$$

2.21. Show that the probability generating function of a constant random variable, that is, one with $P(X = c) = 1$, is $G(z) = (z^c)$.

2.22. Show that the probability generating function of a binomial random variable with parameters n and p is:

$$G(z) = [(1-p) + pz]^n.$$

2.23. Show that the formula given in Exercise 2.22 is the n-fold product of the probability generating function of a Bernoulli random variable with parameter p.

2.24. Suppose a fair die is rolled repeatedly until the third time "1" appears. The probability distribution of the number of numbers other than 1 that had appeared will then be an example of a negative binomial. The *negative binomial distribution* is a discrete probability distribution of the number of successes in a sequence of Bernoulli trials before a specified number of failures (denoted by r) occurs. Its probability mass function is:

$$P(X = k) = \binom{k - r - 1}{k}(1-p)^r p^k, \quad k = 0, 1, 2, \ldots.$$

Show that the probability generating function of a negative binomial random variable, the number of trials until the rth success with probability of success in each trial p, is:

$$G(z) = \left[\frac{pz}{1-(1-p)z}\right]^r, \quad |z| < \frac{1}{1-p}.$$

2.25. Show that the generating function given in Exercise 2.24 is the r-fold product of the probability generating function of a geometric random variable.

2.26. Consider a sequence of independent and identically distributed random variables (iid) $\{X_n, n = 0, 1, 2, \ldots\}$, with each X_n having the *logarithmic distribution* with probability mass function:

$$f_X = \frac{-p^x}{x \ln(1-p)} x = 1, 2, \ldots.$$

Suppose N is a random variable, independent of the sequence $\{X_n\}, n = 0, 1, 2, \ldots$. Suppose also that N has a Poisson distribution with mean $\lambda = -r \ln(1-p)$. Prove that the random sum $Y = \sum_{n=1}^{N} X_n$ has a negative binomial distribution with parameters r and p (which represent a *compound Poisson distribution*).

2.27. Find the generating function $G(z)$ of the sequence $\{a_n\}$:

a. $a_n = (1/5)^n$

b. $a_n = n(n+1)/2$

2.28. Find the sequence $\{a_n\}$ having the generating function $G(z)$:

a. $G(z) = 2/(1-z) + 1/(1-2z)$

b. $G(z) = z/(1-2z)$

CHAPTER THREE

Differential Equations

Historically, German mathematician/philosopher Gottfried Wilhelm Leibniz (1646–1716) and British physicist/mathematician/philosopher Isaac Newton (1643–1727) are credited for the invention of differential and integral calculus. Leibniz introduced several notations we use in mathematics to this day, for instance, the integral sign $\int$ that represents an elongated S, from the Latin word *summa*, and the d for *differential*, from the Latin word *differentia*. Although he did not publish until 1684, he initiated the term "differential equation" in 1676, and it is being universally used to this day. However, Newton is the one who first introduced differential equations in physics. Now, differential equations dominate physics from classical to quantum. Since its inauguration, the subject of differential equations has been considered as a branch in both pure and applied mathematics. Differential equations arise when we want to describe how a physical property changes with regard to another. One of the most important classes of continuous-time systems that arise in a variety of systems and physical phenomena is the one for which the input and output are related through a *linear constant coefficient differential equation*. For instance, in engineering, the response of a simple resistor–capacitor (RC) circuit, in which the patterns of variation over time in the source and capacity voltage can be considered as signals, may be described by a linear constant coefficient differential equation. We start the chapter with basic vocabulary.

3.1. BASIC CONCEPTS AND DEFINITIONS

The most important concept in mathematical analysis is the concept of a function. We have been using this concept in the past two chapters. However, a reminder of the definition at this point could be useful, although the readers might already be familiar with the concept. We need a function not only of one variable, as we have been using, but also a function of more than one variable, in particular, of two variables.

Difference and Differential Equations with Applications in Queueing Theory, First Edition.
Aliakbar Montazer Haghighi and Dimitar P. Mishev.

Definition 3.1.1. *Function of One Variable*
A *function* (of one variable), f, is a relation between two sets, say A and B, such that for each element of the first set, A, there is a unique element in the second set, B. The first set, A, is called the *domain* of f, and the corresponding elements in the second set, B, is called the *range* of f. If a typical element of domain of a function f is denoted by x and its corresponding element in the range is denoted by y, then we can write:

$$y = f(x) \quad \text{or} \quad f : x \rightarrow y \quad \text{or} \quad x \xrightarrow{f} y.$$

The type of elements of a domain of a function, as discrete or continuous, will determine the type of a function as such.

Definition 3.1.2. *Function of Two Variables*
A *function* of two variables, f, is a rule that assigns to each ordered pair of real numbers (x, y) in a set D a unique real number, denoted by $f(x, y)$. The first set, D, is called the *domain* of f and the set of values that f takes is called the *range* of f, that is, $\{f(x, y) : (x, y) \in D\}$.

Example 3.1.1
Consider $f(x, y)$ as a function of two variables x and y such that:

$$f(x, y) = \frac{\sqrt{x + y + 2}}{x - 3}.$$

The domain of this function $f(x, y)$ is $\{f(x, y) : x + y + 2 \geq 0, x \neq 3\}$.

For a function of more than one variable, a derivative with respect to each individual variable is called a *partial derivative* of the function. For $z = f(x, y)$ a function of two variables, the following are notations for partial derivatives (first and second order):

$$\frac{\partial f(x, y)}{\partial x} = \frac{\partial f}{\partial x} = f_x(x, y) = f_x = \frac{\partial z}{\partial x} = D_x f,$$

$$\frac{\partial^2 f(x, y)}{\partial^2 x} = \frac{\partial^2 f}{\partial x^2} = f_{xx}(x, y) = f_{xx} = \frac{\partial^2 z}{\partial x^2} = D_{xx} f,$$

$$\frac{\partial f(x, y)}{\partial y} = \frac{\partial f}{\partial y} = f_y(x, y) = f_y = \frac{\partial z}{\partial y} = D_y f.$$

$$\frac{\partial^2 f(x, y)}{\partial y^2} = \frac{\partial^2 f}{\partial y^2} = f_{yy}(x, y) = f_{yy} = \frac{\partial^2 z}{\partial y^2} = D_{yy} f,$$

$$\frac{\partial^2 f}{\partial y \partial x} = f_{xy}(x, y) = f_{xy} = \frac{\partial^2 z}{\partial y \partial x} = D_{xy} f,$$

$$\frac{\partial^2 f}{\partial x \partial y} = f_{yx}(x, y) = f_{yx} = \frac{\partial^2 z}{\partial x \partial y} = D_{yx} f.$$

Definition 3.1.3

If $f(x, y)$ is a function of two variables x and y, its *partial derivatives* with respect to x and y are defined as:

$$f_x(x, y) = \lim_{h \to 0} \frac{f(x+h, y) - f(x, y)}{h}$$

and

$$f_y(x, y) = \lim_{k \to 0} \frac{f(x, y+k) - f(x, y)}{k}.$$

The rules for finding partial derivatives of a function $z = f(x, y)$ are as follows:

1. To find f_x, regard y as a constant and differentiate $f(x, y)$ with respect to x.
2. To find f_y, regard x as a constant and differentiate $f(x, y)$ with respect to y.

Theorem 3.1.1. *Clairaut's Theorem*

Suppose f is defined on a disk D that contains the point (a, b). If the functions f_{xy} and f_{yx} are both continuous on D, then $f_{xy}(a, b) = f_{yx}(a, b)$.

Proof:

See Stewart (2012, p. A46).

Example 3.1.2

Consider:

$$f(x, y) = x^3 y^2 + \ln(xy) - x \sin x.$$

Then,

$$f_x = 3x^2 y^2 + \frac{1}{x} - \sin x - x \cos x, \quad f_{xy} = 6x^2 y$$

and

$$f_y = 2x^3 y + \frac{1}{y}, \quad f_{yx} = 6x^2 y.$$

Definition 3.1.4

A *differential equation* is an equation that includes derivatives of some unknown functions. If the equation involves derivatives of a function of one variable, it is called an *ordinary differential equation* (denoted by *ODE*).

However, if the equation involves derivatives of a function of more than one variable, it is called a *partial differential equation* (denoted by *PDE*). The *order* of a differential equation is the order of the highest derivative involved.

Example 3.1.3
Here are some examples of differential equations of both types with indicated orders:

(a) $y' + 3xy = 7$. First order, ODE.
(b) $y'' - 5y' + 6y + \sin x + \cos x = 7$. Second order, ODE.
(c) $y''' = (1 + yy')(x^2 + y^2)$. Third order, ODE.
(d) $(\partial^2 u/\partial t^2) - 3(\partial^2 u/\partial x^2) = 0$, or $u_{tt} - 3u_{xx} = 0$. Second order, PDE.

In equations (a), (b), and (c), the unknown function is $y = y(x)$, and y' and y'' are the first and second derivatives with respect to x, respectively. In equation (d), the unknown function is $u = u(t, x)$, and u_{tt} and u_{xx} are the second-order partial derivatives of the function $u(t, x)$ with respect to t and x, respectively.

We note that, generally, differential equations deal with continuous variables. However, in sciences today, in some cases, the time is discretized and differential equations are yet being used with that type of variable. For instance, the motion of atoms within a fluid may be approximated by a continuous process! Moreover, quantum physics is discrete, but uses differential equations in its entirety. Of course, there are those that strongly oppose this type of use of differential equations such as purists who believe "calculus is continuous". Perhaps one should try to use difference equations to avoid this type of controversy that some call the "abuse of differential equations." However, one cannot ignore the vast use of differential equations in signals and systems, calculus of time scales, discrete systems, and so on. We should not ignore such concepts as Simpson's rule!

In general we have the following definition:

Definition 3.1.5
An *ordinary differential equation of order n* is an equation in the form:

$$y^{(n)} = f\left(x, y, y', \ldots, y^{n-1}\right), \tag{3.1.1}$$

where $y = y(x)$ is the unknown function. The independent variable x belongs to some interval I (finite or infinite).

Definition 3.1.6
By an *explicit solution* of the ordinary differential equation, (3.1.1), it is meant a function $\phi(x)$, defined over the interval I, which satisfies equation (3.1.1) identically over the interval I. In other words, $\phi(x)$, is a solution of (3.1.1), if it is n times differentiable and:

$$\phi^{(n)} = f\left(x, \phi(x), \phi'(x), \ldots, \phi^{n-1}(x)\right),$$ (3.1.2)

for all x in some open interval $(a, b) \subset I$. The graph of a solution is called an *integral curve*.

Example 3.1.4
Let us show that the function:

$$\phi(x) = e^{2x}$$ (3.1.3)

is a solution of the second-order ODE:

$$y'' - 4y = 0.$$ (3.1.4)

Answer
We substitute (3.1.3) in (3.1.4) and obtain the following:

$$\phi'' - 4\phi = \left(e^{2x}\right)'' - 4e^{2x} = 4e^{2x} - 4e^{2x} = 0.$$

Hence, (3.1.3) is a solution of (3.1.4) for all $x \in (-\infty, \infty)$.

Example 3.1.5
Let us verify that:

$$\phi = \frac{x^3}{4} - \frac{1}{x}$$ (3.1.5)

is a solution of

$$xy' + y = x^3$$ (3.1.6)

on $(-\infty, 0) \cup (0, \infty)$.

Answer
In order to substitute (3.1.5) into (3.1.6), we need to find the first derivative of it, which is:

$$\phi' = \frac{3x^2}{4} + \frac{1}{x^2}.$$ (3.1.7)

Now substituting (3.1.5) and (3.1.7) into (3.1.6) yields:

$$x\phi' + \phi = x\left(\frac{3x^2}{4} + \frac{1}{x^2}\right) + \left(\frac{x^3}{4} - \frac{1}{x}\right) = \left(\frac{3}{4}x^3 + \frac{1}{x}\right) + \left(\frac{1}{4}x^3 - \frac{1}{x}\right) = x^3$$

for all $x \neq 0$. Therefore, $\phi(x)$, given by (3.1.5), is a solution of (3.1.6) on $(-\infty, 0) \cup (0, \infty)$.

As we will see in the succeeding sections, solving differential equations do not always yield explicit solutions. In such cases, we may have to accept solutions that are implicitly defined. Before we offer some examples, we give a formal definition of implicit solution.

Definition 3.1.7
A relation $F(x, y) = 0$ is said to be an *implicit solution* of the differential equation (3.1.1) on an interval I, if it defines one or more explicit solutions of (3.1.1) on I.

Example 3.1.6
We want to show that the relation

$$y^2 - 2x^3 + 54 = 0 \qquad (3.1.8)$$

is an implicit solution for the given differential equation

$$y' = \frac{3x^2}{y} \qquad (3.1.9)$$

on the interval $(3, \infty)$.

Answer
We solve (3.1.8) for y in terms of x. Hence, we have:

$$y = \pm\sqrt{2x^3 - 54}. \qquad (3.1.10)$$

Suppose one claims either one of the explicit solutions given in (3.1.10) is a solution of (3.1.9). Let us first try the one with the positive square root, that is, $y = \sqrt{2x^3 - 54}$. Then we have:

$$\frac{dy}{dx} = \frac{3x^2}{2\sqrt{2x^3 - 54}}. \qquad (3.1.11)$$

Of course, both y given in (3.1.10) and dy/dx given in (3.1.11) are defined on the interval $(3, \infty)$. Substituting these in (3.1.9), we will have:

$$\frac{3x^2}{2\sqrt{2x^3 - 54}} = \frac{3x^2}{2\sqrt{2x^3 - 54}}.$$

That is, the positive square root in (3.1.10) is indeed a solution of (3.1.9). Similarly, we can see that the negative square root in (3.1.10), that is,

$y = -\sqrt{2x^3 - 54}$, is also a solution. Hence, although two explicit solutions were available, we could have one implicit solution as a combination of the two.

Example 3.1.7

Let us show that the relation:

$$x - y + \sin(xy) = 0 \qquad (3.1.12)$$

is an implicit solution of the differential equation:

$$[1 - x\cos(xy)]\frac{dy}{dx} = 1 + y\cos(xy). \qquad (3.1.13)$$

Solution

Note that despite the case in Example 3.1.6, we are not able to break down (3.1.12) into two explicit solutions. However, from (3.1.12), we see that any change in y requires some change in x for (3.1.13) to hold. It is not easy to show this observation directly. However, the well-known implicit function theorem (see Fitzpatrick, 2009) guarantees that not only such a function exists, but it is differentiable. The knowledge of y being a differentiable function of x clears the way to use implicit differentiation method. Hence, we differentiate (3.1.12) with respect to x and apply the product and chain rule. This will yield:

$$\frac{d}{dx}[x - y + \sin(xy)] = 1 - \frac{dy}{dx} + \cos(xy)\left(y + x\frac{dy}{dx}\right) = 0$$

or

$$[1 - x\cos(xy)]\frac{dy}{dx} = [1 + y\cos(xy)],$$

which is identical to the differential equation (3.1.13). Therefore, relation (3.1.12) is indeed an implicit solution of (3.1.13) on some interval guaranteed by the implicit function theorem.

Differential equations appear in mathematical models that attempt to describe real life situations. Many natural laws and hypotheses can be translated via mathematical language into equations involving derivatives. For example, derivatives appear in physics as velocities and accelerations, in geometry as slopes, and in biology as rates of growth of population. In order to apply a mathematical method to a physical or real-life problem, it is necessary to formulate the problem in mathematical terms, that is, we have to construct a mathematical model for the problems. The mathematical model for an applied problem is almost always simpler than the actual situation being studied, since simplifying assumptions are usually required to obtain a mathematical problem that can be solved.

In the succeeding sections, we offer some examples of such mathematical models.

Example 3.1.8. *Newton's Law of Cooling*
Newton's law of cooling states that the rate of change of the temperature of an object is proportional to the difference between its own temperature and the temperature of its surrounding environment (the so-called ambient temperature). Newton's law makes a statement about an *instantaneous* rate of change of the temperature. Let the body temperature be denoted by $T = T(t)$ and the temperature of the surrounding environment by T_0. Then, by Newton's cooling law, we have:

$$\frac{dT(t)}{dt} = -k(T(t) - T_0), \tag{3.1.14}$$

where $k > 0$ is the constant of proportionality. We made the coefficient negative since we are considering the cooling system in the environment, that is, the decrease in temperature. Equation (3.1.14) is the first-order ODE for $T = T(t)$.

Note that the constant k can be interrelated as the product of αA, where A is the surface area of the body through which heat is transferred and α is the heat transfer coefficient that depends on the geometry of the body, state of surface, heat transfer mode, and other factors.

Example 3.1.9. *Hooke's Law and Vibrating Springs*
Robert Hooke, a British scientist, discovered in 1660 that for relatively small deformation of an object, the displacement or size of the deformation is directly proportional to the deforming force or load. Under these conditions, the object will return to its original shape and size upon removal of the load. This observation is now called the *Hooke's law* of elasticity.

Let us consider the motion of an object with a mass at the end of a spring that is either vertical or horizontal on a level surface. Hooke's law of elasticity states that the applied force F equals a constant k times the displacement or change in length x, $F = -kx$. The value of k depends not only on the kind of elastic material under consideration but also on its dimensions and shape. The negative sign indicates that the force exerted by the spring is in direct opposition to the direction of displacement.

Combine this with Newton's second law $F = m(d^2x/dt^2)$ (force equals mass times acceleration) and we have the following equation for a mass–spring problem:

$$m\frac{d^2x}{dt^2} + kx = 0. \tag{3.1.15}$$

Equation (3.1.15) is a second-order ODE for $x = x(t)$.

Example 3.1.10. *Population Growth and Decay*

Although the number of members of a population (people of a given country, bacteria in a laboratory culture) at any time t is necessarily an integer, in mathematical models the number of members of the population can be regarded as a differentiable function $N = N(t)$. The rate of population growth at any given time can be written as:

$$\frac{dN}{dt} = h(N) \cdot N, \qquad (3.1.16)$$

where $h(N)$ is a continuous function of N that represents the rate of change of population per unit time per individual. Equation (3.1.16) is referred to as the *Malthusian model* or the *Malthusian growth model*. It is sometimes called the *simple exponential growth model*. In this model, it is assumed that $h(N)$ is a constant, denoted by r, so that (3.1.16) becomes:

$$\frac{dN}{dt} = rN. \qquad (3.1.17)$$

The model (3.1.17) assumes that the number of births and deaths per unit time are both proportional to the population, where r ($r \neq 0$) is the constant of proportionality. Hence, (3.1.17) is essentially exponential growth based on a constant rate of compound interest. The model is named after the Reverend Thomas Malthus, who authored *An Essay on the Principle of Population*.

The solution of differential equation (3.1.17) is:

$$N(t) = Ce^{rt}, \qquad (3.1.18)$$

where C is a constant. From this we see that:

$$\lim_{t \to \infty} N(t) = \begin{cases} \infty & \text{if } r > 0, \\ 0, & \text{if } r < 0. \end{cases} \qquad (3.1.19)$$

The Malthusian model (3.1.17) is not that realistic, as it can be seen from (3.1.19) that it either leads to a population explosion (∞) or to extinction (0). Thus, another common *model* of population growth is used, called the *Verhulst* model. It is originally due to Pierre François *Verhulst* in 1838. The Malthusian model was improved by building into differential equation (3.1.18), a way to prevent the runaway growth when $r > 0$. Verhulst proposed the following differential equation as a mathematical model, sometimes called the *logistic differential equation*:

$$\frac{dN(t)}{dt} = rN(t)\left(1 - \frac{N(t)}{\alpha}\right), \qquad (3.1.20)$$

where α is a positive constant.

Note that when N is large so that $(N/\alpha) > 1$ in (3.1.20), the derivative dN/dt will be negative and so the population will decrease.

3.2. EXISTENCE AND UNIQUENESS

Solutions of differential equations usually occur in families, as we have seen in Example 3.1.10. However, real-world situations that we want to model with differential equations often have unique real-world results. For example, if we drop a bowling ball out of a window, the graph of the height of the ball versus time since its release will always be the same. Mathematically, it is often possible to reduce the number of solutions of differential equations to exactly one by adding an additional requirement to the problem, as we explain in the following example.

Example 3.2.1
Consider the following differential equation:

$$\frac{dy}{dx} = 3y. \qquad (3.2.1)$$

All functions of the form $y = Ce^{3x}$ satisfy this equation since $y' = 3Ce^{3x} = 3y$ Now if we pose an additional condition $y(0) = 1$, the problem will have an algebraic and a geometric significance. The algebraic significance of this condition is that it specifies a value of the function $y(x)$ at $x = 0$. That is, $1 = y(0) = Ce^0 = C$. Hence, $C = 1$. Therefore, the problem posed by (3.2.1) will have a unique solution $y = e^{3x}$. The geometric significance of the condition $y(0) = 1$ is that it selects from the family of solutions just the one whose graph passes through the point $(0, 1)$.

Definition 3.2.1
The general form of a first-order differential equation is:

$$y' = F(x, y). \qquad (3.2.2)$$

The *differential* of a function $y = y(x)$ is defined as:

$$dy = y'dx. \qquad (3.2.3)$$

Note that based on (3.2.3), the differential equation (3.2.2) can be written in the differential form as:

$$dy = F(x, y)dx. \qquad (3.2.4)$$

Definition 3.2.2
A problem of the form:

$$\frac{dy}{dx} = F(x, y), \quad y(x_0) = y_0, \tag{3.2.5}$$

where x_0 is any value of the independent variable x, is called an *initial value problem (IVP)* for a first-order differential equation. The given value $y(x_0) = y_0$ is called the *initial condition*.

Theorem 3.2.1.
Consider the IVP (3.2.5). Assume that the functions $F(x, y)$ and $\partial F(x, y)/\partial y$ are continuous in some rectangle $R = \{(x, y): |x - x_0| \leq a, |y - y_0| \leq b\}, a, b > 0$. Then, there is a positive number δ such that the IVP has exactly one solution $y = y(x)$ in the interval $|x - x_0| \leq \delta$.

Proof:
The proof of this theorem can be found in almost any advanced text on differential equations, for example, see Hartman (2002).

Example 3.2.2
Let us show that IVP:

$$y' = x^3 + 5y^2, \quad y(0) = 0 \tag{3.2.6}$$

has a unique solution in some interval of the form $-\delta \leq x \leq \delta$.

Answer
Here in this example, $F(x, y) = x^3 + 5y^2$ and $\partial F(x, y)/\partial y = 10y$ are both continuous in any rectangle R about $(0, 0)$. Now by Theorem 3.2.1, there exists a positive number δ such that the IVP (3.2.6) has a unique solution $y = y(x)$ in the interval $|x - 0| \leq \delta$, that is, $-\delta \leq x \leq \delta$.

Example 3.2.3
By substitution, we can see that the functions $y_1(x) \equiv 0$ and $y_2(x) = x^2$ are two different solutions of the IVP:

$$y' = 4\sqrt{y}, \quad y(0) = 0. \tag{3.2.7}$$

Is this a violation of Theorem 3.2.1?

Answer
Here in this example,

$$F(x, y) = \sqrt{y} \quad \text{and} \quad \frac{\partial F(x, y)}{\partial y} = \frac{1}{2\sqrt{y}}.$$

However, $\partial F(x, y)/\partial y$ is not continuous at (0.0). Hence, one of the assumptions of Theorem 3.2.1 is violated and thus, the theorem does not apply, that is, the theorem has not been violated.

Example 3.2.4
Let us assume that the coefficient $p(x)$ and $q(x)$ of the first-order linear differential equation

$$y' + p(x)y = q(x) \tag{3.2.8}$$

are continuous in some open interval I. We want to show that the differential equation (3.2.8) has a unique solution through any point (x_0, y_0) such that $x_0 \in I$.

Answer
We choose a number a such that the interval $|x - x_0| \le a$ is contained in the interval I. Then,

$$F(x, y) = -p(x)y + q(x) \quad \text{and} \quad \frac{\partial F(x, y)}{\partial y} = p(x)$$

are continuous in the rectangle $R = \{(x, y): x_0 - a \le x \le x_0 + a, -\infty < y < \infty\}$. By Theorem 3.2.1, differential equation (3.2.8) has a unique solution at point (x_0, y_0).

3.3. SEPARABLE EQUATIONS

Definition 3.3.1
A first-order differential equation is said to be *separable* if it can be written in the form:

$$y' = \frac{dy}{dx} = f(x) \cdot g(x). \tag{3.3.1}$$

Example 3.3.1
The equation

$$\frac{dy}{dx} = \frac{3x^6 + x^2 y}{y^2 + 4}$$

is a separable differential equation because by factoring out of x^2 we will have:

$$\frac{3x^6 + x^2 y}{y^2 + 4} = x^2 \cdot \frac{3 + y}{y^2 + 4} = f(x) \cdot g(y),$$

where $f(x) = x^2$ and $g(y) = (3 + y)/(y^2 + 4)$.

Example 3.3.2
The equation

$$\frac{dy}{dx} = 2 - x^2 y$$

is not a separable differential equation because no factorization is possible and hence variables x and y cannot be separated.

3.3.1. Method of Solving Separable Differential Equations

To solve a separable differential equation such as (3.3.1), we multiply both sides of the equation by dx, divide both sides by $g(y)$, and obtain:

$$\frac{dy}{g(y)} = f(x)dx. \tag{3.3.2}$$

Now integrating both sides of (3.3.2), denoting $\int dy/g(y)$ by $G(y)$ and $\int f(x)dx$ by $F(x)$, we will have:

$$\int \frac{dy}{g(y)} = \int f(x)dx$$

or

$$G(x) = F(x) + C, \tag{3.3.3}$$

where C is an arbitrary constant of integration. The relation (3.3.3) gives an implicit solution to the differential equation (3.3.2).

Example 3.3.3
Let us solve the following nonlinear differential equation:

$$\frac{dy}{dx} = 3x(1 + y^2). \tag{3.3.4}$$

Solution
We first separate the variables and rewrite the equation (3.3.4) as:

$$\frac{dy}{1 + y^2} = 3x\,dx. \tag{3.3.5}$$

Integrating both sides of (3.3.5), we will have:

$$\int \frac{dy}{1 + y^2} = \int 3x\,dx$$

or

$$\tan^{-1} y = \frac{3x^2}{2} + C, \tag{3.3.6}$$

and solving (3.3.6) for y gives:

$$y = \tan\left(\frac{3x^2}{2} + C\right), \tag{3.3.7}$$

where C is an arbitrary constant of integration. Relation $y = y(x)$ found in (3.3.7) is the general solution of (3.3.4), that is, a family of solutions due to involving the arbitrary constant C.

Example 3.3.4

Let us solve the initial value problem:

$$\frac{dy}{dx} = -\frac{x}{y}, \quad y(3) = -4. \tag{3.3.8}$$

Solution

Separating the variable in (3.3.8) and putting it in a differential form yields the following:

$$y\, dy = -x\, dx. \tag{3.3.9}$$

Integrating both sides of (3.3.9) yields:

$$\int y\, dy = -\int x\, dx$$

or

$$\frac{y^2}{2} = -\frac{x^2}{2} + C \tag{3.3.10}$$

or

$$y^2 = -x^2 + C_1, \tag{3.3.11}$$

where C_1 is a new arbitrary constant after multiplying both sides of (3.3.10) by 2. Now applying the initial value $y(3) = -4$, we substitute $x = 3$ and $y = -4$ into (3.3.1.10). Hence, we will have $(-4)^2 = -(3)^2 + C_1$ or $C_1 = 25$. Thus, substituting $C_1 = 25$ in (3.3.11), we will have:

$$y^2 = -x^2 + 25. \tag{3.3.12}$$

To find y, we take the square root of (3.3.12) that yields:

$$y = \pm\sqrt{25 - x^2}. \tag{3.3.13}$$

Keeping the initial condition in mind, that is, when $x = 3$, y has to be negative (-4), we see that the unique solution of (3.3.8) is:

$$y = -\sqrt{25 - x^2}. \tag{3.3.14}$$

Example 3.3.5. *Logistic Equation*

Case 1: $\alpha = 1$.
We defined the logistic differential equation by (3.1.20). We now want to attempt to solve it. In this example, we solve (3.1.20) in a special case when $\alpha = 1$. In this case, the equation (3.1.20) reduces to:

$$\frac{dN(t)}{dt} = rN(t)[1 - N(t)]. \tag{3.3.15}$$

Hence, we want to solve (3.3.15) along with the initial condition, $N(0) = 5$, that is, we assume that the initial population size is 5.

Solution
We separate the variables and rewrite equation (3.3.15) as:

$$\frac{dN(t)}{[N(t)(1 - N(t))]} = r\,dt. \tag{3.3.16}$$

Integrating both sides of (3.3.16), using the partial fraction method, yields:

$$\int \frac{dN(t)}{N(t)(1 - N(t))} = \int \left(\frac{1}{N(t)} - \frac{1}{N(t) - 1} \right) dN(t) = \int r\,dt,$$

or

$$\ln|N(t)| - \ln|N(t) - 1| = rt + C,$$

or

$$\ln\left| \frac{N(t)}{N(t) - 1} \right| = rt + C,$$

or

$$\frac{N(t)}{N(t) - 1} = e^C \cdot e^{rt} = C_1 e^{rt}, \tag{3.3.17}$$

where C_1 is an arbitrary constant of integration.

Now using the initial condition, we can find C_1. Thus, substituting 0 for t in (3.3.17), we will have $N(0)/(N(0) - 1) = C_1 e^{r \cdot 0}$ from which $C_1 = 5/4$. Thus, (3.3.17) can be written as:

$$\frac{N(t)}{N(t)-1} = \frac{5}{4}e^{rt}. \tag{3.3.18}$$

From (3.3.18), solving for $N(t)$, we will have:

$$N(t) = \frac{5}{5 - 4e^{-rt}}, \tag{3.3.19}$$

which is the population size at time t. If $r > 0$, then as t increases indefinitely, we will have:

$$\lim_{t \to \infty} N(t) = \lim_{t \to \infty} \frac{5}{5 - 4e^{-rt}} = 1. \tag{3.3.20}$$

Therefore, as (3.3.20) suggests, the population tends toward the equilibrium $N(t) = 1$, which is typical behavior in many biological problems.

Figure 3.3.1 shows the behavior of the population growth in time. Each graph represents one solution according to the initial condition chosen. The first initial population, which must be positive, is chosen as $N(0) = 0.5$ and with increase of increments of 0.5 for other initial values. We have cut off the time units depending on how fast the population reaches its steady state. We also have chosen the initial time as $t = 1$. This time unit could very well represent the time 0 as the starting point.

Figure 3.3.2 shows as r increases how fast the convergence to equilibrium occurs, almost half of the previous case.

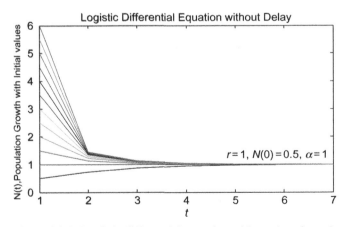

Figure 3.3.1. Logistic differential equation with $r = 1$ and $\alpha = 1$.

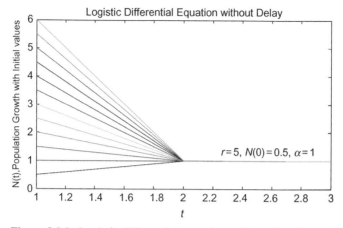

Figure 3.3.2. Logistic differential equation with $r = 5$ and $\alpha = 1$.

Case 2: $\alpha \neq 1$.

In the previous example, we solved logistic equation (3.1.20) when $\alpha = 1$. We leave it as an exercise to show that, similar to the case when $\alpha = 1$, we can find the solution of (3.1.20) with an initial value to be:

$$N(t) = \frac{\alpha N_0}{N_0 + (\alpha - N_0)e^{-rt}}, \tag{3.3.21}$$

where $N_0 = N(0)$ is the initial condition.

We note that, in this case, as the previous one, for $N_0 > 0$, $N(t) \to \alpha$ as $t \to \infty$ since $e^{-rt} \to 0$. Moreover, be reminded that when N is large so that $N/\alpha > 1$ in (3.1.20), the derivative dN/dt will be negative and so the population will decrease. This could very well be the case when $0 < \alpha < 1$.

Now through numerical examples, we want to concentrate on the behavior of the standard logistic differential equation (3.1.20), where r is the growth rate and α is the carrying capacity when the equation is being used to model population growth.

For the programming pursues, the mathematical software MATLAB is used. We will use an essential feature of the MATLAB code, which is the plotting of multiple solutions curves. We present different solutions, starting with an initial condition. We should expect that each solution curve, which is a unique solution, approaches the carrying capacity, α, as $t \to \infty$.

It should be noted that the population increases and approaches the equilibrium when the initial condition is less than the carrying capacity, α, and it decreases and approaches the equilibrium when the initial value is greater than α. If the initial condition is the same as α, the population stays steady as time increases. The role of r is the rate of convergence to the equilibrium position; the larger it is, the faster it will reach.

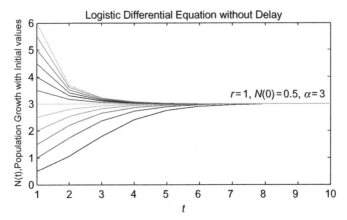

Figure 3.3.3. Logistic differential equation with $r = 1$ and $\alpha = 3$.

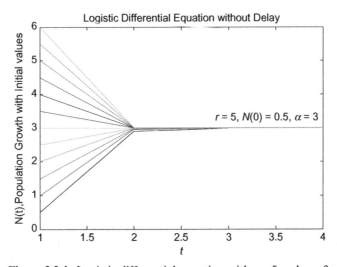

Figure 3.3.4. Logistic differential equation with $r = 5$ and $\alpha = 3$.

Now we generate the graphs of solutions for different values of α and r and different initial values $N(0)$. The results are shown in Figure 3.3.3, Figure 3.3.4, Figure 3.3.5, and Figure 3.3.6.

Introducing a *delay time*, say $\tau > 0$, in the time response of the logistic differential equation, the equation becomes:

$$\frac{dN(t)}{dt} = rN(t)\left[1 - \frac{N(t-\tau)}{\alpha}\right], \tag{3.3.22}$$

and it is called the *standard logistic differential equation with delay*, where $N(t)$ is the population size at a given time t and r and α are as before.

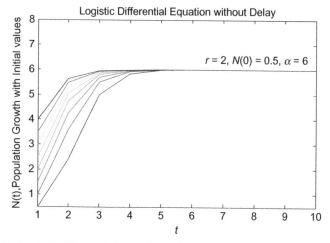

Figure 3.3.5. Logistic differential equation with $r = 2$ and $\alpha = 6$. All initial values are less than α. All solutions are increasing.

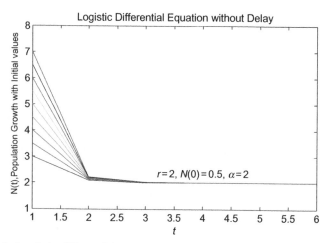

Figure 3.3.6. Logistic differential equation with $r = 2$ and $\alpha = 6$. All initial values are greater than α. All solutions are decreasing.

Equation (3.3.22) was first introduced by Hutchinson (1948). Since this paper was published, the delay logistic equation has been studied by many mathematicians. A particular feature factor of equation (3.3.22) is that its solutions exhibit oscillatory behaviors. The term $rN(t)$ describes the rise of the population to the level determined by the carrying capacity. As the population size increases, the growth is delayed by the term proportional to $N(t - \tau)$.

Gyori and Ladas (1991) established the necessary and sufficient conditions for the oscillation of every positive solution of the delay logistic equation (3.3.22). They first proved that every positive solution of (3.3.22) oscillates

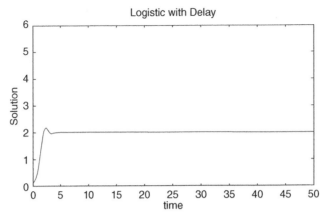

Figure 3.3.7. $r\tau > 1/e, r = 1, \alpha = 1, \tau = 1$.

about α as $(dy/dt) + r[e^{y(t-\tau)} - 1] = 0$ oscillates about 0. Then, they proved the following theorem:

Theorem 3.3.1
Suppose $r, \alpha, \tau \in (0, \infty)$. Then, every positive solution of (3.3.22) oscillates about the steady-state α if and only if $r\tau > 1/e$.

    We note from Theorem 3.3.1 that if $r\tau$ is less than or equal to $1/e$, then the solution would be a nonoscillating one.

Example 3.3.6. *Numerical Solution for Logistic Differential Equation with Delay*
In this example, we consider the logistic differential equation with delay (3.3.22). The analytic solution for a logistic equation with delay is not known. Hence, it is solved numerically. The mathematical software MATLAB does have some built-in programs that may be used and that is what we have done. Keeping Theorem 3.3.1 in mind, we have Figure 3.3.7 and Figure 3.3.8 showing behaviors of the solutions under different initial conditions. The vertical axis shows the initial conditions and the graph shows the population growth in time.

3.4. LINEAR DIFFERENTIAL EQUATIONS

Definition 3.4.1
A first-order differential equation is said to be *linear* if it can be written as:

$$y' + p(x)y = f(x). \tag{3.4.1}$$

A linear differential equation of the form (3.4.1) is called *homogeneous* if $f(x) = 0$; otherwise, it is said to be *nonhomogeneous*.

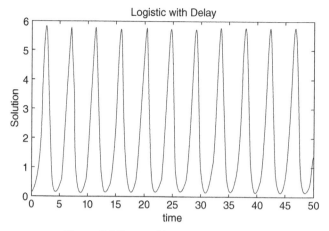

Figure 3.3.8. $r\tau > 1/e, r = 1, \alpha = 1, \tau = 2.$

Example 3.4.1
The equation:

$$\cos x \frac{dy}{dx} + (\sin x) y = 2$$

is a linear differential equation because it can be rewritten in the form (3.4.1)
as follows:

$$y' + (\tan x) y = 2 \sec x.$$

Example 3.4.2
The differential equation

$$\frac{dy}{dx} + (\sin x) y^2 = e^x$$

is not linear because of the term y^2.

3.4.1. Method of Solving a Linear First-Order Differential Equation

To solve the linear nonhomogeneous differential equation (3.4.1), we will
determine a function $\mu(x)$, called *the integrating factor*, such that after multiplying both sides of (3.4.1) by $\mu(x)$, that is,

$$\mu(x) y' + \mu(x) p(x) y = \mu(x) f(x), \tag{3.4.2}$$

the left-hand side of (3.4.2) is the derivative of the product $\mu(x) y$, that is,

$$\mu(x) y' + \mu(x) p(x) y = [\mu(x) y]' = \mu'(x) y + \mu(x) y'. \tag{3.4.3}$$

Therefore, comparison of the left-hand side and right-hand side of equalities in (3.4.3) yields:

$$\mu'(x) = \mu(x)p(x). \qquad (3.4.4)$$

To find such a function, $\mu(x)$, we rewrite (3.4.4) in the separable differential equations form and solve it with the method discussed in the previous section. Hence, we have:

$$\frac{d\mu(x)}{dx} = \mu(x) \cdot p(x)$$

or

$$\frac{d\mu(x)}{\mu(x)} = p(x)dx,$$

from which we have

$$\mu(x) = e^{\int p(x)dx}. \qquad (3.4.5)$$

With this choice for $\mu(x)$, equation (3.4.2) becomes:

$$[\mu(x)y]' = \mu(x)f(x), \qquad (3.4.6)$$

which has the solution

$$y(x) = \frac{1}{\mu(x)}\left[C + \int \mu(x)f(x)dx\right], \qquad (3.4.7)$$

where C is an arbitrary integration constant. The solution (3.4.7) is the general solution of linear differential equation (3.4.2).

We now summarize the method in the following steps:

Step 1. Write the given equation in the form (3.4.1).

Step 2. Calculate the integrating factor $\mu(x)$ using (3.4.5).

Step 3. Multiply the equation (3.4.2) by $\mu(x)$ and rewrite the equation obtained as in (3.4.6).

Step 4. Integrate the equation obtained in Step 3 and solve for $y = y(x)$ to obtain the general solution through formula (3.4.7).

Note: If $f(x) = 0$, then the equation (3.4.2) is homogeneous and from (3.4.7), the general solution will be:

$$y(x) = Ce^{-\int p(x)dx}. \qquad (3.4.8)$$

Example 3.4.3
Let us find the general solution of

$$(x^2 - 7)y' + 2y = 0. \tag{3.4.9}$$

Solution
Following the steps mentioned, we write given the differential equation (3.4.9) in the standard form (3.4.1) as:

$$y' + \frac{2x}{x^2 - 7}y = 0. \tag{3.4.10}$$

We see from (3.4.10) that $f(x) = 0$. Hence, the equation under consideration is a linear homogenous ODE of the first order. From (3.4.10), $p(x) = 2x/(x^2 - 7)$ and, hence, from (3.4.5), the integrating factor $\mu(x)$ is:

$$\mu(x) = e^{\int p(x)dx} = e^{\int \frac{2x}{x^2-7}dx} = e^{\ln(x^2-7)} = x^2 - 7. \tag{3.4.11}$$

Note that we have dropped the integration constant in the exponent in (4.4.11). This is because that constant can be combined with the constant we will have after another integration.
    We now multiply both sides of (3.4.10) by the integrating factor found in (3.4.11) and obtain the form (3.4.6) as:

$$[(x^2 - 7)y]' = 0. \tag{3.4.12}$$

Integrating both sides of (3.4.12) and solving for y, we will have the general solution of (3.4.9), following the form (3.4.8), as:

$$y(x) = \frac{C}{x^2 - 7}, \tag{3.4.13}$$

where C is an arbitrary constant.

Example 3.4.4
Let us find the general solution of:

$$\left(\frac{1}{x^2}\right)y' - \left(\frac{3}{x^3}\right)y = x\sin x, \quad x > 0. \tag{3.4.14}$$

Solution
To write the given equation in standard form (3.4.1), we multiply both side of (3.4.14) by x^2 and obtain the following:

$$y' - \frac{3}{x}y = x^3 \sin x. \tag{3.4.15}$$

As we see from (3.4.15), $p(x) = (-3/x)$ and $f(x) = x^3 \sin x$. Hence, $\int p(x)dx = -\int(3/x)dx = -3\ln x = \ln x^{-3}$. Thus, the integrating factor is $\mu(x) = e^{\int p(x)dx} = e^{\ln x^{-3}} = x^{-3}$. Multiplying both sides of equation (3.4.15) by $\mu(x)$, just found, we will have $[x^{-3}y]' = \sin x$. Now integrating both sides of this, we obtain:

$$x^{-3}y = \int \sin x \, dx = -\cos x + C, \qquad (3.4.16)$$

where C is an arbitrary constant. Solving (3.4.16) for y, we find the general solution as:

$$y = Cx^3 - x^3 \cos x. \qquad (3.4.17)$$

Example 3.4.5

Let us now solve the initial value problem:

$$y' - (\tan x)y = x^2 \sec x, \quad y(0) = 1. \qquad (3.4.18)$$

Solution

The given equation, (3.4.18), is already the standard form (3.4.1). Thus, $p(x) = -\tan x$ and $f(x) = x^2 \sec x$. Hence,

$$\int p(x)dx = -\int \tan x \, dx = -\int \frac{\sin x}{\cos x} dx = \ln(\cos x). \qquad (3.4.19)$$

Thus, the integrating factor is $\mu(x) = e^{\int p(x)dx} = e^{\ln(\cos x)} \cos x$. Following the aforementioned steps, we will have $[\cos xy]' = x^2$ and, hence, the general solution is:

$$y(x) = \frac{1}{\cos x}\left(C + \frac{x^3}{3}\right). \qquad (3.4.20)$$

Now applying the initial condition, $y(0) = 1$ on (3.4.20), we will have $C = 1$, and the particular solution to the given initial value problem (3.4.18) is thus:

$$y(x) = \frac{1}{\cos x}\left(1 + \frac{x^3}{3}\right) = \frac{x^3 + 3}{3\cos x}. \qquad (3.4.21)$$

3.5. EXACT DIFFERENTIAL EQUATIONS

In this section, we continue studying the first-order differential equations. However, we will write them differently so that we can solve them differently. In other words, we will try to find a special integrating factor to solve a certain type of first-order ODE. More specifically, we will try to write a given ODE in the form:

$$M(x, y)dx + N(x, y)dy = 0. \tag{3.5.1}$$

We may rewrite (3.5.1) so that one of x and y can be considered as the independent variable and the other as dependent variable. Hence,

$$M(x, y) + N(x, y)\frac{dy}{dx} = 0 \tag{3.5.2}$$

is the form in which x is the independent and y is the dependent variable, that is, $y = y(x)$. In contrast, we consider the form

$$M(x, y)\frac{dx}{dy} + N(x, y) = 0, \tag{3.5.3}$$

in which y is the independent and x is the dependent variable, that is, $x = x(y)$.

Definition 3.5.1
Let $x = f(x, y)$ be a function of two variables x and y with continuous first partial derivatives in a region R of the x–y plane. Then, the *differential of z* is:

$$dz = \frac{\partial f}{\partial x}dx + \frac{\partial f}{\partial y}dy. \tag{3.5.4}$$

We note that in a special case, if $f(x, y) = C$ (a constant), then:

$$\frac{\partial f}{\partial x}dx + \frac{\partial f}{\partial y}dy = 0. \tag{3.5.5}$$

Let us regard y as a function of x. We differentiate $f(x, y) = C$ with respect to x. This will yield the following equation:

$$\frac{\partial f(x, y)}{\partial x} + \frac{\partial f(x, y)}{\partial y}\frac{dy}{dx} = 0. \tag{3.5.6}$$

Now regard x as a function of y and differentiate $f(x, y) = C$ implicitly with respect to y. Hence, we will have:

$$\frac{\partial f(x, y)}{\partial x}\frac{dx}{dy} + \frac{\partial f(x, y)}{\partial y} = 0. \tag{3.5.7}$$

This means that $f(x, y) = C$ is an implicit solution of the differential equation (3.5.5) in either of its two forms (3.5.6) and (3.5.7).

Example 3.5.1

The function $f(x, y) = x^3 - 7xy + y^2 = C$ is an implicit solution of the equation

$$(3x^2 - 7y)dx + (-7x + 2y)dy = 0.$$

This is because

$$\frac{\partial f}{\partial x} = 3x^2 - 7y \quad \text{and} \quad \frac{\partial f}{\partial y} = (-7x + 2y).$$

Definition 3.5.2

A differential expression

$$M(x, y)dx + N(x, y)dy \qquad (3.5.8)$$

is called an *exact differential* in a region R of the plane if there is a function $f(x, y)$ such that:

$$\frac{\partial f}{\partial x} = M(x, y) \quad \text{and} \quad \frac{\partial f}{\partial y} = N(x, y), \quad \forall (x, y) \in R. \qquad (3.5.9)$$

Definition 3.5.3

A first-order differential equation of the form (3.5.1) is said to be an *exact differential equation* if the expression (3.5.8) is an exact differential.

Example 3.5.2

We want to show that the differential equation

$$2x \sin y \, dx + x^2 \cos y \, dy = 0 \qquad (3.5.10)$$

is an exact equation.

Answer

The left-hand side of (3.5.10) is an exact differential because

$$d(x^2 \sin y) = 2x \sin y \, dy + x^2 \cos y \, dy.$$

Let:

$$M(x, y) = 2x \sin y \quad \text{and} \quad N(x, y) = x^2 \cos y. \qquad (3.5.11)$$

Then, from (3.5.11), we have:

$$\frac{\partial M}{\partial y} = 2x \cos y = \frac{\partial N}{\partial x}.$$

Theorem 3.5.1. *Test for Exactness*

Suppose that the first partial derivatives of $M(x, y)$ and $N(x, y)$ are continuous in a rectangle R. Then, a necessary and sufficient condition that ODE (3.5.1) is exact in R is:

$$\frac{\partial M(x, y)}{\partial y} = \frac{\partial N(x, y)}{\partial x}. \tag{3.5.12}$$

Proof:

(a) *Necessity.* Let us assume that $M(x, y)$ and $N(x, y)$ have continuous first partial derivatives for all points (x, y) in the rectangle R. Now, if the equation (3.5.1) is an exact equation, then by definition (3.5.1) there exists some function $f(x, y)$ such that for all x in R, we have:

$$M(x, y)dx + N(x, y)dy = \frac{\partial f}{\partial x}dx = \frac{\partial f}{\partial y}dy.$$

Hence,

$$M(x, y) = \frac{\partial f}{\partial y} \quad \text{and} \quad N(x, y) = \frac{\partial f}{\partial x}. \tag{3.5.13}$$

It is a fact that for a function $f(x, y)$,

$$\frac{\partial^2 f}{\partial x \partial y} = \frac{\partial^2 f}{\partial y \partial x}. \tag{3.5.14}$$

See Theorem 3.1.1. Thus, differentiating the first equation of (3.5.13) with respect to y and the second with respect to x, we will have:

$$\frac{\partial^2 f}{\partial y \partial x} = \frac{\partial M}{\partial y} \quad \text{and} \quad \frac{\partial^2 f}{\partial x \partial y} = \frac{\partial N}{\partial x}.$$

Then, the necessity of exactness of (3.5.1) follows from (3.5.14).

(b) *Sufficiency.* To show the sufficiency of the Theorem (3.5.1), we have to show that there exists a function $f(x, y)$ for which (3.5.11) holds, that is,

$$\frac{\partial f}{\partial x} = M(x, y) \quad \text{and} \quad \frac{\partial f}{\partial y} = N(x, y),$$

whenever (3.5.12) holds. The construction of the function $f(x, y)$ is a basic procedure for solving exact equations. Therefore, integrating the first equation of (3.5.13) with respect to x yields:

$$f(x, y) = \int M(x, y)dx + g(y), \qquad (3.5.15)$$

where the arbitrary function $g(y)$ is the "constant" of integration. To determine $g(y)$, we differentiate both sides of (3.5.15) with respect to y and assume that $\partial f/\partial y = N(x, y)$. Hence, we have:

$$\frac{\partial f}{\partial y} = \frac{\partial}{\partial y} \int M(x, y)dx + g'(y) = N(x, y). \qquad (3.5.16)$$

Solving (3.5.16) for $g'(y)$ gives:

$$g'(y) = N(x, y) - \frac{\partial}{\partial y} \int M(x, y)dx. \qquad (3.5.17)$$

Finally, integrate (3.5.17) with respect to y and substitute the result in (3.5.15). The implicit solution of the equation is $f(x, y) = C$, where C is an arbitrary constant.

Now, there are two important questions. First, how can one be sure that $g'(x)$, given in (3.5.17), is really a function of just y alone? The second question is, how does one verify if $f(x, y)$ satisfies the condition (3.5.13)? We leave responses to these questions as exercises.

We now present an algorithm for solving an exact equation.

Step 1. Check that the equation (3.5.1) satisfies the exactness condition (3.5.12).

Step 2. Integrate $\partial f/\partial x = M(x, y)$ with respect to x to obtain:

$$f(x, y) = \int M(x, y)dx + g(y), \qquad (3.5.18)$$

where $g(y)$ is an unknown function.

Step 3. Take the partial derivative of both sides of (3.5.18) with respect to y and substitute $N(x, y)$ for $\partial f/\partial y$. We can now solve for $g'(y)$.

Step 4. Integrate $g'(y)$ to obtain $g(y)$ up to a constant. Substituting $g(y)$ into equation (3.5.17) gives $f(x, y)$.

Step 5. The solution of equation (3.5.1) is given implicitly by:

$$f(x, y) = C. \qquad (3.5.19)$$

Note that we can alternatively start with $\partial f/\partial y = N(x, y)$. The implicit solution, then, can be found by first integrating with respect to y.

Example 3.5.3
Let us solve the differential equation:

$$\left(3x^2 y - 2\right)dx + \left(4y^3 + x^3\right)dy = 0. \qquad (3.5.20)$$

Solution

In this case, we have:

$$M(x, y) = 3x^2 y - 2 \quad \text{and} \quad N(x, y) = 4y^3 + x^3. \tag{3.5.21}$$

Thus, from (3.1.21) we have:

$$\frac{\partial M(x, y)}{\partial y} = 3x^2 = \frac{\partial N(x, y)}{\partial x}.$$

Hence, equation (3.5.20) is exact. We now integrate $M(x, y)$ with respect to x and follow Step 2 to find $f(x, y)$. Thus:

$$f(x, y) = \int (3x^2 y - 2) dx + g(y) = x^3 y - 2x + g(y). \tag{3.5.22}$$

Now we take partial derivative of (3.5.22) with respect to y and substitute $N(x, y) = 4y^3 + x^3$ for $\partial f(x, y)/\partial y$ as follows:

$$\frac{\partial f(x, y)}{\partial y} = x^3 + g'(y) = 4y^3 + x^3. \tag{3.5.23}$$

Thus, from (3.5.23), comparing corresponding terms in equivalent polynomials, we have:

$$g'(y) = 4y^3. \tag{3.5.24}$$

Integrating both sides of (3.5.24), we will have:

$$g(y) = y^4. \tag{3.5.25}$$

Note that since the choice of constant of integration will not matter, we choose it to be equal to zero. Therefore, from (3.5.22) and (3.5.25), we have the solution as:

$$f(x, y) = x^3 y - 2x + y^4. \tag{3.5.26}$$

Hence, from (3.5.26), the solution to equation (3.5.20) is given implicitly by the relation:

$$x^3 y - 2x + y^4 = C,$$

where C is an arbitrary constant.

Example 3.5.4

We now want to solve the initial value problem:

$$\left(2x + e^{\frac{x}{y}}\right) dx + e^{\frac{x}{y}}\left(1 - \frac{x}{y}\right) dy = 0, \quad y(0) = 1. \tag{3.5.27}$$

Solution

From (3.5.27), we have:

$$M(x, y) = 2x + e^{\frac{x}{y}} \quad \text{and} \quad N(x, y) = e^{\frac{x}{y}}\left(1 - \frac{x}{y}\right). \tag{3.5.28}$$

Thus, we have:

$$\frac{\partial M(x, y)}{\partial y} = -\frac{x}{y^2}e^{\frac{x}{y}} = \frac{\partial N(x, y)}{\partial x},$$

and, hence, equation (3.5.27) is exact. Now following the rest of steps, to find $f(x, y)$, we integrate $M(x, y)$ with respect to x and follow Step 2 to find $f(x, y)$. Thus:

$$f(x, y) = \int\left(2x + e^{\frac{x}{y}}\right)dx + g(y) = x^2 + ye^{\frac{x}{y}} + g(y). \tag{3.5.29}$$

Now we take the partial derivative of (3.5.29) with respect to y, and setting the result equal to $N(x, y)$ gives:

$$e^{\frac{x}{y}} - \frac{x}{y}e^{\frac{x}{y}} + g'(y) = e^{\frac{x}{y}}\left(1 - \frac{x}{y}\right).$$

Thus:

$$g'(y) = 0, \quad \text{and then } g(y) = \int 0\,dy = C = 0. \tag{3.5.30}$$

From (3.5.29) and (3.5.30) we will have:

$$f(x, y) = x^2 + ye^{\frac{x}{y}}.$$

Therefore, the general solution of the equation (3.5.27) is given by the relation:

$$x^2 + ye^{\frac{x}{y}} = C, \tag{3.5.31}$$

where C is an arbitrary constant. Now apply the initial condition $y_0 = 1$, when $x_0 = 0$ demands that $0 + e^0 = C$ and, thus, $C = 1$. An implicit solution of the IVP is, then, $x^2 + ye^{x/y}$.

So far, in this section, we have been discussing exact differential equations. Now let us see what happens if a given differential equation is not exact.

Definition 3.5.4. *Integrating Factor*
If the differential equation given in (3.5.1) is not exact, but the following equation

$$\mu(x, y)M(x, y)dx + \mu(x, y)N(x, y)dy = 0, \qquad (3.5.32)$$

which results from multiplying equation (3.5.1) by the function $\mu(x, y)$ is exact, then $\mu(x, y)$ is called an *integrating factor* of the differential equation (3.5.1).
    By Theorem 3.5.1, equation (3.5.32) is exact if and only if:

$$\frac{\partial[\mu(x, y)M(x, y)]}{\partial y} = \frac{\partial[\mu(x, y)N(x, y)]}{\partial x}. \qquad (3.5.33)$$

By the product rule of differentiation (using subscript for partial differentiation and dropping x and y from M, N, and μ, for ease of writing), the equation (3.5.33) will be:

$$\mu M_y + \mu_y M = \mu N_x + \mu_x N$$

or

$$\mu_x N - \mu_y M = (M_y - N_x)\mu. \qquad (3.5.34)$$

If μ is a function of only one variable, say $\mu = \mu(x)$, then:

$$\mu_x = \frac{d\mu}{dx} = \mu', \quad \mu_y = 0.$$

Hence, (3.5.34) can be written as:

$$\mu' = \frac{M_y - N_x}{N}\mu. \qquad (3.5.35)$$

If the quotient in (3.5.35) is just a function of x, then since (3.5.35) is a homogeneous linear equation and by formula (3.4.8), μ can determined as:

$$\mu(x) = e^{\int \frac{M_y - N_x}{N} dx}. \qquad (3.5.36)$$

Likewise, it follows from (3.5.34) that if the quotient $Nx - My/M$ is just a function of y, then the integrating factor is

$$\mu(y) = e^{\int \frac{N_x - M_y}{M} dy}. \qquad (3.5.37)$$

We summarize these special cases in the following theorem.

Theorem 3.5.2
If $(My - Nx)/M$ is continuous and depend only on x, then the integrating factor $\mu = \mu(x)$ is given by (3.5.36). If $(Nx - My)/M$ is continuous and depends only on y, then the integrating factor is given by (3.5.37).

Example 3.5.5
We want to solve the IVP:

$$\left(\frac{\cos x}{x^2} + 3y\right)dx + x\,dy = 0, \quad y(\pi) = 2 \tag{3.5.38}$$

by finding an appropriate integrating factor.

Solution
From (3.5.38), we have:

$$M(x, y) = \frac{\cos x}{x^2} + 3y \text{ and } N(x, y) = x. \tag{3.5.39}$$

Thus, we have:

$$\frac{\partial M(x, y)}{\partial y} = 3 \text{ and } \frac{\partial N(x, y)}{\partial x} = 1.$$

Hence, equation (3.5.38) is not exact. Now the quotient:

$$\frac{M_y - N_x}{N} = \frac{3 - 1}{x} = \frac{2}{x}$$

depends only on x. Thus, by Theorem 3.5.2, the integrating actor $\mu = \mu(x)$ is:

$$\mu(x) = e^{\int \frac{2}{x}dx} = e^{\ln x^2} = x^2. \tag{3.5.40}$$

After we multiply the given differential equation (3.5.38) by $\mu = x^2$, the resulting equation is:

$$\left(\cos x + 3x^2 y\right)dx + x^3 dy = 0. \tag{3.5.41}$$

Equation (3.5.41) is exact because:

$$M_1(x, y) = \cos x + 3x^2 y, \, N_1(x, y) = x^3, \quad \text{and} \quad \frac{\partial M_1(x, y)}{\partial y} = 3x^2 = \frac{\partial N_1(x, y)}{\partial x}.$$

To find $f(x, y)$, we integrate $N_1(x, y)$ with respect to y. Thus,

$$f(x, y) = \int x^3 dy + g(y) = x^3 y + g(x), \tag{3.5.42}$$

where $g(x)$ is an arbitrary function. Taking the partial derivatives of (3.5.42) with respect to x and setting the result equal to $M_1(x, y)$ gives:

$$3x^2 y + g'(y) = \cos x + 3x^2 y. \tag{3.5.43}$$

Thus, from (3.5.43) we have:

$$g'(x) = \cos x, \text{ and then } g(x) = \int \cos x \, dx = \sin x. \tag{3.5.44}$$

From (3.5.42) and (3.5.44), we have:

$$f(x, y) = x^3 y + \sin x. \tag{3.5.45}$$

Hence, the general solution is:

$$x^3 y + \sin x = C, \tag{3.5.46}$$

where C is an arbitrary constant. If we apply initial condition $y = 2$ when $x = \pi$ to (3.5.46), we will have $C = 2\pi^3$. Hence, the implicit solution to the IVP (3.5.36) is:

$$x^3 y + \sin x - 2\pi^3 = 0.$$

3.6. SOLUTION OF THE FIRST ODE BY SUBSTITUTION METHOD

Let us go back to the differential equation (3.5.1) when it is not separable, linear, or exact. However, it may be transformed to a form that can be solved. In this section, we sill study three types of equations that can be transformed into either separable or linear equations.

We recall that in Definition 3.4.1 we defined a linear differential equation that is "homogenous." We want to caution the reader as we are about to define "homogenous" differential equation in a different sense.

Definition 3.6.1
Consider the differential equation:

$$\frac{dy}{dx} = f(x, y). \tag{3.6.1}$$

If $f(x, y)$ in (3.6.1) can be expressed as a function of the ratio y/x, say $F(y/x)$, then equation (3.6.1) is called *homogeneous*.

Example 3.6.1

Let us investigate if the equation

$$(x^2 + 3y^2)dx + x^2 dy = 0 \qquad (3.6.2)$$

is homogeneous.

Answer

Equation (3.6.2) can be written as:

$$\frac{dy}{dx} = -\frac{x^2 + 3y^2}{x^2} = -\left[1 + 3\left(\frac{y}{x}\right)^2\right] \equiv F\left(\frac{y}{x}\right), \qquad (3.6.3)$$

where $F(u) = (1 + 3u^2)$. Thus, through (3.6.3), we were able to write (3.6.2) in the form of (3.6.1), in which $f(x, y)$ is of the form $F(y/x)$. Hence, (3.6.2) is homogeneous.

Example 3.6.2

Let us investigate if the equation:

$$(2x + 3y + 1)dx + (x + y)dy = 0 \qquad (3.6.4)$$

is homogeneous.

Answer

Equation (3.6.4) can be written as:

$$\frac{dy}{dx} = -\frac{2x + 3y + 1}{x + y} = -\frac{2 + 3\dfrac{y}{x} + \dfrac{1}{x}}{1 + \dfrac{y}{x}}. \qquad (3.6.5)$$

We can see that (3.6.5) cannot be expressed as a function of y/x alone because of the term $1/x$ in the numerator. Hence, equation (3.6.4) is not homogeneous.

3.6.1. Substitution Method

Expressing a differential equation as a function of y/x alone gives us the idea of the solution of a homogeneous equation using a substitution

$$u(x) = \frac{y(x)}{x}, \qquad (3.6.6)$$

where we may use $u = u(x)$ and $y = y(x)$, where ever there is no ambiguity. Hence, the substitution method is as follows. From $u = y/x$, we have $y = ux$. Differentiating this product, we will have:

$$\frac{dy}{dx} = u + x\frac{du}{dx}. \qquad (3.6.7)$$

Substitute dy/dx from (3.6.7) in (3.6.1), we will obtain:

$$u + x\frac{du}{dx} = F(u). \qquad (3.6.8)$$

Equation (3.6.8) is separable and hence an implicit solution can be found through:

$$\int \frac{1}{F(u) - u}\, du = \int \frac{1}{x}\, dx. \qquad (3.6.9)$$

Therefore, the solution of (3.6.1) will be obtained by back substitution.

Alternatively, the solution of (3.6.1) can be obtained by rewriting the differential equation as:

$$\frac{dx}{dy} = \frac{1}{f(x, y)} = G\left(\frac{x}{y}\right). \qquad (3.6.10)$$

Now, we substitute

$$v = \frac{x}{y} \text{ or } x = vy, \text{ where } v = v(y) \text{ and } x = x(y), \qquad (3.6.11)$$

and the corresponding derivative

$$\frac{dx}{dy} = v + y\frac{dv}{dy} \qquad (3.6.12)$$

into equation (3.6.10). We leave it as an exercise for the reader to see that after some simplification, the resulting differential equation is separable.

Example 3.6.3
Let us solve the equation

$$\left(4x^2 + y^2 + xy^2\right)dx - x\, dy = 0. \qquad (3.6.13)$$

Solution
Equation (3.6.13) is none of the forms separable, linear, or exact. It may be rewritten in the following derivative form:

$$\frac{dy}{dx} = \frac{4x^2 + y^2 + xy}{x^2} = 4 + \left(\frac{y}{x}\right)^2 + \frac{y}{x}. \tag{3.6.14}$$

Since the right-hand side of equation (3.6.14) is a function of just y/x, the equation (3.6.13) is homogeneous. Substituting equation (3.6.6) and equation (3.6.7) into equation (3.6.14), we obtain:

$$u + x\frac{du}{dx} = 4 + u^2 + u. \tag{3.6.15}$$

Equation (3.6.15) is separable and, hence, upon separating the variables and using the integrating method, we obtain:

$$\int \frac{1}{4+u^2} du = \int \frac{1}{x} dx$$

or

$$\frac{1}{2}\tan^{-1}\frac{u}{2} = \ln|x| + C$$

or

$$u = 2\tan(2\ln|x| + C_1), \quad C_1 = 2C. \tag{3.6.16}$$

Substituting $u = y/x$ into (3.6.16), we find the explicit solution of the given equation (3.6.13), as desired, which is:

$$y = 2x\tan(\ln x^2 + C_1).$$

Definition 3.6.2. *Bernoulli Equation*
The differential equation:

$$\frac{dy}{dx} + p(u)y = f(x)y^n, \tag{3.6.17}$$

where $p(x)$ and $f(x)$ are continuous on an interval I and n is a real number, is called a *Bernoulli equation*.

Note that for $n = 0$ and $n = 1$, equation (3.6.17) is linear. For $n \neq 0$, and $n \neq 1$, the substitution

$$u = y^{1-n} \tag{3.6.18}$$

reduces any equation of the form (3.6.17) to a linear equation, as we will show later.

Dividing equation (3.6.17) by y^n yields:

$$y^{-n}\frac{dy}{dx}+p(x)y^{1-n}=f(x).\qquad(3.6.19)$$

We find from (3.6.18) and the chain rule that:

$$\frac{du}{dx}=(1-n)y^{-n}\frac{dy}{dx},\qquad(3.6.20)$$

and hence equation (3.6.19) becomes:

$$\frac{1}{1-n}\frac{du}{dx}+p(x)u=f(x).$$

Equation (3.6.20) is linear because $1/1-n$ is a constant.

Example 3.6.4
Let us solve the initial value problem:

$$\frac{dy}{dx}-y=xy^3,\quad y(0)=1.\qquad(3.6.21)$$

Solution
The equation (3.6.21) is a Bernoulli equation with $n=3$. We divide both side of (3.6.21) by y^3 to obtain:

$$y^{-3}\frac{dy}{dx}-y^{-2}=-x.\qquad(3.6.22)$$

Next, we make the substitution $u=y^{1-n}=y^{1-3}=y^{-2}$. Since

$$\frac{du}{dx}=-2y^{-3}\frac{dy}{dx},$$

the equation (3.6.22) becomes:

$$-\frac{1}{2}\frac{du}{dx}-u=-x$$

or

$$\frac{du}{dx}+2u=2x.\qquad(3.6.23)$$

Equation (3.6.23) is linear, so we can solve it for u using the method mentioned in Section 3.4 and formulas (3.4.5) and (3.4.7). Thus:

$$u = Ce^{-2x} + x - \frac{1}{2}. \qquad (3.6.24)$$

Substituting $u = y^{-2} = 1/y^2$, relation (3.6.24) gives the solution of (3.6.21) as:

$$y = \pm \frac{1}{\sqrt{Ce^{-2x} + x - \frac{1}{2}}}. \qquad (3.6.25)$$

Keeping the initial condition in mind, that is, when $x = 0$, y has to be positive 1. Hence, $y(0) = 1$, we obtain $1 = 1/\sqrt{Ce^0 + 0 - 1/2}$, from which we will have $C = 3/2$. Therefore, the explicit solution of the IVP (3.9.21) is:

$$y = \frac{1}{\sqrt{\frac{3}{2}e^{-2x} + x - \frac{1}{2}}}. \qquad (3.6.26)$$

3.6.2. Reduction to Separation of Variables

A differential equation of the form

$$\frac{dy}{dx} = f(ax + by) \qquad (3.6.27)$$

can always be reduced to an equation with separable variables by substitution:

$$u = ax + by. \qquad (3.6.28)$$

The method is illustrated in the next example.

Example 3.6.5
Let us solve the equation:

$$\frac{dy}{dx} = 6x - 2y + 1 + (y - 3x + 2)^2. \qquad (3.6.29)$$

Solution
The right-hand side of (3.6.29) can be expressed as a function of $y - 3x$, that is,

$$6x - 2y + 1 + (y - 3x + 2)^2 = -2(y - 3x) + 1 + [(y - 3x) + 2]^2.$$

If we let $u = y - 3x$, then $du/dx = (dy/dx) - 3$, and hence the differential equation (3.6.29) is transformed into:

$$\frac{du}{dx} + 3 = -2u + 1 + (u+2)^2$$

or

$$\frac{du}{dx} = u^2 + 2u + 2. \tag{3.6.30}$$

Equation (3.6.30) is separable. To solve it, we write:

$$\frac{du}{u^2 + 2u + 2} = dx,$$

and integrate, that is,

$$\int \frac{1}{u^2 + 2u + 2} du = \int dx.$$

Thus:

$$\tan^{-1}(u+1) = x + C.$$

Finally, replacing u by $y - 3x$ yields:

$$\tan^{-1}(y - 3x + 1) = x + C$$

or

$$y = 3x - 1 + \tan(x + C). \tag{3.6.31}$$

Therefore, relation (3.6.31) is the explicit solution of (3.6.29).

3.7. APPLICATIONS OF THE FIRST-ORDER ODEs

In this section, we offer some examples where the first-order differential equations are used. Since these examples somehow deal with time, the independent variable is denoted by the letter t.

Example 3.7.1. *Growth and Decay*
A model for diverse phenomena involving either growth or decay is:

$$\frac{dy(t)}{dt} = ky(t), \quad y(t_o) = y_0, \tag{3.7.1}$$

where k is a constant of proportionality. When $k > 0$, the model represents "growth" and when $k < 0$, it represents "decay." Of course, $k = 0$ is not of our concern in this model since there is no change in that case. We saw in Example 3.1.10 that in biological applications, the rate of growth of certain populations such as bacteria, small animals, and people over a short period of time is proportional to the population present at time t. Knowing the population size at a given point t_0 will allow us to use model (3.7.1) to predict the future population size.

We give two particular cases for the model (3.7.1) as follows:

Example 3.7.1a
Let us apply the Malthusian model given in (3.1.17) to the demographic history of the United States of America. It is known that in 1900 the population of the United States was about 76,094,000. The census shows that it grew to about 248,464,000 in 1990. We want to use the Malthusian model to estimate the population of the United States as a function of time.

Answer
By comparing differential equation (3.7.1) and equation (3.1.17), we see that in (3.7.1), if we replace $y(t)$ by $N(t)$ and taking the values of k in mind, as discussed earlier, then $dy(t)/dt = ky(t)$ becomes (3.1.17). Hence, we set $t_0 = 1900$, $y_0 = N(1900) = 76,094,000$, and $y_1 = N(1900) = 249,464,000$. The constant of proportionality k may be calculated given these values. Now, from (3.1.17), we can write:

$$\frac{dN(t)}{dt} - kN(t) = 0, \tag{3.7.2}$$

which is a linear homogeneous or separable differential equation. The solution of (3.7.2) can be obtained from (3.4.8) as:

$$N(t) = Ce^{\int k \, dt} = Ce^{kt}, \tag{3.7.3}$$

where C is a constant. Note that the solution given in (3.1.18) is now confirmed by (3.7.3). To find C, if we choose the year 1900 as the base year and let it be 0, then 1990 becomes 90. Hence, by given information, we have $N(0) = 76,094,000 = Ce^0 = C$. Thus:

$$N(t) = 76094000e^{kt}, \tag{3.7.4}$$

where $N(t)$ is the population at any time t. We now calculate k using the additional information given to us, simply $N(90) = 249,464,000$. Thus:

$$N(90) = 249,464,000 = 76,094,000e^{90k},$$

from which we have:

$$\ln(249464000) = \ln(76094000) + 90k$$

or

$$k = \frac{\ln(249464000) - \ln(76094000)}{90} = 0.0131927244. \tag{3.7.5}$$

Therefore, from (3.7.4) and (3.7.5), we have:

$$N(t) = 76094000e^{0.0131927244t}. \tag{3.7.6}$$

For instance, the population of one more year after 1990, will be $N(91) \approx 252,780,000$. This is very close to the population announced as about 252,150,000.

Example 3.7.1b
As our second particular growth example, we consider the following biological case. A culture initially starts with $B(0) = B_0$ bacteria. After an hour, the number of bacteria increases to $B(1) = 5/4B_0$. Let us assume that the rate of growth of bacteria is proportional to the number of bacteria $B(t)$ at time t. We want to determine the time it will take for the number of bacteria to increase to three times as the original number.

Answer
We replace y in (3.7.1) by $B(t)$ and solve for it. Similar to Example 3.7.1a, we will have the homogeneous linear or separable differential equation:

$$\frac{dB(t)}{dt} - kB(t) = 0, \tag{3.7.7}$$

whose solution is

$$B(t) = Ce^{\int k\,dt} = Ce^{kt}. \tag{3.7.8}$$

Hence, using the initial number of bacteria, we will have the solution as:

$$B(t) = B_0 e^{kt}. \tag{3.7.9}$$

Using other information given, we will have $B(1) = 5/4B_0 = B_0 e^k$, or $e^k = 5/4$, from which we have $k = \ln 5/4 = 0.223$ and from (3.7.9), we will have:

$$B(t) = B_0 e^{0.223t}. \tag{3.7.10}$$

Now, to have the number of bacteria to be three times as much as the initial number, we will have:

$$3B_0 = B_0 e^{0.223t}. \tag{3.7.11}$$

Solving t from (3.7.11), we will have $t = \ln 3/0.223 \approx 4.93$ hours.

Example 3.7.2. *Newton's Law of Cooling*
A pot of liquid is put on the stove to boil. The temperature of the liquid reaches 180°F and then the pot is taken off the burner and placed on a counter in the kitchen. The temperature of the air in the kitchen is 75°F. After 2 minutes, the temperature of the liquid in the pot is 132°F. How long before the temperature of the liquid in the pot will be 85°F?

Solution
Let us denote by $T(t)$ the temperature of the liquid at time t (in minutes). According to the Newton's law of cooling (3.1.14), we have:

$$\frac{dT}{dt} = -k(T - T_0),$$

where $T_0 = 75°F$ (the temperature of the air in the kitchen), with initial conditions $T(0) = 180°F$. Thus:

$$\frac{dT}{dt} = -k(T - 75), \quad T(0) = 180. \tag{3.7.12}$$

Solving the initial value problem (3.7.12), we will have:

$$\frac{dT}{T - 75} = -k\,dt,$$

from which we will have:

$$\ln|T - 75| = -kt + C$$

or

$$|T - 75| = e^{-kt} e^c,$$

or

$$T(t) = 75 + C_1 e^{-kt}. \tag{3.7.13}$$

Now applying the initial condition to (3.7.13), we will have:

$$180 = T(0) = 75 + C_1 e^{-k \cdot 0}.$$

This implies that $C_1 = 105$ and, hence,

$$T(t) = 75 + 105 e^{-kt}. \tag{3.7.14}$$

To find k, we use the fact that after 2 minutes the temperature of the liquid is 132°F. Thus, from (3.7.14), we have:

$$132 = T(2) = 75 + 105 e^{-2k},$$

from which we will have $k = -1/2 \ln(19/35) \approx 0.3054545412$, and hence from (3.7.14) we will have:

$$T(t) = 75 + 105 e^{-0.3054545412t}. \tag{3.7.15}$$

Finally, if $T(t) = 85$, then from (3.7.15) we find the time t as follows:

$$85 = 75 + 105 e^{-0.3054545412t}$$

or

$$t \approx 7.7 \text{ minutes.}$$

Example 3.7.3. *Newton's Law of Cooling*
Suppose that a cup of hot coffee at 93°C is left in a room with a temperature of 22°C. After 5 minutes, the coffee's temperature drops to 78°C. Using the Newton's cooling law, we want to find the time it will take for the coffee's temperature to drop to 48°C.

Solution
Let us denote the coffee's initial temperature, coffee's temperature at time t, and the room temperature by $T(0)$, $T(t)$, and T_0, respectively. Hence, from the assumption, we have $T(0) = 93$, $T_0 = 22$, and the initial value problem is:

$$\frac{dT}{dt} = -k(T - 22), \quad T(0) = 93. \tag{3.7.16}$$

To solve (3.7.16), we rewrite it as:

$$\frac{dT}{T - 22} = -k \, dt$$

and integrate this with respect to t to obtain:

$$\ln|T - 22| = -kt + C. \tag{3.7.17}$$

Thus, from (3.7.17), we have:

$$T(t) = 22 + C_1 e^{-kt}. \tag{3.7.18}$$

Using the initial condition to find the constant of integration (rewriting it as $C_1 = e^C$), we will have:

$$93 = T(0) = 22 + C_1 e^{-k},$$

from which we will have $C_1 = 71$ and, consequently,

$$T(t) = 22 + 71e^{-5k}. \tag{3.7.19}$$

Now using the fact that in 5 minutes the coffee's temperature dropped 15 degrees from 93°C to 78°C, from (3.7.19) we will have:

$$78 = T(5) = 22 + 71e^{-5k},$$

from which we will have $k = -1/5 \ln \frac{56}{71} \approx 0.0474656373$, and hence from (3.7.19), we will have the temperature function as:

$$T(t) = 22 + 71e^{-0.0474656373t}. \tag{3.7.20}$$

For the time required for the temperature to get to 48°C, we will have:

$$48 = 22 + 71e^{-0.0474656373t}$$

or

$$t \approx 21 \text{ minutes and 27 seconds.}$$

3.8. SECOND-ORDER HOMOGENEOUS ODE

Because of their many applications in science, technology, and engineering, the second-order differential equations are very important. The second-order differential equations has historically been the most studied class of differential equations and has kept their importance in research today in such areas as Newton's second law (force equals mass time acceleration, $F = ma$), which is the most commonly encountered differential equation in practice. The formula $F = ma$ is a second-order differential equation because acceleration is the second derivative of position function, denoted by y, with respect to time, that is, $a = d^2y/dt^2$. See, for instance, Example 3.1.9, when the second law was applied to a mass–spring oscillator and the resulting motion experienced energy for life. We will study this class of differential equations in this section.

Definition 3.8.1

A *second-order differential equation* is called *linear* if it can be written as:

$$y'' + p(x)y' + q(x)y = f(x). \tag{3.8.1}$$

It is called *homogeneous* if $f(x) \equiv 0$ and *nonhomogeneous* if $f(x) \not\equiv 0$

Note that terms defined in Definition 3.8.1 for second-order ODE are similar to those in Definition 3.4.1 for the first-order ones, such as equation (3.4.1). Hence, it is natural to expect similarities in properties for (3.4.1) and (3.8.1). However, solving (3.8.1) is more difficult than (3.4.1). For instance, we found a general solution for a nonhomogeneous linear first-order equation given by (3.4.7) and for a homogenous one by (3.4.8). However, there is no explicit solution formula for the second linear one. Thus, we are bind to solve (3.8.1) in special cases.

Definition 3.8.2

A *linear second-order differential equation with constant coefficient* is an equation of the form:

$$ay'' + by' + cy = f(x), \tag{3.8.2}$$

where $a \neq 0$, and b and c are constants. It is called *homogeneous* if $f(x) \equiv 0$ and *nonhomogeneous* if $f(x) \not\equiv 0$

We note that the linear homogenous second-order ODE with constant coefficient arises in studying the mass–spring oscillators vibrating freely, that is, without external forces applied. See Example 3.1.9. We will study this case in more detail later.

3.8.1. Solving a Linear Homogeneous Second-Order Differential Equation

$$ay'' + by' + cy = 0. \tag{3.8.3}$$

Solution

We start by trying a solution of the form:

$$y = e^{rx}, \quad r = \text{constant}. \tag{3.8.4}$$

Substituting (3.8.4) into (3.8.3) yields:

$$ar^2 e^{rx} + bre^{rx} + ce^{rx} = e^{rx}\left(ar^2 + br + c\right) = 0. \tag{3.8.5}$$

Since, $e^{rx} > 0$, $\forall x$, dividing both sides of (3.8.5) by e^{rx} will yield:

$$ar^2 + br + c = 0. \tag{3.8.6}$$

Hence, $y = e^{rx}$ is a solution of (3.8.1) if and only if r satisfies (3.8.6).

Definition 3.8.3
Equation (3.8.6) is called the *characteristic equation* (or auxiliary equation) associated with the second-order linear homogeneous equation (3.8.3).

Example 3.8.1
We want to find a pair of solutions for:

$$y'' - 5y' + 6y = 0. \tag{3.8.7}$$

Solution
The characteristic equation for the equation (3.8.7) is:

$$r^2 - 5r + 6 = (r - 2)(r - 3) = 0,$$

which has two solutions: $r_1 = 2$ and $r_2 = 3$. Thus, $y = e^{2x}$ and $y = e^{3x}$ are solutions of (3.8.7).

Theorem 3.8.1
If $y_1(x)$ and $y_2(x)$ are two solutions of the second-order linear homogeneous differential equation (3.8.3), then the linear combination of these solutions, that is,

$$y(x) = c_1 y_1(x) + c_2 y_2(x), \tag{3.8.8}$$

where c_1 and c_2 are arbitrary constants, is also a solution.

Proof:
Substituting (3.8.8) in (3.8.3) yields the following:

$$\begin{aligned}
ay'' + by' + cy &= a(c_1 y_1 + c_2 y_2)'' + b(c_1 y_1 + c_2 y_2)' + c(c_1 y_1 + c_2 y_2) \\
&= a(c_1 y_1'' + c_2 y_2'') + b(c_1 y_1' + c_2 y_2') + c(c_1 y_1 + c_2 y_2) \\
&= c_1(ay_1'' + by_1' + cy_1) + c_2(ay_2'' + by_2' + cy_2) \\
&= c_1 \cdot 0 + c_2 \cdot 0 = 0.
\end{aligned}$$

Thus, $y(x) = c_1 y_1(x) + c_2 y_2(x)$ is a solution of (3.8.3), as was to prove.

Definition 3.8.4
A problem of the form:

$$ay'' + by' + cy = 0, \quad y(x_0) = k_0,\, y'(x_0) = h_0, \tag{3.8.9}$$

where x_0 is any value of the independent variable x, and k_0 and h_0 are given numbers, is called an *initial value problem (IVP)* for a homogeneous second-order differential equation. The given values k_0 and h_0 are called the *initial conditions*.

Theorem 3.8.2. *Existence and Uniqueness Theorem*
Consider the initial value problem (3.8.9). For real constants a, bc, k_0, and h_0, there exists a unique solution $y(x)$, $x \in (-\infty, \infty)$, for the initial value problem given in (3.8.9).

Proof:
See Hartman (2002, p. 326).
   Note that if $y_0 = y_0' = 0$, then $y(x) \equiv 0$ because the trivial solution is unique.

Definition 3.8.5
A pair of solutions $y_1(x)$ and $y_2(x)$ for (3.8.3) are called *linearly independent* on an interval I if and only if neither $y_1(x)$ nor $y_2(x)$ is a constant multiple of the other on I. Otherwise, they are called *linearly dependent* on I.

Theorem 3.8.3. *Linearity Condition*
Let $y_1(x)$ and $y_2(x)$ be a pair of solutions for (3.8.3). Let $W(x)$ be defined as:

$$W(x) = \begin{vmatrix} y_1(x) & y_2(x) \\ y_1'(x) & y_2'(x) \end{vmatrix} = y_1(x)\,y_2'(x) - y_1'(x)\,y_2(x). \tag{3.8.10}$$

Let also x_0 be any number on $(-\infty\ \infty)$. Then, if:

$$W(x_0) = y_1(x_0)\,y_2'(x_0) - y_1'(x_0)\,y_2(x_0) = 0, \tag{3.8.11}$$

then $y_1(x)$ and $y_2(x)$ are linearly dependent on $(-\infty\ \infty)$.
   Note that the function $W(x)$ given by (3.8.10) is called the *Wronskian* of $y_1(x)$ and $y_2(x)$.

Proof:
We consider three cases:

Case 1: $y_1(x_0) \neq 0$.
In this case, we let $\tau = y_2(x_0)/y_1(x_0) = $ constant. We, then, choose a solution of (3.8.3) as:

$$y(x) = \tau y_1(x). \tag{3.8.12}$$

The chosen solution (3.8.12) satisfies the same initial conditions at $x = x_0$ as $y_2(x)$ does, that is,

$$y(x_0) = \tau y_1(x_0) = \frac{y_2(x_0)}{y_1(x_0)} \cdot y_1(x_0) = y_2(x_0)$$

and by (3.8.11):

$$y'(x_0) = \tau y_1'(x_0) = \frac{y_2(x_0)}{y_1(x_0)} \cdot y_1'(x_0) = \frac{y_1(x_0)}{y_1(x_0)} \cdot y_2'(x_0) = y_2'(x_0).$$

By the uniqueness of the solution of the initial value problem (3.8.9), Theorem 3.8.2, $y_2(x) \equiv y(x)$ or $y_2(x) = \tau y_1(x)$. Thus, $y_1(x)$ and $y_2(x)$ are linearly dependent solutions of (3.8.3) on $(-\infty\ \infty)$.

Case 2: $y_1(x_0) = 0$ but $y_1'(x_0) \neq 0$.
For this case, from (3.8.11) we will have $y_2(x_0) = 0$. We now let $\tau = y_2'(x_0)/y_1'(x_0) = $ constant. Then, the solution of (3.8.3) given by (3.8.12) satisfies the same initial conditions at x_0 as $y_2(x)$ does, that is,

$$y(x_0) = \frac{y_2'(x_0)}{y_1'(x_0)} \cdot y_1(x_0) = 0 = y_2(x_0)$$

and

$$y'(x_0) = \frac{y_2'(x_0)}{y_1'(x_0)} \cdot y_1'(x_0) = y_2'(x_0).$$

By the uniqueness of the solution of initial value problem (3.8.9), Theorem 3.8.2, we have $y_2(x) = \tau y_1(x)$. Thus, $y_1(x)$ and $y_2(x)$ are linearly dependent solutions of (3.8.3) on $(-\infty, \infty)$.

Case 3: $y_1(x_0) = 0$ and $y_1'(x_0) = 0$.
In this case, as we mentioned earlier, $y_1(x) \equiv 0$ and $0 = y_1(x) = 0 \cdot y_2(x)$, that is, $y_1(x)$ and $y_2(x)$ are linearly dependent solutions of (3.8.3) on $(-\infty, \infty)$.

Theorem 3.8.4
If $y_1(x)$ and $y_2(x)$ are the only two linearly independent solutions of (3.8.3) on $(-\infty, \infty)$, and $y(x)$ is a solution of the initial value problem (3.8.9), then the unique constants c_1 and c_2 can be found such that $y(x) = c_1 y_1(x) + c_2 y_2(x)$.

Proof:
By Theorem 3.8.1 we have that $y(x) = c_1 y_1(x) + c_2 y_2(x)$ is a solution of (3.8.3). To find c_1 and c_2 we have to solve the system of two equations:

$$\begin{cases} y(x_0) = c_1 y_1(x_0) + c_2 y_2(x_0) = k_0 \\ y'(x_0) = c_1 y_1'(x_0) + c_2 y_2'(x_0) = h_0 \end{cases} \tag{3.8.13}$$

for c_1 and c_2. After some algebra and simplification, we can see that the solutions of (3.8.13) are

$$c_1 = \frac{k_0 y_2'(x_0) - h_0 y_2(x_0)}{y_1(x_0) y_2'(x_0) - y_1'(x_0) y_2(x_0)} \text{ and } c_2 = \frac{h_0 y_1(x_0) - k_0 y_1'(x_0)}{y_1(x_0) y_2'(x_0) - y_1(x_0) y_2'(x_0)}.$$
$$\tag{3.8.14}$$

Since $y_1(x)$ and $y_2(x)$ are linearly independent on $(-\infty, \infty)$, then by Theorem 3.8.3:

$$W(x_0) = y_1(x_0) y_2'(x_0) - y_1'(x_0) y_2(x_0) \neq 0$$

and the denominator in (3.8.14) is nonzero.

Theorem 3.8.5
If $y_1(x)$ and $y_2(x)$ are the only two linearly independent solutions of (3.8.3) on $(-\infty, \infty)$, then $y(x) = c_1 y_1(x) + c_2 y_2(x)$ (formula 3.8.8) is the general solution of (3.8.3).

Proof:
Let $Y(x)$ be any solution of (3.8.3), $x_0 \in (-\infty, \infty)$, $Y(x_0) = k_0$, and $Y'(x_0) = h_0$. By Theorem 3.8.4, we can find c_1 and c_2 such that the function $c_1 y_1(x) + c_2 y_2(x) = y(x)$ is a solution of the initial value problem (3.8.9). By the uniqueness of the solution, Theorem 3.8.2, we have $Y(x) \equiv y(x)$, that is, formula (3.8.8) is the general solution of equation (3.8.3).

Now the question is, how do we find a general solution of (3.8.3)? According to our previous discussions, we first find the roots of the characteristic equation (3.8.6) associated with the homogeneous equation (3.8.3). These roots are:

$$r = \frac{-b \pm \sqrt{b^2 - 4ac}}{2a}. \tag{3.8.15}$$

There are three cases based on the value of the discriminate in (3.8.15). These are: (1) $b^2 - 4ac > 0$, (2) $b^2 - 4ac = 0$, and (3) $b^2 - 4ac < 0$. Roots of the characteristic equation in case (1) will be two distinct reals, in case (2) will be a repeated root, and in case (3) will be two complex conjugates. We will consider these cases in the following examples.

Example 3.8.2

Case 1: $b^2 - 4ac = 0$.
Consider the linear homogeneous second-order differential equation

$$y'' + 5y' + 6y = 0. \tag{3.8.16}$$

We want to find:

(a) the general solution of (3.8.16) and

(b) the particular solution of (3.8.16) under the initial conditions $y(0) = 1$ and $y'(0) = 2$, given that $y(x)$ is the general solution of (3.8.16) found in part (a).

Solutions

(a) The characteristic equation of (3.8.16) is:

$$r^2 = 5r + 6 = (r+2)(r+3) = 0. \tag{3.8.17}$$

There are two distinct real roots $r_1 = -2$ and $r_2 = -3$. Hence, $y_1(x) = e^{-2x}$ and $y_2(x) = e^{-3x}$ are a pair of solutions for (3.8.16). Since the ratio $y_2(x)/y_1(x) = e^{-3x}/e^{-2x} = e^{-x}$ is not a constant, the solutions $y_1(x)$ and $y_2(x)$ are linearly independent. Therefore, by Theorem 3.8.4, the general solution of (3.8.16) is:

$$y(x) = c_1 e^{-2x} + c_2 e^{-3x}, \tag{3.8.18}$$

where c_1 and c_2 are arbitrary constants.

(b) To find a particular solution, we will have to find c_1 and c_2 that satisfy the initial conditions given by the problem. Hence, differentiating (3.8.18) yields:

$$y'(x) = -2c_1 e^{-2x} - 3c_2 e^{-3x}. \tag{3.8.19}$$

Applying the initial conditions on (3.8.19), we will have the following system of two equations with two unknowns c_1 and c_2:

$$\begin{cases} y(0) = 1 = c_1 + c_2 \\ y'(0) = 2 = -2c_1 - 3c_2. \end{cases} \tag{3.8.20}$$

Clearly, the solution of the system (3.8.20) is $c_1 = 5$ and $c_2 = -4$, two distinct reals. Therefore, the particular solution satisfying the given initial condition is:

$$y(x) = 5e^{-2x} - 4e^{-3x}.$$

Theorem 3.8.6

If the characteristic equation (3.8.6) has two distinct real roots, say r_1 and r_2, then each of $y_1(x) = e^{r_1 x}$ and $y_2(x) = e^{r_2 x}$ is a solution of (3.8.3) and the general solution of (3.8.3) is:

$$y(x) = c_1 e^{r_1 x} + c_2 e^{r_2 x}, \tag{3.8.21}$$

where c_1 and c_2 are arbitrary constants.

Proof:
Since the ratio $y_2(x)/y_1(x) = e^{r_2 x}/e^{r_1 x} = e^{(r_2 - r_1)x}$ is not a constant, the solutions $y_1(x)$ and $y_2(x)$ are linearly independent. Therefore, by Theorem 3.8.4, the statement of the theorem is true.

Example 3.8.3

Case 2: $b^2 - 4ac = 0$.
Consider the linear homogeneous second-order differential equation

$$y'' - 6y' + 9y = 0. \tag{3.8.22}$$

We want to find:

(a) the general solution of (3.8.22) and
(b) the particular solution of (3.8.22) under the initial conditions $y(0) = -3$, and $y'(0) = 5$, given that $y(x)$ is the general solution of (3.8.22) found in part (a).

Solutions
(a) The characteristic equation of (3.8.22) is:

$$r^2 - 3r + 9 = (r + 3)^2 = 0. \tag{3.8.23}$$

Equation (3.8.23) has a double root, $r = 3$. Hence, $y_1(x) = e^{3x}$ is a solution for (3.8.22). Since the characteristic equation has no other root than 1, the solution $y(x)$, $y(x) = e^{rx}$, is the only form of solution for (3.8.23). Thus, we will look at a solution of the form:

$$y(x) = u(x) y_1(x) = u(x) e^{3x}. \tag{3.8.24}$$

Later (in Section 3.9) we will discuss the method of variation of parameters to explain the choice of $y(x)$ given in (3.8.24). The function $u(x)$ in (3.8.24) must be determined. To do this, we differentiate (3.8.24) twice with respect to x. Hence, we will have:

$$y'(x) = u'(x) e^{3x} + 3u(x) e^{3x} \tag{3.8.25}$$

and

$$y''(x) = u''(x) e^{3x} + 6u'(x) e^{3x} + 9u(x) e^{3x}. \tag{3.8.26}$$

Hence (dropping the x's for simplicity) substituting (3.8.25) and (3.8.26) in (3.8.22), we will have:

$$y'' - 6y' + 9y = u''e^{3x} + 6u'e^{3x} + 9ue^{3x} - 6\left(u'e^{3x} + 3ue^{3x}\right) + ue^{3x}$$

$$= u''e^{3x}. \tag{3.8.27}$$

Therefore, $y = ue^{3x}$ is a solution of (3.8.22), if and only if (3.8.27) is zero, that is, $u''(x) = 0$, which is equivalent to $u(x) = c_1 x + c_2$, where c_1 and c_2 are arbitrary constants. If we choose $c_1 = 0$ and $c_2 = 1$, then $y_2(x) = ue^{3x}$ is a solution of (3.8.22). Since the ratio $y_2(x)/y_1(x) = xe^{3x}/e^{3x} = x$ is not a constant, the solutions $y_1(x)$ and $y_2(x)$ are linearly independent. Thus, by Theorem 3.8.4, the general solution of (3.8.22) is:

$$y(x) = c_1 e^{3x} + c_2 x e^{3x} = e^{3x}(c_1 + c_2 x). \tag{3.8.28}$$

(b) Differentiating (3.8.28) yields:

$$y'(x) = 3c_1 e^{3x} + c_2 e^{3x} + 3c_2 x e^{3x} = 3e^{3x}(c_1 + c_2 x) + c_2 e^{3x}. \tag{3.8.29}$$

Now applying the initial conditions on (3.8.28) and (3.8.29), we will have the following system of two equations with two unknowns c_1 and c_2:

$$\begin{cases} y(0) = -3 = c_1 \\ y'(0) = 5 = 3c_1 + c_2. \end{cases} \tag{3.8.30}$$

The solution of the system (3.8.30) is $c_1 = -3$ and $c_2 = 14$; two distinct reals. Therefore, the particular solution satisfying the given initial conditions is:

$$y(x) = -3e^{3x} + 14 + xe^{3x} = e^{3x}(14x - 3).$$

Theorem 3.8.7
If the characteristic equation (3.8.6) has a repeated root, say r_0, then each of $y_1(x) = e^{r_0 x}$ and $y_2(x) = xe^{r_0 x}$ is a solution of (3.8.3), and

$$y(x) = c_1 e^{r_0 x} + c_2 x e^{r_0 x} = e^{r_0 x}(c_1 + c_2 x). \tag{3.8.31}$$

is the general solution of (3.8.3), where c_1 and c_2 are arbitrary constants.

Proof:
From the assumption of the theorem, we have:

$$ar^2 + br + c = a(r - r_0)^2 = a(r^2 - 2rr_0 + r_0^2) = 0. \tag{3.8.32}$$

Since $a \neq 0$, dividing both sides of (3.8.32) by a yields:

$$r^2 - 2rr_0 + r_0^2 = 0. \tag{3.8.33}$$

From (3.8.33), it follows that the equation (3.8.3) can be written as:

$$y'' - 2r_0 y' + r_0^2 y = 0. \tag{3.8.34}$$

Proceeding as in Example 3.8.3, we will look for other solutions of (3.8.34) of the form:

$$y_2(x) = u(x)e^{r_0 x}. \tag{3.8.35}$$

As mentioned in Example 3.8.3, we must determine the function $u(x)$. To do this, we differentiate (3.8.35) twice with respect to x. Hence, we will have:

$$y_2'(x) = u'(x)e^{r_0 x} + r_0 u(x)e^{r_0 x} \tag{3.8.36}$$

and

$$y_2''(x) = u''(x)e^{r_0 x} + 2r_0 u'(x)e^{r_0 x} + r_0^2 u(x)e^{r_0 x}. \tag{3.8.37}$$

Hence, substituting (3.8.36) and (3.8.37) in (3.8.34) (dropping the x's for convenience), we will have:

$$\begin{aligned} y_2'' - 2r_0 y' + r_0^2 y &= u''e^{r_0 x} + 2r_0 u'e^{r_0 x} + r_0^2 u e^{r_0 x} \\ &- 2r_0 \left(u'e^{r_0 x} + r_0 u e^{r_0 x} \right) + r_0^2 u e^{r_0 x} = u''e^{r_0 x}. \end{aligned} \tag{3.8.38}$$

Therefore, $y_2 = u e^{r_0 x}$ is a solution of (3.8.34) if and only if $u''(x) = 0$, since it will make (3.8.38) zero. Thus, $u(x) = c_1 + c_2 x$. If we choose $c_1 = 0$ and $c_2 = 1$, then $y_2(x) = x e^{rx}$ is a solution of (3.8.3). Since the ratio $y_2(x)/y_1(x) = x e^{rx}/e^{rx} = x$ is not a constant, the solutions $y_1(x)$ and $y_2(x)$ are linearly independent. Thus, by Theorem 3.8.4, the general solution of (3.8.3) is:

$$y(x) = c_1 e^{r_0 x} + c_2 x e^{r_0 x} = e^{r_0 x}(c_1 + c_2 x).$$

Example 3.8.4

Case 3: $b^2 - 4ac < 0$.

Theorem 3.8.8
If the characteristic equation (3.8.6) has two complex conjugate roots $r_1 = \alpha + i\beta$ and $r_2 = \alpha - i\beta$, then each of $y_1(x) = e^{\alpha x} \cos \beta x$ and $y_2(x) = e^{\alpha x} \sin \beta x$ is a solution of (3.8.3) and

$$y(x) = e^{\alpha x}(c_1 \cos \beta x + c_2 \sin \beta x) \tag{3.8.39}$$

is the general solution of (3.8.3), where c_1 and c_2 are arbitrary constants.

***Proof*:**
Formally, there is no difference between this case and Case 1. Thus,

$$y(x) = c_1 e^{(\alpha+i\beta)x} + c_2 e^{(\alpha-i\beta x)} \tag{3.8.40}$$

is the general solution of (3.8.3), where c_1 and c_2 are arbitrary constants. However, in practice, we prefer to work with real functions rather than complex functions. Hence, we use a consequence of Euler's rule, that is, $e^{i\beta x} + e^{-i\beta x} = 2 \cos \beta x$, and $e^{i\beta x} - e^{-i\beta x} = 2i \sin \beta x$, and two choices of c_1 and c_2 as $c_1 = c_2 = 1/2$ and $c_1 = 1/2i$, $c_2 = -1/2i$ (since the constants in the general solution are arbitrary). Hence, using these quantities from (3.8.40) we have:

$$y_1(x) = \frac{1}{2} e^{\alpha x} \left(e^{i\beta x} + e^{-i\beta x} \right) = e^{\alpha x} \cos \beta x \tag{3.8.41}$$

and

$$y_2(x) = \frac{1}{2i} e^{\alpha x} \left(e^{i\beta x} - e^{-i\beta x} \right) = e^{\alpha x} \sin \beta x. \tag{3.8.42}$$

Thus, $y_1(x)$ and $y_2(x)$, given by (3.8.41) and (3.8.42), respectively, are two linearly independent real solutions of (3.8.3) and, therefore, by Theorem 3.8.4, the general solution is (3.8.39).

Example 3.8.5

Case 4: $b^2 - 4ac < 0$.
Consider the linear homogeneous second-order differential equation

$$y'' - 2y' + 2y = 0. \tag{3.8.43}$$

We want to find:

(a) the general solution of (3.8.43) and
(b) the particular solution of (3.8.43) under the initial conditions $y(0) = 2$, and $y'(0) = -1$, given that $y(x)$ is the general solution of (3.8.43) found in part (a).

Solutions
(a) The characteristic equation of (3.8.43) is:

$$r^2 - 2r + 2 = 0. \tag{3.8.44}$$

Roots of (3.8.44) are $r_1 = 1 + i$ and $r_2 = 1 - i$. Hence, $\alpha = \beta = 1$, in this case. Now, by (3.8.39), the general solution is:

$$y(x) = e^x(c_1 \cos x + c_2 \sin x). \tag{3.8.45}$$

(b) To find a particular solution, we will have to find c_1 and c_2 that satisfy the initial conditions given by the problem. Hence, differentiating (3.8.45) yields:

$$y'(x) = e^x(c_1 \cos x + c_2 \sin x) + e^x(-c_1 \sin x + c_2 \sin x) \tag{3.8.46}$$

or

$$y'(x) = e^x[c_1(\cos x - \sin x) + c_2(\sin x + \cos x)]. \tag{3.8.47}$$

Applying the initial conditions on (3.8.45) and (3.8.46), we will have the following system of two equations with two unknowns c_1 and c_2:

$$\begin{cases} y(0) = 2 = c_1 \\ y'(0) = -1 = c_1 + c_2. \end{cases} \tag{3.8.48}$$

From the system (3.8.47), we have $c_1 = 2$ and $c_2 = -3$. Therefore, the particular solution satisfying the given initial conditions is:

$$y(x) = e^x(2\cos x - 3\sin x).$$

3.9. THE SECOND-ORDER NONHOMOGENEOUS LINEAR ODE WITH CONSTANT COEFFICIENTS

We will now consider the nonhomogeneous linear second-order equation with constant coefficients that is of the form

$$ay'' + by' + cy = f(x), \tag{3.9.1}$$

where the function $f(x)$ is not identically zero. The following theorem gives sufficient conditions for the existence and uniqueness of solutions of an initial value problem for (3.9.1).

Theorem 3.9.1. *Sufficient Condition*
Consider the following initial value problem:

$$ay'' + by' + cy = f(x), \quad y(x_0) = k_0, \ y'(x_0) = h_0, \tag{3.9.2}$$

where $f(x)$ is a continuous function on an open interval I, c_1 and c_2 are arbitrary constants, and $y(x_0) = k_0$ and $y'(x_0) = h_0$ are given initial values that are real numbers. Then there exists a unique solution $y(x), x \in I$, of this problem.

Proof:
See Hartman (2002, p. 326).

In order to find the general solution of (3.9.1) where $f(x)$ is continuous on I, it is necessary to find the general solution of the associated homogeneous equation:

$$ay'' + by' + cy = 0. \tag{3.9.3}$$

The homogenous equation (3.9.3) is part of (3.9.1) and is called the *complementary equation* for (3.9.1). The following theorem shows how to find the general solution of (3.9.1) if (1) we know one solution, say $y_p(x)$, of (3.9.1) and (2) two linearly independent solutions $y_1(x)$ and $y_2(x)$ of the homogeneous equation (3.9.3). $y_p(x)$ is called a *particular solution* of (3.9.1).

Theorem 3.9.2
Suppose $f(x)$ is a continuous function on an interval I. Let $y_p(x)$ be a particular solution of (3.9.1) and $y_1(x)$ and $y_2(x)$ be two linear independent solutions of complementary equations of (3.9.3). Denote by $y_c(x)$ the general solution of the complimentary equation of (3.9.1), that is,

$$y_c(x) = c_1 y_1(x) + c_2 y_2(x). \tag{3.9.4}$$

Then, a continuous function $y(x)$ is a solution of (3.9.1) on I if and only if:

$$y(x) = y_p(x) + y_c(x) = y_p(x) + c_1 y_1(x) + c_2 y_2(x). \tag{3.9.5}$$

Proof:
Let us first show that $y(x)$ is a solution of (3.9.1) for any choice of the constants c_1 and c_2. Differentiating (3.9.5) twice (dropping x for convenience) yields the followings:

$$y' = y_p' + c_1 y_1' + c_2 y_2' \quad \text{and} \quad y'' = y_p'' + c_1 y_1'' + c_2 y_2''. \tag{3.9.6}$$

Hence,

$$\begin{aligned}
ay'' + by' + cy &= a\left(y_p'' + c_1 y_1'' + c_2 y_2''\right) + b\left(y_p' + c_1 y_1' + c_2 y_2'\right) + c\left(y_p + c_1 y_1 + c_2 y_2\right) \\
&= \left(ay_p'' + by_p' + cy_p\right) + c_1\left(ay_1'' + by_1' + cy_1\right) + c_2\left(ay_2'' + by_2' + cy_2\right) \\
&= f(x) + c_1 \cdot 0 + c_2 \cdot 0 \\
&= f(x).
\end{aligned}$$

Now, we will show that every solution of (3.9.1) is of the form (3.9.5). Suppose that $y(x)$ is a solution of (3.9.1). We define:

$$Y(x) = y(x) - y_p(x). \tag{3.9.7}$$

Then (again dropping the x), substituting (3.9.7) in (3.9.1) we will have:

$$
\begin{aligned}
aY'' + bY + cY &= a(y - y_p)'' + b(y - y_p)' + c(y - y_p) \\
&= (ay'' + by' + cy) - (ay_p'' + by_p' + cy_p) \\
&= f(x) - f(x) = 0,
\end{aligned}
$$

since $y(x)$ and $y_p(x)$ both satisfy (3.9.1). This shows that $Y(x)$ is a solution of the homogeneous equation (3.9.3). Hence, for $y_1(x)$ and $y_2(x)$ to be two linear independent solutions of complementary equations (3.9.3), by Theorem 3.8.4,

$$
Y(x) = c_1 y_1(x) + c_2 y_2(x) = y_c(x). \tag{3.9.8}
$$

Then, from (3.9.6) and (3.9.8) we obtain (3.9.5). Equation (3.9.5) is called the *general solution* of (3.9.1) on I.

The following theorem enables us to break a nonhomogeneous equation into simpler parts, find a particular solution for each part, and then combine these particular solutions to obtain a particular solution of the original problem.

Theorem 3.9.3. *The Principle of Superposition*

Let us consider the following second-order nonhomogeneous differential equations with constant coefficients:

$$
ay'' + by' + cy = f_i(x), \quad i = 1, 2. \tag{3.9.9}
$$

Suppose that two particular solutions y_{p_i}, $i = 1, 2$, on I are found. Then,

$$
y_p(x) = y_{p_1} + y_{p_2} \tag{3.9.10}
$$

is a particular solution of:

$$
ay'' + by' + cy = f_1(x) + f_2(x) \tag{3.9.11}
$$

on I.

Proof:

Substituting (3.9.10) into (3.9.11) yields:

$$
\begin{aligned}
ay_p'' + by_p' + cy_p &= a(y_{p_1} + y_{p_2})'' + b(y_{p_1} + y_{p_2})' + c(y_{p_1} + y_{p_2}) \\
&= a(y_{p_1}'' + by_{p_1}' + cy_{p_1}) + a(y_{p_2}'' + by_{p_2}' + cy_{p_2}) \\
&= f_1(x) + f_2(x).
\end{aligned}
$$

In this section, we examine two methods for obtaining a particular solution of the nonhomogeneous equation (3.9.1).

3.9.1. Method of Undetermined Coefficients

The underlying idea in undetermined coefficients method is guessing the form of a particular solution, $y_p(x)$, that is motivated by the kinds of functions that compare the nonhomogeneous part $f(x)$. The method of undetermined coefficients is not applicable to equations of the form (3.9.1) when $f(x)$ has such forms as $f(x) = 1/x, \ln x, \cos^{-1}x, \tan x$, and so on. Differential equations, in which $f(x)$ is a function of these kinds, will be considered in the next subsection. However, it does apply when $f(x)$ has such forms as $f(x) = 17, 2x^2 - 3x + 1,$ $(3x - 1)e^{2x}, \sin 3x - 4x \cos 3x, e^{-x}(\cos 2x - 5x^2 \sin 2x)$, and so on, that is, $f(x)$ is a linear combination of functions of the type $p(x) = a_n x^n + a_{n-1}x^{n-1} + \ldots + a_1 x + a_0, p(x)e^{\alpha x}, p(x)e^{\alpha x} \cos \beta x, p(x)e^{\alpha x} \sin \beta x$, where n is a nonnegative integer and α and β are real numbers.

The set of functions that consist of constants, polynomials, exponentials $e^{\alpha x}$, sines, and cosines has the property that derivatives of their sums and products are again sums and products of constants, polynomials, exponentials $e^{\alpha x}$, sines, and cosines. Since the linear combination of derivatives $ay_p'' + by_p' + cy_p$ must be identical to $f(x)$, it seems reasonable to assume that y_p has the same form as $f(x)$.

The next four examples illustrate the basic method.

Example 3.9.1
Let us solve the equation:

$$y'' - y' - 2y = 3x^2 - 2x + 6. \qquad (3.9.12)$$

Solution
We will show steps to solve the problem.

Step 1
Solve the associated homogeneous equation:

$$y'' - y' - 2y = 0. \qquad (3.9.13)$$

The characteristic equation for (3.9.13) is a quadratic equation $r^2 - r - 2 = 0$ that has two distinct real roots $r_1 = -1$ and $r_2 = 2$. Hence, the general solution for the complimentary equation (3.9.13) is:

$$y_c(x) = c_1 e^{-x} + c_2 e^{2x}. \qquad (3.9.14)$$

Step 2
Now, since the function $f(x)$ is a quadratic polynomial, we assume that a particular solution is also in the form of a quadratic polynomial

$$y_p(x) = Ax^2 + Bx + C. \qquad (3.9.15)$$

We need to determine the constants A, B, and C such that y_p, defined in (3.9.15), is a solution of (3.9.12). Thus, from (3.9.15), substituting y_p, its first derivative $y'_p = 2A + B$ and its second derivative $y''_p = 2A$ into the equation (3.9.12), we get:

$$\begin{aligned} y''_p - y'_p - 2y_p &= 2A - (2Ax + B) - 2(Ax^2 + Bx + C) \\ &= -2Ax^2 - (2A + 2B)x + (2A - B - 2C) \quad\quad (3.9.16) \\ &\equiv 3x^2 - 2x + 6. \end{aligned}$$

Equating the corresponding coefficient of like powers of the two sides of (3.9.16), we will have the following system of three equations with three unknowns A, B, and C:

$$\begin{cases} -2A = 3 \\ -(2A + 2B) = -2 \\ 2A - B - 2C = 6. \end{cases} \quad\quad (3.9.17)$$

Solving the system of equations (3.9.17) yields $A = -3/2$, $B = 5/2$, and $C = '-23/4$. Thus, a particular solution will be:

$$y_p(x) = -\frac{3}{2}x^2 + \frac{5}{2}x - \frac{23}{4}. \quad\quad (3.9.18)$$

Step 3
Therefore, from (3.9.14) and (3.9.18), we can write the general solution of (3.9.12) as:

$$y(x) = y_c(x) + y_p(x) = c_1 e^{-x} + c_2 e^{2x} - \frac{3}{2}x^2 + \frac{5}{2}x - \frac{23}{4}.$$

Example 3.9.2
Let us solve the equation:

$$y'' + y' - 2y = 3\sin 2x. \quad\quad (3.9.19)$$

Solution
To solve (3.9.19), we follow the steps in Example 3.9.1. Thus, we start with solving the associated homogeneous equation:

$$y'' + y' - 2y = 0. \qu\quad (3.9.20)$$

The characteristic equation of (3.9.20) is $r^2 + r - 2 = 0$ that can be factored as $r^2 + r - 2 = (r - 1)(r + 2) = 0$. The roots of this polynomial equation are $r_1 = 1$ and $r_2 = -2$. Hence, a natural first guess for a particular solution would be $A \sin 2x$. But, since the derivative of $\sin 2x$ is $\cos 2x$, we assume that a particular solution would include both $\sin 2x$ and $\cos 2x$. Thus, we choose our particular solution as:

$$y_p = A\cos 2x + B\sin 2x. \qquad (3.9.21)$$

Twice differentiating y_p given by (3.9.21) and substituting the results into equation (3.9.20) gives the following:

$$y_p' = -2A\cos 2x + 2B\sin 2x,$$
$$y_p'' = -4A\cos 2x - 4B\sin 2x,$$

and

$$\begin{aligned}
y_p'' + y_p' - 2y_p &= (-4A\cos 2x - 4B\sin 2x) \\
&\quad + (-2A\cos 2x + 2B\sin 2x) \\
&\quad - 2(A\cos 2x + B\sin 2x) \\
&\equiv 0 \cdot \cos 2x + 3\sin 2x.
\end{aligned} \qquad (3.9.22)$$

Again, equating the corresponding coefficient of both sides, we will have the following system of equations:

$$\begin{cases} -6A + 2B = 0 \\ -2A - 6B = 3. \end{cases} \qquad (3.9.23)$$

Solving the system (3.9.23) we will obtain $A = -3/20$ and $B = 9/20$. Thus, a particular solution of the equation (3.9.19) will be:

$$y_p = -\frac{3}{20}\cos 2x - \frac{9}{20}\sin 2x. \qquad (3.9.24)$$

Therefore, the general solution of (3.9.20) would be:

$$y(x) = y_c(x) + y_p(x) = c_1 e^x + c_2 e^{-2x} - \frac{3}{20}\cos 2x - \frac{9}{20}\sin 2x.$$

Example 3.9.3
Let us solve the equation:

$$y'' - 4y' + 5y = 2x + 1 + 3xe^x. \qquad (3.9.25)$$

Solution
We start by solving the associated homogeneous equation:

$$y'' - 4y' + 5y = 0. \qquad (3.9.26)$$

The roots of the characteristic equation $r^2 - 4r + 5 = 0$ are $r_1 = 2 + i$ and $r_2 = 2 - i$. Hence,

$$y_c = e^{2x}(c_1 \cos x + c_2 \sin x). \tag{3.9.27}$$

The presence of $2x + 1$ on the right-hand side of (3.9.25), which is $f(x)$, suggests that the particular solution includes a linear polynomial. Moreover, since the derivative of the product xe^x produces xe^x and e^x, we assume that the particular solution would include both xe^x and e^x. In other words, $f(x)$ is the sum of the polynomial and exponential functions as follows:

$$f(x) = f_1(x) + f_2(x) = (2x + 1) + xe^x. \tag{3.9.28}$$

Theorem 3.9.3 (the principle of superposition) suggests that we seek a particular solution of the form:

$$y_p(x) = y_{p_1} + y_{p_2} = Ax + B + Cxe^x + De^x, \tag{3.9.29}$$

where we are using $y_{p_1}(x) = Ax + B$ and $y_{p_2}(x) = (Cx + D)e^x$. Now, substituting (3.9.29) in the given equation (3.9.25) and grouping like terms gives the following result:

$$y_p'' - 4y_p' + 5y_p = 5Ax + (5B - 4A) + 2Cxe^x + (-2C + 2D)e^x$$
$$\equiv 2x + 1 + xe^x + 0 \cdot e^x. \tag{3.9.30}$$

From the identity (3.9.30), we obtain four equations with four unknowns:

$$\begin{cases} 5A = 2 \\ 5B - 4A = 1 \\ 2C = 3 \\ -2C + 2D = 0. \end{cases} \tag{3.9.31}$$

Solving the system (3.9.31), we find $A = 2/5$, $B = 13/25$, and $C = D = 3/2$. Thus, a particular solution of the equation (3.9.25) would be:

$$y_p(x) = \frac{2}{5}x + \frac{13}{25} + \frac{3}{2}xe^x + \frac{3}{2}e^x. \tag{3.9.32}$$

The general solution of (3.9.25), therefore, is:

$$y(x) = y_c(x) + y_p(x) = e^{2x}(c_1 \cos x + c_2 \sin x) + \frac{2}{5}x + \frac{13}{25} + \frac{3}{2}xe^x + \frac{3}{2}e^x.$$

Example 3.9.4

Let us solve the equation:

$$y'' - 3y' + 2y = 5e^x. \qquad (3.9.33)$$

Solution

The associated homogeneous equation

$$y'' - 3y' + 2y = 0 \qquad (3.9.34)$$

has the characteristic equation $r^2 - 3r + 2 = 0$ that has two roots $r_1 = 1$ and $r_2 = 2$. Hence,

$$y_c = c_1 e^x + c_2 e^{2x}. \qquad (3.9.35)$$

A natural first guess for a particular solution would be $y_p = Ae^x$. However, substituting this expression into the equation (3.9.33) yields the untrue statement $0 = 5e^x$. The difficulty here is that e^x is a solution of the associated homogeneous differential equation (3.9.34), and $y_p = Ae^x$ when substituted in (3.9.34) necessarily produces zero. Hence, we will try something else. Let us try a particular solution of the form $y_p = Axe^x$. Substituting derivatives $y'_p = Axe^x + Ae^x$ and $y''_p = Axe^x + 2Ae^x$ into the equation (3.9.33) and simplifying gives the following result:

$$y''_p - 3y'_p + 2y_p = Axe^x \equiv 5e^x,$$

from which we will have $A = -5$. Therefore, a particular solution is:

$$y_p = -5xe^x. \qquad (3.9.36)$$

The general solution of the (3.9.33), therefore, is:

$$y(x) = y_c(x) + y_p(x) = c_1 e^x + c_2 e^{2x} - 5xe^x.$$

The following theorem gives the method to find a particular solution of equation (3.9.1) in two cases for $f(x)$ (see Nagle et al., 2012):

Theorem 3.9.4

Case 1

For a particular solution of the differential equation

$$ay'' + by' + cy = f_n(x)e^{kx}, \qquad (3.9.37)$$

where $f_n(x)$ is a polynomial of degree n and k is a constant, choose $y_p(x)$ as:

$$y_p(x) = x^s \left(A_n x^n + A_{n-1} x^{n-1} + \cdots + A_1 x + A_0 \right) e^{kx}, \qquad (3.9.38)$$

and $s = 0, 1$, or 2 as follows:

$s = 0$, if k is not a root of the associated characteristic equation,
$s = 1$, if k is a simple root of the associated characteristic equation, and
$s = 2$, if k is the double root of the associated characteristic equation.

Case 2
For a particular solution of the differential equation

$$ay'' + by' + cy = e^{\alpha x} [f_{n_1}(x) \cos \beta x + f_{n_2}(x) \sin \beta x], \qquad (3.9.39)$$

where $f_{n_i}, i = 1, 2$, are polynomials of degrees $n_i, i = 1, 2$, and α and β are constants, choose $y_p(x)$ as:

$$y_p(x) = x^s e^{\alpha x} [(A_n x^n + \cdots + A_0) \cos \beta x + (B_m x^n + \cdots + B_0) \sin \beta x], \qquad (3.9.40)$$

where $n = \max(n_i, i = 1, 2)$ and $s = 0$ or 1 as follows:

$s = 0$, if $\alpha + i\beta$ is not a root of the associated characteristic equation and
$s = 1$, if $\alpha + i\beta$ is a root of the associated characteristic equation.

Example 3.9.5
Let us solve the equation:

$$y'' + 9y = 2\cos 3x - 5\sin 3x. \qquad (3.9.41)$$

Solution
The associated homogeneous equation:

$$y'' + 9y = 0 \qquad (3.9.42)$$

has the characteristic equation $r^2 + 9 = 0$ that has two complex roots $r_1 = 3i$ and $r_2 = -3i$. Hence,

$$y_c = c_1 \cos 3x + c_2 \sin 3x. \qquad (3.9.43)$$

The right-hand side of the equation (3.9.41) may be written as:

$$f(x) = e^{0 \cdot x} (2\cos 3x - 5\sin 3x). \qquad (3.9.44)$$

This is Case 2 of Theorem 3.9.4 with $\alpha + i\beta = 0 + 3i = 3i$ as a root of the characteristic equation. Therefore, the form of the particular solution $y_p(x)$ is:

$$y_p(x) = x(A\cos 3x + B\sin 3x).$$ (3.9.45)

Differentiating (3.9.45) twice gives:

$$y_p'(x) = (3Bx + A)\cos 3x + (-3Ax + B)\sin 3x$$ (3.9.46)

and

$$y_p''(x) = (-9Ax + 6B)\cos 3x + (-9Bx - 6A)\sin 3x.$$ (3.9.47)

Substituting y_p, y_p', and y_p'' from (3.9.45), (3.9.46), and (3.9.47) into (3.9.41) and grouping like terms, we obtain:

$$6B\cos 3x - 6A\sin 3x \equiv 2\cos 3x - 5\sin 3x.$$ (3.9.48)

From the identity in (3.9.48), we will have $6B = 2$ and $-6A = -5$. Hence, $A = 5/6$ and $B = 1/3$. A particular solution of equation (3.9.41), therefore, is:

$$y_p = x\left(\frac{5}{6}\cos 3x + \frac{1}{3}\sin 3x\right).$$ (3.9.49)

From (3.9.43) and (3.9.49), the general solution of (3.9.41) will be:

$$y(x) = y_c(x) + y_p(x) = c_1\cos 3x + c_2\sin 3x + x\left(\frac{5}{6}\cos 3x + \frac{1}{3}\sin 3x\right)$$

or

$$y(x) = \left(c_1 + \frac{5}{6}x\right)\cos 3x + \left(c_2 + \frac{1}{3}x\right)\sin 3x.$$

3.9.2. Variation of Parameters Method

To find a particular solution, we found that the undetermined coefficient was a simple method to do the task we needed. However, that was when the coefficients were constant and the nonhomogeneous part of the equation had special form. We now present a more general method of finding a particular solution of the nonhomogeneous equation (3.8.2), and that is called *variation of parameters*.

Suppose $y_1(x)$ and $y_2(x)$ are two linearly independent solutions for the corresponding homogeneous equation (3.8.3). Then, by Theorem 3.8.5, the general solution is given by formula (3.8.8) as:

$$y(x) = c_1 y_1(x) + c_2 y_2(x),$$

where c_1 and c_2 are arbitrary constants. To find a particular solution, we replace the constants c_1 and c_2 by functions of x, say, $u_1(x)$ and $u_2(x)$. Thus, we seek a solution of equation (3.8.2) as:

$$y_p(x) = u_1(x)y_1(x) + u_2(x)y_2(x), \tag{3.9.50}$$

where $u_1(x)$ and $u_2(x)$ are unknown functions. Differentiating (3.9.50) with respect to x twice, and dropping x for convenience, we will have:

$$y'_p = u'_1 y_1 + u_1 y'_1 + u'_2 y_2 + u_2 y'_2 \tag{3.9.51}$$

and

$$y''_p = u''_1 y_1 + u'_1 y'_1 + u'_1 y'_1 + u_1 y''_1 + u''_2 y_2 + u'_2 y'_2 + u'_2 y'_2 + u_2 y''_2. \tag{3.9.52}$$

Substituting (3.9.50), (3.9.51), and (3.9.52) into (3.8.2) and grouping terms, we obtain:

$$\begin{aligned}
ay''_p + by'_p + cy_p(x) &= a(u''_1 y_1 + 2u'_1 y'_1 + u_1 y''_1 + u''_2 y_2 + 2u'_2 y'_2 + u_2 y''_2) \\
&\quad + b(u'_1 y_1 + u_1 y'_1 + u'_2 y_2 + u_2 y'_2) + c(u_1 y_1 + u_2 y_2) \\
&= u_1(ay''_1 + by'_1 + cy_1) + u_2(ay''_2 + by'_2 + cy_2) \\
&\quad + a(u''_1 y_1 + 2u'_1 y'_1 + u''_2 y_2 + u_2 y''_2) + a(u'_1 y'_1 + u'_2 y'_2) \\
&\quad + b(u'_1 y_1 + u'_2 y_2) \\
&= u_1 \cdot 0 + u_2 \cdot 0 + a \cdot \frac{d}{dx}(u'_1 y_1 + u'_2 y_2) \\
&\quad + b(u'_1 y_1 + u'_2 y_2) + a(u'_1 y'_1 + u'_2 y'_2)
\end{aligned}$$

or

$$\begin{aligned}
ay''_p + by'_p + cy_p(x) &= a \cdot \frac{d}{dx}(u'_1 y_1 + u'_2 y_2) \\
&\quad + b(u'_1 y_1 + u'_2 y_2) + a(u'_1 y'_1 + u'_2 y'_2) = f(x).
\end{aligned} \tag{3.9.53}$$

Since we are looking for two unknown functions $u_1(x)$ and $u_2(x)$, we need two equations. Thus, we assume that $u_1(x)$ and $u_2(x)$ satisfy:

$$u'_1 y_1 + u'_2 y_2 = 0. \tag{3.9.54}$$

Hence, from equation (3.9.53) and equation (3.9.54), we will have:

$$a(u'_1 y'_1 + u'_2 y'_2) = f(x)$$

or

$$u_1' y_1' + u_2' y_2' = \frac{f(x)}{a}. \tag{3.9.55}$$

Equation (3.9.54) and equation (3.9.55) are the ones we were looking for. In other words, we have the system of two equations with two unknowns as $u_1' = u_1'(x)$ and $u_2' = u_2'(x)$, that is,

$$\begin{cases} y_1 u_1' + y_2 u_2' = 0 \\ y_1' u_1' + y_2' u_2' = \dfrac{f(x)}{a}. \end{cases} \tag{3.9.56}$$

Using Cramer's rule, to solve (3.9.56), denoting the Wronskian of $y_1(x)$ and $y_2(x)$ as $D(x)$, we will have:

$$u_1'(x) = \frac{D_1(x)}{D(x)} \quad \text{and} \quad u_2'(x) = \frac{D_2(x)}{D(x)}, \tag{3.9.57}$$

where

$$D(x) = \begin{vmatrix} y_1(x) & y_2(x) \\ y_1'(x) & y_2'(x) \end{vmatrix},$$

$$D_1(x) = \begin{vmatrix} 0 & y_2(x) \\ \dfrac{f(x)}{a} & y_2'(x) \end{vmatrix}, \quad \text{and} \quad D_2(x) = \begin{vmatrix} y_1(x) & 0 \\ y_1'(x) & \dfrac{f(x)}{a} \end{vmatrix}. \tag{3.9.58}$$

The functions $u_1(x)$ and $u_2(x)$ can now be found by integrating (3.9.57). We know that $D(x) \neq 0$, $\forall x \in I$ because of $y_1(x)$ and $y_2(x)$ being linearly independent and the Theorem 3.8.3.

Let us now summarize the variation of the parameters (or of coefficients) method in steps as follows:

Step 1. Find two linearly independent solutions $y_1(x)$ and $y_2(x)$ to the corresponding homogeneous equation and define a particular solution, $y_p(x)$, as in (3.9.50).

Step 2. Determine $u_1(x)$ and $u_2(x)$ by solving the system (3.9.56) for $u_1'(x)$ and $u_2'(x)$, and integrating. This may be done using several methods, one of which is Cramer's rule, (3.9.57).

Step 3. Substitute $u_1(x)$ and $u_2(x)$ into (3.9.50) to obtain $y_p(x)$.

Example 3.9.6
Let us solve the following nonhomogeneous linear differential equation with constant coefficient:

$$9y'' + 36y = \sec 2x. \tag{3.9.59}$$

Solution

The characteristic equation of the homogenous part has two complex roots $r_1 = 2i$ and $r_2 = -2i$. Thus, the complementary function will be:

$$y_c(x) = c_1 \cos 2x + c_2 \sin 2x. \tag{3.9.60}$$

Hence,

$$y_1(x) = \cos 2x \text{ and } y_2(x) = \sin 2x \tag{3.9.61}$$

are the two linearly independent solution of the homogenous part of (3.9.58). Thus, we let:

$$y_p(x) = u_1(x)\cos 2x + u_2(x)\sin 2x. \tag{3.9.62}$$

From (3.8.2) and (3.9.59), we see that $a = 9$ and $f(x) = \sec 2x$. Hence, $f(x)/a = 1/9 \sec 2x$.

Now using Cramer's rule, from (3.9.58), we have:

$$D(x) = \begin{vmatrix} \cos 2x & \sin 2x \\ -\sin 2x & 2\cos 2x \end{vmatrix} = 2,$$

$$D_1(x) = \begin{vmatrix} 0 & \sin 2x \\ \dfrac{1}{9}\sec 2x & 2\cos 2x \end{vmatrix} = -\dfrac{1}{9}\tan 2x,$$

and

$$D_2(x) = \begin{vmatrix} \cos 2x & 0 \\ -2\sin 2x & \dfrac{1}{9}\sec 2x \end{vmatrix} = -\dfrac{1}{9}.$$

Thus, from (3.9.57), we have:

$$u_1' = -\dfrac{1}{18}\tan 2x \text{ and } u_2' = -\dfrac{1}{18}. \tag{3.9.63}$$

Integrating (3.9.63), we will have:

$$u_1 = \dfrac{1}{36}\ln|\cos 2x| \text{ and } u_2 = -\dfrac{1}{18}x. \tag{3.9.64}$$

Substitute $u_1(x)$ and $u_2(x)$ from (3.9.64) into (3.9.62), we obtain:

$$y_p = \frac{1}{36}\ln|\cos 2x| \cdot \cos 2x - \frac{1}{18}x \cdot \sin 2x. \tag{3.9.65}$$

Therefore, from (3.9.60) and (3.9.65), the general solution of (3.9.59) is:

$$y(x) = y_c(x) + y_p(x) = c_1 \cos 2x + c_2 \sin 2x$$
$$+ \frac{1}{36}\ln|\cos 2x| \cdot \cos 2x - \frac{1}{18}x \cdot \sin 2x$$

or after combining terms:

$$y(x) = \left(c_1 + \frac{1}{36}\ln|\cos 2x|\right)\cos 2x + \left(c_2 - \frac{1}{18}x\right)\sin 2x.$$

3.10. MISCELLANEOUS METHODS FOR SOLVING ODE

3.10.1. Cauchy–Euler Equation

Definition 3.10.1

A Cauchy–Euler equation is a linear ordinary differential equation of the form:

$$ax^2 y'' + bxy' + cy = f(x), \tag{3.10.1}$$

where a, b, and c are constants and $a \neq 0$.

The coefficient of y'', ax^2, is zero when $x = 0$. Thus, if we rewrite (3.10.1) as:

$$y'' + \frac{b}{x}y' + \frac{c}{x^2}y = \frac{f(x)}{x^2}, \tag{3.10.2}$$

it has discontinuous coefficients at $x = 0$. Thus, we may expect the solution to be located in intervals $(-\infty, 0)$ or $(0, \infty)$.

It is possible that (3.10.1) be reduced to an equation with constant coefficient. That, of course, is the importance of this equation and why it is worth considering it as a special case.

First, let us consider the homogeneous part of (3.10.1), that is,

$$ax^2 y'' + bxy' + cy = 0. \tag{3.10.3}$$

Equation (3.10.3) can be converted to an equation with constant coefficient by assuming $x > 0$ and letting

$$x = e^t. \tag{3.10.4}$$

With the change of variable (3.10.4), we can rewrite $y(x)$ as:

$$Y(t) \equiv y(x) = y(e^t). \tag{3.10.5}$$

Using the chain rule, we will have:

$$\frac{dY}{dt} = \frac{dy}{dx}\frac{dx}{dt} = y'e^t = xy'. \tag{3.10.6}$$

Now, differentiating (3.10.6) with respect to t, we obtain:

$$\frac{d^2Y}{dt^2} = \frac{dx}{dt}y' + x\frac{dy'}{dx}\frac{dx}{dt} = xy' + x^2y'' = \frac{dY}{dt} + x^2y''. \tag{3.10.7}$$

From (3.10.7), we will have:

$$x^2y'' = \frac{d^2Y}{dt^2} - \frac{dY}{dt}. \tag{3.10.8}$$

Substituting (3.10.6) and (3.10.8) into (3.10.3) gives the following:

$$a\left(\frac{d^2Y}{dt^2} - \frac{dY}{dt}\right) + b\frac{dY}{dt} + cY = 0$$

or

$$a\frac{d^2Y}{dt^2} + (b-a)\frac{dY}{dt} + cY = 0. \tag{3.10.9}$$

But (3.10.9) is a second-order differential equation with constant coefficients and we can solve it with the method offered in Section 3.8 and Section 3.9.

Example 3.10.1
Let us solve the Cauchy–Euler equation

$$x^2y'' + 6xy' + 6y = \ln x, \quad x > 0. \tag{3.10.10}$$

Solution
From (3.10.4), (3.10.6), and (3.10.8), we have:

$$\left(\frac{d^2Y}{dt^2} - \frac{dY}{dt}\right) + 6\frac{dY}{dt} + 6Y = \ln(e^t) = t$$

or

$$\frac{d^2Y}{dt^2} + 5\frac{dY}{dt} + 6Y = t, \tag{3.10.11}$$

where $Y = Y(t)$ that is defined in (3.10.5). Equation (3.10.11) being with the constant coefficient has its characteristic equation as:

$$r^2 + 5r + 6 = (r+2)(r+3) = 0,$$

with two real roots $r_1 = -2$ and $r_2 = -3$. Thus, the complimentary solution is:

$$Y_c = c_1 e^{-2t} + c_2 e^{-3t}. \tag{3.10.12}$$

Applying the method of undetermined coefficients on (3.10.11), we choose a particular solution as:

$$Y_p = At + B. \tag{3.10.13}$$

Substituting (3.10.13) into (3.10.11) yields $5A + 6(At + B) = t$, from which we obtain $A = 1/6$ and $B = -5/36$. Therefore, the general solution for Y will be:

$$Y = Y_c + Y_p = c_1 e^{-2t} + c_2 e^{-3t} + \frac{1}{6}t - \frac{5}{36}.$$

Hence, the general solution of the original problem, equation (3.10.10), on the interval $(0, \infty)$ is:

$$y(x) = c_1 \frac{1}{x^2} + c_2 \frac{1}{x^3} + \frac{1}{6}\ln x - \frac{5}{36}.$$

3.10.2. Elimination Method to Solve Differential Equations

In mathematics, an *operator* is a transformation that transforms a function into another function. It is usually denoted and describes a mathematical procedure. In differential calculus, symbols such as d/dx, D_x, D, $\partial/\partial x$, and ∂_x are symbols that play the role of differential operators. What they do is take a derivative of the function under consideration. Of course, partials are used for the derivative of functions of more than one variable, as we have seen earlier in this chapter. Derivatives of higher orders are denoted by powers of these symbols, such as d^n/dx^n, D_x^n, D^n, $\partial^n/\partial x^n$ and ∂_x^n to stand for derivatives of n order. The function that the operator is to operate on will follow the operator symbol. For instance, dy/dx, $D_x f(x, y)$, Dy, $\partial z/\partial x$, and $\partial_x z$ to represent derivatives of y, $f(x, y)$, y, and z. Hence, for example, $D(\sin 3x) = 3\cos 3x$, and $D(Dy) = D^2 y = d^2/dx^2 = y''$. Since differentiation is linear, we have $D(f+g) = Df + Dg$ and $D(af) = aDf$, where a is a constant. A *differential operator* may appear as a polynomial expression to express a differential form. For instance, $2D - 3$, $D^2 - 5D + 6$, $2x^2 D^2 - 2x + 7$, and so on. Thus, in such cases we will have $(2D - 3)y = 2y' - 3y$, $(D^2 - 5D + 6)y = y'' - 5y' + 6$, and $(2x^2 D^2 - 2xD + 7)y = 2x^2 y'' - 2xy' + 7y$. In general, a second-order differential operator is defined as:

$$L = p(x)D^2 + q(x)D + r(x). \tag{3.10.14}$$

Example 3.10.2
Let us show that the operator $(D - 2)(D + 3)$ is the same as the operator $D^2 + D - 6$.

Solution
Suppose $y = y(t)$ is twice differentiable. Then, we have:

$$\begin{aligned}
[(D-2)(D+3)]y &= (D-2)[(D+3)y] \\
&= (D-2)(y'+3y) = D(y'+3y) - 2(y'+3y) \\
&= y'' + 3y' - 2y' - 6y = y'' + y' - 6y \\
&= (D^2 + D - 6)y.
\end{aligned}$$

As noticed, we manipulate the polynomial expressions of the operator the same as a polynomial function if coefficients in (3.10.14) are constants. Thus, as we mentioned, since a differential operator is linear, L in (3.10.14) enjoys such a property. Hence, for example, $(D^2 + 7D - 13)y = \sin x$ is the operator form of the differential equation $y'' + 7y' - 13y = \sin x$.

We now want to use the idea of differential operator to solve a system of differential equations. Once again, to see how we apply the differential operator, let us consider the system of nonhomogeneous differential equations:

$$\begin{cases} a_1 x'(t) + b_1 x(t) + c_1 y'(t) + d_1 y(t) = f_1(t), \\ a_2 x'(t) + b_2 x(t) + c_2 y'(t) + d_2 y(t) = f_2(t), \end{cases} \tag{3.10.15}$$

where a_i, b_i, c_i, d_i, $i = 1, 2$, are constants and $x(t)$ and $y(t)$ are the unknown functions to be determined. The operator notation equivalence of (3.10.15), using D as the derivative operator, is as follows:

$$\begin{cases} (a_1 D + b_1)x + (c_1 D + d_1)y = f_1(t), \\ (a_2 D + b_2)x + (c_2 D + d_2)y = f_2(t). \end{cases} \tag{3.10.16}$$

Example 3.10.3
Let us solve the following system of differential equations using the elimination method:

$$\begin{cases} x'(t) - 4x(t) + y'(t) = 0, \\ x'(t) + x(t) + y(t) = 3t^2. \end{cases} \tag{3.10.17}$$

Solution
We first write (3.10.17) in terms of the differential operator as follows:

$$\begin{cases} (D-4)x + Dy = 0, \\ (D+1)x + y = 3t^2. \end{cases} \tag{3.10.18}$$

Applying the operator D on the second equation of (3.10.18) and subtracting the result from the first equation will eliminate y and results in:

$$(D-4)x+Dy-D(D+1)x-Dy=-3D(t^2),$$

or

$$[(D-4)-D(D+1)]x=-3D(t^2),$$

or

$$(D^2+4)x=6t,$$

or

$$x''+4x=6t. \tag{3.10.19}$$

The characteristic equation of (3.10.19) is $r^2+4=0$ that has two complex roots $r_1=2i$ and $r_2=-2i$. Thus, we will have:

$$x_c(t)=c_1\cos 2t+c_2\sin 2t. \tag{3.10.20}$$

To determine a particular solution, $x_p(t)$, we use the undetermined coefficient method. Hence, we choose $x_p(t)$ as:

$$x_p=At+B. \tag{3.10.21}$$

Substituting (3.10.21) into (3.10.19), we will have:

$$x_p''+4x_p=4At+4B\equiv 6t. \tag{3.10.22}$$

From equivalence in (3.10.22), we obtain $A=3/2$ and $B=0$. Thus, from (3.10.21), we will have:

$$x_p(t)=\frac{3}{2}t. \tag{3.10.23}$$

Therefore, from (3.10.20) and (3.10.23), we have:

$$x(t)=x_c(t)+x_p(t)=c_1\cos 2t+c_2\sin 2t+\frac{3}{2}t. \tag{3.10.24}$$

Now substituting (3.10.24) into the second equation of (3.10.17), we will have:

$$y(t)=3t^2-x'(t)-x(t)$$
$$=3t^2-\left(-2c_1\sin 2t+2c_2\cos 2t+\frac{3}{2}\right)-\left(c_1\cos 2t+c_2\sin 2t+\frac{3}{2}t\right)$$

or

$$y(t) = -(c_1 + 2c_2)\cos 2t + (2c_1 - c_2)\sin 2t + 3t^2 - \frac{3}{2}t - \frac{3}{2}. \qquad (3.10.25)$$

Thus, a pair solution of (3.10.17) has been found by (3.10.24) and (3.10.25).

3.10.3. Application of Laplace Transform to Solve ODE

We have already discussed the Laplace transform and some of properties and usefulness of it in Chapter 2. It is particularly useful in solving linear ordinary differential equations with constant coefficients. Here is an example:

Example 3.10.4
Let us solve the following initial value problem using the Laplace transform:

$$ay''(x) + by'(x) + cy(x) = f(x), \quad y(x_0) = k_0, \, y'(x_0) = h_0, \quad (3.10.26)$$

where a $(\neq 0)$, b, c, k_0, and h_0 are constants.

Solution
Now applying the Laplace transform (operator $\mathcal{L}$) along with its linearity property on (3.10.26), we will have:

$$a\mathcal{L}[y''(x)] + b\mathcal{L}[y'(x)] + c\mathcal{L}[y(x)] = \mathcal{L}[f(x)]. \qquad (3.10.27)$$

Let us denote the following: $Y(s) = \mathcal{L}[y(x)]$ and $F(s) = \mathcal{L}[f(x)]$. Then, applying formulas 2 and 3 in Table 2.2.1 of Chapter 2 on (3.10.27) and the initial values given in (3.10.26), we will have:

$$a\left[s^2 Y(s) - sk_0 - h_0\right] + b\left[sY(s) - k_0\right] + cY(s) = F(s). \qquad (3.10.28)$$

Therefore, we have transformed the initial value problem to an algebraic equation. Solving (3.10.28) for $Y(s)$, we will have:

$$(as^2 + bs + c)Y(s) = ask_0 + ah_0 + bk_0 + F(s)$$

or

$$Y(s) = \frac{ask_0 + ah_0 + bk_0}{as^2 + bs + c} + \frac{F(s)}{as^2 + bs + c}. \qquad (3.10.29)$$

Thus, by inversion, we can find $y(x)$, that is, $y(x) = \mathcal{L}^{-1}[Y(s)]$. To do that, we need to apply the partial fraction method to split the rational functions on the right-hand side of (3.10.29).

Example 3.10.5

Let us solve the following initial value problem using the Laplace transform:

$$y''(x) - 5y'(x) + 6y(x) = 10e^x \cos x, \quad y(0) = 2, \ y'(0) = -1. \qquad (3.10.30)$$

Solution

As in the previous example, applying the Laplace transform on (3.10.30) will result in:

$$\mathcal{L}[y''(x)] - 5\mathcal{L}[y'(x)] + 6\mathcal{L}[y(x)] = \mathcal{L}[10e^x \cos x]. \qquad (3.10.31)$$

Let us denote $Y(s) = \mathcal{L}[y(x)]$. Then, applying formulas 2, 3 and 11 in Table 2.2.1 of Chapter 2 on (3.10.31) and the initial values given in (3.10.30), that is, $\mathcal{L}[y'(x)] = sY(s) - 2$ and $\mathcal{L}[y''(x)] = s^2 Y(s) - 2s + 1$, we will have:

$$s^2 Y(s) - 2s + 1 - 5[sY(s) - 2] + 6Y(s) = \frac{10s - 10}{s^2 - 2s + 2}$$

or

$$(s^2 - 5s + 6) Y(s) = 2s - 11 + \frac{10s - 10}{s^2 - 2s + 2}.$$

Hence:

$$Y(s) = \frac{2s - 11}{(s-2)(s-3)} + \frac{10s - 10}{(s-2)(s-3)(s^2 - 2s + 2)}. \qquad (3.10.32)$$

Applying the partial fraction method on (3.10.32), we will have:

$$Y(s) = 2\frac{1}{s-2} - \frac{1}{s-3} + \frac{s-1}{(s-1)^2 + 1} - 3\frac{1}{(s-1)^2 + 1}. \qquad (3.10.33)$$

Now, by inversion of (3.10.33), we find $y(x)$ as:

$$y(x) = \mathcal{L}^{-1}[Y(s)]$$

$$= 2\mathcal{L}^{-1}\left[\frac{1}{s-2}\right] - \mathcal{L}^{-1}\left[\frac{1}{s-3}\right] + \mathcal{L}^{-1}\left[\frac{s-1}{(s-1)^2 + 1}\right] - 3\mathcal{L}^{-1}\left[\frac{1}{(s-1)^2 + 1}\right]$$

or

$$y(x) = 2e^{2x} - e^{3x} + e^x \cos x - 3e^x \sin x.$$

3.10.4. Solution of Linear ODE Using Power Series

There are many important and interesting differential equations that cannot be solved using standard methods for exact solutions. However, approximation (numerical approach) is a tool to settle for qualitative information. Approximation sometimes goes through an analytic process to give a closed form approximation formula. The power series centered on the point that initial condition is one of the approximation methods available. The succeeding sections describe how it works.

Let us consider the linear second-order differential equation

$$y'' + p(x)y' + q(x)y = 0 \qquad (3.10.34)$$

in an open interval containing the point x_0. The idea is to find conditions under which the general solution of (3.10.34) can be represented by a power series that converges in a neighborhood of the point x_0. Since coefficients of (3.10.34) are not constants, there is a possibility of having singularities. We start by defining the basic terms from analysis.

Definition 3.10.2
A function $f(x)$ is called *analytic* at a point x_0 if it can be represented by the Taylor series as:

$$f(x) = \sum_{n=0}^{\infty} \frac{f^{(n)}(x_0)}{n!}(x - x_0)^n, \qquad (3.10.35)$$

with the radius of convergence $R > 0$.

Note that if a function $f(x)$ is represented by a power series as (3.10.35), its derivative of all order at the point x_0 exits. Hence, if at least one order fails to exit, $f(x)$ is not analytic at the point x_0.

Definition 3.10.3
If both coefficients $p(x)$ and $q(x)$ in (3.10.34) are analytic at the point x_0, then x_0 is called an *ordinary point* of (3.10.34). If $p(x)$ or $q(x)$ is not analytic at the point x_0, then x_0 is called a singular point.

Example 3.10.6
Let us consider the equation:

$$y'' + \frac{x}{x^2 - 1}y' + (\tan x)y = 0 \qquad (3.10.36)$$

in the open interval $-2 < x < 2$. We want to determine if coefficients are analytic in the interval and find the points of singularities.

Answer

From (3.10.36), the coefficients are:

$$p(x) = \frac{x}{x^2 - 1} \text{ and } q(x) = \tan x.$$

The roots of the denominator of $p(x)$ are $x = \pm 1$ that are within the interval $-2 < x < 2$, at which $p(x)$ fails to be analytic and, hence, are singular points. The function $q(x)$, on the other hand, is undefined at $x = \pm \pi/2$ that are also within the interval and, hence, they are singular points of $q(x)$. Thus, the four points $x = \pm 1$ and $x = \pm \pi/2$ are singular points and all other points of the interval are ordinary points for equation (3.10.36).

The following theorem shows that we can present the general solution of (3.10.34) in a neighborhood of an ordinary point in terms of a convergent series.

Theorem 3.10.1

Let $p(x)$ and $q(x)$ be analytic at the point x_0. Let also R denote the smallest radii of convergence of their respective Taylor series representations. The, the initial value problem

$$y'' + p(x)y' + q(x)y = 0, \quad y(x_0) = h_0, \; y'(x_0) = k_0, \qquad (3.10.37)$$

has a unique solution that is analytic in the interval $|x - x_0| < R$, that is, $(x_0 - R, x_0 + R)$.

Proof:

See Birkhoff and Rota (1989).

What the Theorem 3.10.1 says is that if x_0 is an ordinary point, then the initial value problem (3.10.37) has a power series of the form:

$$y(x) = \sum_{n=0}^{\infty} a_n (x - x_0)^n = a_0 + a_1 (x - x_0) + a_2 (x - x_0)^2 + \cdots + a_n (x - x_0)^n + \cdots.$$

$$(3.10.38)$$

The first two coefficients in (3.10.38), that is, a_0 and a_1, will be determined by the initial conditions given in (3.10.37), since a_0 and a_1 are, indeed, $y(x_0)$ and $y'(x_0)$, respectively.

To find a particular solution satisfying the initial value condition, in this case, is similar to what we did in Section 3.9 in finding a particular solution of nonhomogeneous differential equations by the method of undetermined coefficients. That is, we substitute $y(x)$ from (3.10.38) into the differential equation (3.10.34), combine the terms of series, and then equate all corresponding coefficients to the right-hand side of the equation to determine the coefficient a_n. However, because the right-hand side is zero, it implies that all coefficients of x must be equated to zero. We know we can find the solution because of Theorem 3.10.1.

Example 3.10.7

Let us find a power series representation at $x_0 = 0$ for the general solution of the following equation:

$$y'' + 2xy' + 2y = 0. \tag{3.10.39}$$

Solution

Since there are no finite singular points for equation (3.10.39), Theorem 3.10.1 guarantees the existence of a power series solution centered on $x = x_0 = 0$ that converges for all x. Hence, let:

$$y(x) = \sum_{n=0}^{\infty} a_n x^n, \tag{3.10.40}$$

where $a_n, n = 0, 1, 2, \ldots$, are constants. Differentiating the series (3.10.40) twice yields:

$$y'(x) = \sum_{n=1}^{\infty} n a_n x^{n-1} \tag{3.10.41}$$

and

$$y''(x) = \sum_{n=2}^{\infty} n(n-1) a_n x^{n-2} = \sum_{n=0}^{\infty} (n+2)(n+1) a_{n+2} x^n. \tag{3.10.42}$$

Substituting (3.10.40), (3.10.41), and (3.10.42) into (3.10.39), we will have:

$$\sum_{n=0}^{\infty} (n+2)(n+1) a_{n+2} x^n + \sum_{n=1}^{\infty} 2n a_n x^n + \sum_{n=0}^{\infty} 2a_n x^n = 0$$

or

$$a_0 + 2a_2 + \sum_{n=1}^{\infty} [(n+2)(n+1) a_{n+2} + 2(n+1) a_n] x^n = 0. \tag{3.10.43}$$

Since (3.10.43) is identically zero, it is necessary that the coefficient of each power of x be equal to zero. That is,

$$a_0 + 2a_2 = 0 \tag{3.10.44}$$

and

$$a_{n+2} = -\frac{2}{n+2} a_n, \quad n = 1, 2, \ldots. \tag{3.10.45}$$

As it can be seen from (3.10.45), all coefficients can be found and hence (3.10.45) is called a *recursive formula for coefficients*. It generates consecutive coefficients of the assumed solutions one at a time for each n in (3.10.45). The goal is to obtain a closed form solution for coefficients. Therefore, we set $a_0 = c_1$ and $a_1 = c_2$. (These are the constants in the general solution.) Thus, from these assumptions and (3.10.44) we have $a_2 = -1/2c_1$. Therefore, we are to calculate several of the coefficients as follows:

$$n = 1, a_3 = -\frac{2}{3}a_1 = -\frac{2}{3}c_2,$$

$$n = 2, a_4 = -\frac{2}{4}a_2 = \left(-\frac{2}{4}\right)\left(-\frac{1}{2}c_1\right) = \frac{2}{4\cdot2}c_1,$$

$$n = 3, a_5 = -\frac{2}{5}a_3 = \left(-\frac{2}{5}\right)\left(-\frac{2}{3}c_2\right) = \frac{2^2}{5\cdot3}c_2,$$

$$n = 4, a_6 = -\frac{2}{6}a_4 = \left(-\frac{2}{6}\right)\left(\frac{2}{4\cdot2}c_1\right) = -\frac{2^2}{6\cdot4\cdot2}c_1,$$

$$n = 5, a_7 = -\frac{2}{7}a_5 = \left(-\frac{2}{7}\right)\left(\frac{2^2}{5\cdot3}c_2\right) = -\frac{2^3}{7\cdot5\cdot3}c_2,$$

$$n = 6, a_8 = -\frac{2}{8}a_6 = \left(-\frac{2}{8}\right)\left(-\frac{2^2}{6\cdot4\cdot2}c_1\right) = \frac{2^3}{8\cdot6\cdot4\cdot2}c_1,$$

and so on. We can see a pattern from these few cases. It can be seen that when n is an even number, say $n = 2k, k = 0, 1, 2, \ldots,$

$$a_{2k} = (-1)^k \frac{2^k}{2k(2k-2)\ldots2}c_1 = (-1)^k \frac{2^k}{(2k)!!}c_1, \qquad (3.10.46)$$

where

$$(2k)!! = 2k\cdot(2k-2)(2k-4)\ldots2 \qquad (3.10.47)$$

is the *even number factorial*. In contrast, if n is an odd number, say $n = 2k + 1,$ $k = 0, 1, 2, \ldots,$ then

$$a_{2k+1} = (-1)^k \frac{2^k}{(2k+1)(2k-1)\ldots1}c_2 = (-1)^k \frac{2^k}{(2k+1)!!}c_2, \qquad (3.10.48)$$

where

$$(2k+1)!! = (2k+1)(2k-1)(2k-3)\cdot(2k-2)(2k-4)\ldots1 \qquad (3.10.49)$$

is the *odd number factorial*.

Now that we have formulated the coefficients, we can write two linearly independent solutions $y_1(x)$ and $y_2(x)$ of (3.10.39) by choosing specific values for c_1 and c_2. So let us choose $c_1 = 1$ and $c_2 = 0$. Then, as one particular solution, from (3.10.46) we have:

$$y_1(x) = \sum_{k=0}^{\infty} (-1)^k \frac{2^k}{(2k)!!} x^{2k}. \tag{3.10.50}$$

From (3.10.47), factoring out of 2, we have $(2k)!! = 2^k \cdot k!$. Hence, from (3.10.50) we have:

$$y_1(x) = \sum_{k=0}^{\infty} (-1)^k \frac{x^{2k}}{(k)!} = e^{-x^2}. \tag{3.10.51}$$

Now choosing specific values for c_1 and c_2 as $c_1 = 0$ and $c_2 = 1$, from (3.10.48) we can write the second particular solution as:

$$y_2(x) = \sum_{k=0}^{\infty} (-1)^k \frac{2^k}{(2k+1)!!} x^{2k+1}. \tag{3.10.52}$$

However, (3.10.52) cannot be written in terms of elementary functions as we did for the first solution. Therefore, the general solution can be written as:

$$y(x) = c_1 y_1(x) + c_2 y_2(x), \tag{3.10.53}$$

where c_1 and c_2 are arbitrary constants and $y_1(x)$ and $y_2(x)$ are given in (3.10.51) and (3.10.52), respectively.

3.11. APPLICATIONS OF THE SECOND-ORDER ODE

In this section we offer some applications of ordinary differential equations that we have discussed in this chapter.

3.11.1. Spring–Mass System: Free Undamped Motion

In physics and mechanics *damping* is an effect that reduces the amplitude of oscillations in an oscillatory system. Suppose that a flexible spring is suspended vertically from a rigid support and then a mass m is attached to its free end. The amount of stretch of the spring will depend on the mass. If there is no external force and no damping, then the ordinary differential equation for a mass–spring problem is given by (3.1.15), where $k > 0$ is the spring constant.

Example 3.11.1
A weight whose mass is 4 kg is attached to a linear spring suspended vertically from a support. A weight is attached to the spring and it causes a stretch of 0.2 m. The weight is held stationary 0.5 m below this equilibrium position and released. We want to describe the motion of the system.

Solution
To determine the stiffness k, we note that the weight exerts a force of $mg = 39.2$ N on the spring and stretches it 0.2 m. Thus, $k = 39.2/0.2 = 196$ N/m. Now let $x(t)$ denote the displacement of the weight below the equilibrium position t seconds after release. The velocity of the weight is then $v(t) = x'(t)$ and the acceleration is $a(t) = x''(t)$.

The net force on the weight is $F = -196x$. By Newton's second law of motion, $F = ma(t)$, where $m = 4$, is the mass (in kilograms). Hence, $x(t)$ is a solution of the ODE $4x''(t) = -196x(t)$ or $x''(t) = -49x(t)$.

The initial conditions given are $x(0) = 0.5$ and $x'(0) = v(0) = 0$. This is because the weight was stationary before release. Therefore, $x(t)$ is the solution of the initial value problem

$$x''(t) + 49x(t) = 0, \quad x(0) = 0.5, x'(0) = 0. \tag{3.11.1}$$

The characteristic equation is $r^2 + 49 = 0$, which has two complex solutions $r_1 = 7i$ and $r_2 = -7i$. Therefore, the general solution of equation (3.11.1) is:

$$x(t) = c_1 \cos(7t) + c_2 \sin(7t), \tag{3.11.2}$$

where c_1 and c_2 are arbitrary constants.

To determine c_1 and c_2, we use the initial conditions as follows:

$$x(0) = c_1 \cos(0) + c_2 \sin(0) = c_1 = 0.5$$

and

$$x'(t) = -7c_1 \sin(7t) + 7c_2 \cos(7t),$$
$$x'(0) = -7c_1 \sin(0) + 7c_2 \cos(0) = 7c_2 = 0.$$

Hence, $c_2 = 0$. Thus, from (3.11.2) we will have $x(t) = 0.5 \cos(7t)$.

3.11.2. Damped-Free Vibration

If we include the effect of damping, the differential equation of the motion is:

$$mx''(t) + bx'(t) + kx(t) = 0, \tag{3.11.3}$$

where m, b, and k are positive constants. The corresponding characteristic equation is $mr^2 + br + k = 0$ with the roots:

$$r_1 = -\frac{b}{2m} - \frac{\sqrt{b^2 - 4ac}}{2m} \text{ and } r_1 = -\frac{b}{2m} + \frac{\sqrt{b^2 - 4ac}}{2m}. \tag{3.11.4}$$

We can consider three possible cases depending upon the discriminant $b^2 - 4ac$. Since each solution contains the *damping factor* $e^{-(b/2m)t}$, the displacements of the mass become negligible as t approaches infinity.

Case 1: $b^2 - 4ac > 0$.
The system in this case is called *overdamped*. The general solution of equation (3.11.3) will be:

$$x(t) = c_1 e^{r_1 t} + c_2 e^{r_2 t}, \tag{3.11.5}$$

where r_1 ($r_1 < 0$) and r_2 ($r_2 < 0$) are given in (3.11.4).

Case 2: $b^2 - 4ac = 0$.
The system in this case is called *critically damped*. The general solution of the equation (3.11.3) will be:

$$x(t) = (c_1 + c_2 t) e^{-\frac{b}{2m}t}, \tag{3.11.6}$$

where r_1 ($r_1 < 0$) and r_2 ($r_2 < 0$) are given in (3.11.4).

Case 3: $b^2 - 4ac < 0$.
The system in this case is called *underdamped*. The general solution of the equation (3.11.3) will be:

$$x(t) = (c_1 \cos \mu t + c_2 \sin \mu t) e^{-\frac{b}{2m}t}, \tag{3.11.7}$$

where $\mu = \sqrt{4km - b^2}/2m > 0$.

Example 3.11.2
Let us consider the mechanical system discussed in Example 3.11.1 when it is damped. If the initial conditions are the same as in Example 3.11.1, we want to determine the subsequent motion of the weight, where the damping is:

i. $b = 40$ kg/s,
ii. $b = 14$ kg/s, and
iii. $b = 4$ kg/s.

Solution
Let $x(t)$ denote the distance of the weight from its equilibrium position. Then, $x(t)$ satisfies the following initial value problem:

$$x''(t) + bx'(t) + 49x(t) = 0, \quad x(0) = 0.5, x'(0) = 0. \tag{3.11.8}$$

i. If $b = 40$, then $b^2 - 4ac = 40^2 - 4 \cdot 1 \cdot 49 = 1404 > 0$ and the system is overdamped. The characteristic equation is:

$$r^2 + 40r + 49 = 0 \tag{3.11.9}$$

with roots

$$r_1 = -20 - 3\sqrt{39} \text{ and } r_2 = -20 + 3\sqrt{39}. \tag{3.11.10}$$

Thus, the general solution of (3.11.9) is:

$$x(t) = c_1 e^{r_1 t} + c_2 e^{r_2 t},$$

where r_1 and r_2 are given in (3.11.10). Since

$$x(0) = c_1 e^0 + c_2 e^0 = 0.5$$

and

$$x'(0) = c_1 r_1 e^0 + c_2 r_2 e^0 = 0,$$

we will have:

$$\begin{cases} c_1 + c_2 = 0.5 \\ r_1 c_1 + r_2 c_2 = 0 \end{cases}$$

or $c_1 = -0.01688$ and $c_2 = 0.51688$. The equation of the motion is:

$$x(t) = -0.01688 e^{(-20 - 3\sqrt{39})} + 0.51688 e^{(-20 + 3\sqrt{39})}.$$

ii. If $b = 14$, then $b^2 - 4ac = b^2 - 4mk = 14^2 - 4 \cdot 1 \cdot 49 = 0$ and the system is critically damped. The roots of the characteristic equation in this case are $r_1 = r_2 = 7$. Hence, the general solution of (3.11.9) is:

$$x(t) = (c_1 + c_2)e^{-7t}$$

and

$$x'(t) = (c_2 - 7c_1 - 7c_2 t)e^{-7t}.$$

Thus,

$$\begin{cases} x(0) = c_1 e^0 = 0.5 \\ x'(t) = -7c_1 + c_2 = 0 \end{cases}$$

or $c_1 = 0.5$ and $c_2 = 3.5$. Thus, the equation of the motion is:

$$x(t) = (0.5 + 3.5t)e^{-7t}.$$

iii. If $b = 4$, then $b^2 - 4ac = b^2 - 4mk = 4^2 - 4 \cdot 1 \cdot 19 = -80 < 0$ and the system is underdamped. The roots of the characteristic equation in this case are $r_1 = -2 - 3i\sqrt{5}$ and $r_2 = -2 + 3i\sqrt{5}$. Hence, the general solution of (3.11.9) is:

$$x(t) = e^{-2t}\left[c_1 \cos\left(3\sqrt{5}t\right) + c_2 \sin\left(3\sqrt{5}t\right)\right].$$

Thus, $x(0) = c_1 = 0.5$ and

$$x'(t) = -2e^{-2t}\left[c_1 \cos\left(3\sqrt{5}t\right) + c_2 \sin\left(3\sqrt{5}t\right)\right]$$
$$+ e^{-2t}\left[-3\sqrt{5}c_1 \sin\left(3\sqrt{5}t\right) + 3\sqrt{5}c_2 \cos\left(3\sqrt{5}t\right)\right].$$

Hence, $x'(0) = -2c_1 + 3\sqrt{5}c_2 = 0$ or $c_2 = 2c_1/3\sqrt{5} = \sqrt{5}/15$. Thus, the equation of the motion is:

$$x(t) = e^{-2t}\left[0.5\cos\left(3\sqrt{5}t\right) + \frac{\sqrt{5}}{15}\sin\left(3\sqrt{5}t\right)\right].$$

3.12. INTRODUCTION TO PDE: BASIC CONCEPTS

Definition 3.12.1
A *partial differential equation (PDE)* is an equation that involves one or more partial derivatives of an unknown function, say u, that depends on two or more variables. The *order* of a PDE is the order of highest derivative involved. Thus, an nth-order partial differential equation is an equation of the form:

$$F\left(x, y, \ldots, u, \frac{\partial u}{\partial x}, \frac{\partial u}{\partial y}, \ldots, \frac{\partial^n u}{\partial x^n}, \ldots\right) = 0. \qquad (3.12.1)$$

There are two common notations for partial derivatives and we shall use them interchangeably. For instance, if $u = u(x, y, z)$ is a function of three independent variables, partial derivatives of different orders of u with respect to the each of the individual variables x, y, and z will be denoted by (1) $\partial u/\partial x$, $\partial^2 u/\partial x^2$, $\partial u/\partial y$, $\partial^3 u/\partial z^3$, $\partial^2 u/\partial x \partial y$, $\ldots$, (the symbol ∂ is read "partial") and (2) u_x, u_{xx}, u_y, u_{zzz}, u_{xy}, $\ldots$, respectively.

Very well-known and commonly discussed partial differential equations are (1) simple first-order equations that arise as models of wave phenomena and (2) linear second-order partial differential equations of (a) the heat, modeling thermodynamics in a continuous medium; (b) the wave, modeling vibrations of bars, strings, plates, solid bodies, and electromagnetic vibrations; and (c) the Laplace and its inhomogeneous counterpart, the Poisson, governing the mechanical and thermal equilibriums of bodies, as well as fluid mechanical and electromagnetic potentials.

Definition 3.12.2
A partial differential equation is said to be *linear* if it is of the first degree in the unknown function, say u, and its partial derivatives. Otherwise, it is said to be *nonlinear*. A linear partial differential equation is called *homogeneous* if each term of the equation contains either the function u or one of its partial derivatives. Otherwise, it is called *nonhomogeneous* or *inhomogeneous*. Thus, a linear first-order partial differential equation on two variables x and y, non-homogeneous and homogeneous, are, respectively, of the type:

$$A(x, y)\frac{\partial u}{\partial x} + B(x, y)\frac{\partial u}{\partial y} = C(x, y)u(x, y), \qquad (3.12.2)$$

where $u = u(x, y)$ and

$$A(x, y)\frac{\partial u}{\partial x} + B(x, y)\frac{\partial u}{\partial y} = 0. \qquad (3.12.3)$$

Example 3.12.1
Here are some examples of partial differential equations with their orders listed:

(a) $u_t + cu_x = 0$, linear homogeneous first-order *transport equation.*
(b) $u_t + uu_x = 0$, nonlinear first-order *Poisson/Riemann transport equation.*
(c) $\partial^2 u/\partial t^2 = c^2(\partial^2 u/\partial t^2)$, linear, second-order, homogeneous, one-dimensional *wave equation.*
(d) $\partial u/\partial t = k(\partial^2 u/\partial x^2 + \partial^2 u/\partial y^2)$, linear, second-order, homogeneous, two-dimensional *heat equation.*
(e) $\partial^2 u/\partial x^2 + \partial^2 u/\partial y^2 = f(x, y)$, linear, second-order, inhomogeneous two-dimensional *Poisson equation.*
(f) $\partial^2 u/\partial x^2 + \partial^2 u/\partial y^2 + \partial^2 u/\partial z^2 = 0$, linear, second order, homogeneous, three-dimensional *Laplace equation.*
(g) $u_t + uu_x + u_{xxx} = 0$, *KdV equation.*

Definition 3.12.3
By a *solution* of a partial differential equation in some region D of the space of independent variables, it is meant a sufficiently smooth function, say u, of the independent variables that possess all partial derivatives appearing in the equation under consideration in the space D and satisfies the given partial differential equation at every point of its domain D.

Notes:

(a) It is not necessarily required that the solution be defined for all possible values of the independent variables.

(b) In general, the domain D is an open set, usually connected and often bounded with a reasonably nice boundary.

(c) Unlike the theory of ordinary differential equations, which relies on the "existence and uniqueness theorem," there is no single theorem, which is central to the partial differential equations. Instead, there are separate theories used for each of the major types of partial differential equations that commonly arise, although there are several basic skills, which are essential for studying all types of partial differential equations.

(d) The technique of linear algebra proves to be effective for linear partial differential equations.

Example 3.12.2
Let us solve the PDE:

$$u_{yy} - 9u = 0. \tag{3.12.4}$$

Solution
Since the given equation, (3.12.4), involves a derivative with respect to only one variable, that is, y, it can be solved by methods discussed in Section 3.8 as an ordinary differential equation:

$$\frac{d^2u}{dy^2} - 9u = 0. \tag{3.12.5}$$

Again, from Section 3.8, the general solution of (3.12.5) is $u = c_1 e^{3y} + c_2 e^{-3y}$, where c_1 and c_2 are constants. These constants may be functions of the second variable, x. That is, the general solution of (3.12.4) and, thus, of (3.12.4), is:

$$u(x, y) = c_1(x)e^{3y} + c_2(x)e^{-3y},$$

where $c_1(x)$ and $c_2(x)$ are arbitrary functions of x.

3.12.1. First-Order Partial Differential Equations

First-order partial differential equations arise most naturally from the simplest models of exchange processes and other physical conservation laws. By the "simplest model" of physical processes, it is referring to those that neglect the dispersive effects of heat conduction, diffusion, or viscosity. The first-order PDE also arises in many problems from chemical engineering and mathematics.

Example 3.12.3. *Cauchy–Riemann Equations*
Suppose $z = x + iy$ is a complex number and $f(z)$ is a function of z. We want to know when $f(z)$ is differentiable.

Answer
In the theory of functions of complex variables, it has been proved that if

$$f(z) = u(x, y) + iv(x, y) \tag{3.12.6}$$

is a function of a complex variable defined in a domain D of the complex plan, a necessary and sufficient condition for $f(z)$ to be differentiable at $z = z_0 = x_0 + iy_0$ is that $u(x, y)$ and $v(x, y)$ should be differentiable functions of the real variables x and y and that the first derivatives should satisfy the *Cauchy–Riemann conditions*:

$$\frac{\partial u}{\partial x} = \frac{\partial v}{\partial y} \quad \text{and} \quad \frac{\partial v}{\partial x} = -\frac{\partial u}{\partial y}. \tag{3.12.7}$$

The two equations in (3.12.7) represent the system of first-order PDE.

Definition 3.12.4
A *quasilinear* partial differential equation is a slightly more general form of linear, and with two variables x and y, it is of the form:

$$A(x, y, u)\frac{\partial u}{\partial x} + B(x, y, u)\frac{\partial u}{\partial y} = C(x, y, u), \tag{3.12.8}$$

Example 3.12.4
The following are examples of quasilinear equations:

(a) $x(\partial u/\partial x) - y(\partial u/\partial y) = x - y$.
(b) $x(\partial u/\partial y) - y(\partial u/\partial y) = 2xy$.
(c) $(3y - 2u)(\partial u/\partial x) + (u - 3x)(\partial u/\partial y) = 2x - y$.

Now, let us consider the following quasilinear first-order partial equation:

$$P(x, y, z)\frac{\partial z}{\partial x} + Q(x, y, z)\frac{\partial z}{\partial y} = R(x, y, z), \tag{3.12.9}$$

where P, Q, and R are arbitrary functions of x, y, and z.

Theorem 3.12.1

The general solution of the quasilinear first-order partial equation (3.12.9) is:

$$F(u, v) = 0, \tag{3.12.10}$$

where F is an arbitrary differentiable function of u and v, and

$$u(x, y, z) = c_1 \text{ and } v(x, y, z) = c_2 \tag{3.12.11}$$

are two independent solutions of the so-called *characteristic equation* (due to Lagrange):

$$\frac{dx}{P} = \frac{dy}{Q} = \frac{dz}{R}. \tag{3.12.12}$$

Proof:

See Rhee et al. (2001, p. 102).

We note that by the method of characteristics, the equation reduces a quasilinear first-order partial differential equation to a system of ordinary differential equations (3.12.12). The disadvantage of this method is that it gives the solution u as a function on a characteristic curve, not as a function on the points of a rectangular grid in $\mathbb{R}^2$

To find the general solution of (3.12.9), we need to solve the system of equations (3.12.12) to obtain (3.12.11). The general solution, then, will be found by (3.12.10).

Example 3.12.5

Let us solve the equation:

$$x^2 \frac{\partial z}{\partial x} - y^2 \frac{\partial z}{\partial y} = (x - y)z. \tag{3.12.13}$$

Solution

From (3.12.12), the characteristic equation of (3.12.13) is:

$$\frac{dx}{x^2} = \frac{dy}{-y^2} = \frac{dz}{(x - y)z}. \tag{3.12.14}$$

Integrating the left equality of (3.12.14), that is, $dx/x^2 = dy/-y^2$, we have:

$$\frac{1}{x} + \frac{1}{y} = c_1. \tag{3.12.15}$$

From (3.12.14), by adding the numerator and denominator of the left equality, that is, $(dx + dy/x^2 - y^2) = (dz/(x - y)z)$, we have:

$$\frac{d(x+y)}{x+y} = \frac{dz}{z}$$

or

$$\frac{z}{x+y} = c_2. \tag{3.12.16}$$

Hence, from (3.12.10), the general solution of the equation under consideration, (3.12.13), is:

$$F\left(\frac{1}{x}+\frac{1}{y}, \frac{z}{x+y}\right) = 0$$

or

$$\frac{z}{x+y} = G\left(\frac{1}{x}+\frac{1}{y}\right),$$

where G is an arbitrary function.

3.12.2. Second-Order Partial Differential Equations

Definition 3.12.5
The most general form of a *second-order quasilinear* partial differential equation is:

$$Au_{xx} + 2Bu_{xy} + Cu_{yy} = F(x, y, u, u_x, u_y), \tag{3.12.17}$$

where $u = u(x, y)$.

We note that from (3.12.17), depending upon the type of the equation based on $AC - B^2$, we obtain equations of wave, heat, and Laplace, mentioned in Example 3.12.1, as follows:

$AC - B^2$	Type	Name in Example 3.12.1
<0	Hyperbolic	(c) Wave
$=0$	Parabolic	(d) Heat
>0	Elliptic	(f) Laplace

In other words, there are three basic types of quasilinear second-order partial differential equations that commonly arise in applications. They are: (1) *hyperbolic* (wave), (2) *parabolic* (heat), and (3) *elliptic* (Laplace). The solutions of the equations pertaining to each of the types have their own characteristic qualitative differences. Equations for these three types are:

1. **Wave or Vibration of a String (Hyperbolic) Equation:**

$$\frac{\partial^2 u}{\partial t^2} - \frac{\partial^2 u}{\partial t^2} = 0, \tag{3.12.18}$$

where the variables t and x denote the time and the spatial coordinate, respectively. This equation is often encountered in elasticity, aerodynamics, acoustics, and electrodynamics. The general solution of (3.12.18) is:

$$w = \theta(x+t) + \phi(x-t), \tag{3.12.19}$$

where $\theta(x)$ and $\phi(x)$ are arbitrary twice continuously differentiable functions. The physical interpretation of the solution (3.12.19) is two traveling waves of arbitrary shape that propagate to the right and to the left along the x-axis with a constant speed equal to 1.

2. **Heat (Parabolic) Equation:**

$$\frac{\partial u}{\partial t} - \frac{\partial^2 u}{\partial x^2} = 0, \tag{3.12.20}$$

where the variables t and x denote the time and the spatial coordinate, respectively. Equation (3.12.20) often represents the theory of heat and mass transfer. As in many partial differential equations, equation (3.12.20) has infinitely many particular solutions.

3. **Laplace (Elliptic) Equation:**

$$\frac{\partial^2 u}{\partial x^2} + \frac{\partial^2 u}{\partial y^2} = 0, \tag{3.12.21}$$

where x and y denote the spatial coordinates. Equation (3.12.21) often represents heat and mass transfer theory, fluid mechanics, elasticity, electrostatics, and other areas of mechanics and physics. A solution to the Laplace equation (3.12.21) is called a *harmonic function*.

3.12.2a Fourier Series and Partial Differential Equations

Definition 3.12.6

The *gradient* or *gradient vector* is a (row) vector operator, denoted by ∇, of the form:

$$\nabla f(x_1, x_2, \ldots, x_n) = \text{grad}(f) = \left(\frac{\partial f}{\partial x_1}, \frac{\partial f}{\partial x_2}, \ldots, \frac{\partial f}{\partial x_n} \right). \tag{3.12.22}$$

Definition 3.12.7

The *one-dimensional heat equation* for $u(x, t)$ is a partial differential equation, which describes the evolution of temperature $u(x, t)$ inside a media (such as a homogeneous metal rod). It is formulated as:

$$u_t = k u_{xx}, \quad x \in [a, b] \subset \mathbb{R}. \tag{3.12.23}$$

The *three-dimensional heat equation* is a (parabolic) partial differential equation that describes the heat distribution $u(x, y, z, t)$ on a region $D \subset \mathbb{R}^3$ over time t. It is formulated as:

$$\frac{\partial u}{\partial t} = k \left(\frac{\partial^2 u}{\partial x^2} + \frac{\partial^2 u}{\partial y^2} + \frac{\partial^2 u}{\partial z^2} \right) = c^2 \nabla^2 u, \quad (x, y, z) \in D \subset \mathbb{R}^3, \tag{3.12.24}$$

where $k = K/\sigma p$ is a positive constant, with K the thermal conductivity, σ the specific heat, and ρ the density of the material.

We note that the heat equation is useful, among other things, for the study of Brownian motion in stochastic processes.

The one-dimensional heat equation (3.12.23) can be solved using the Fourier series, if appropriate boundary conditions are provided.

Example 3.12.6

(a) The one-dimensional heat equation with initial condition (ID) and the *Dirichlet-type boundary conditions* (BC) is as follows:

$$u_t(x, t) = k u_{xx}(x, t), \quad x \in (0, L), \quad t > 0,$$
$$\text{IC:} \quad u(x, 0) = \varphi(x), \quad x \in (0, L), \tag{3.12.25}$$
$$\text{BC:} \quad u(0, t) = f_1(x), \quad u(L, t) = f_2(t), \quad t > 0.$$

(b) The one-dimensional heat equation with initial condition (ID) and the *Neumann-type boundary conditions* (BC) is as follows:

$$u_t(x, t) = k u_{xx}(x, t), \quad x \in (0, L), \quad t > 0,$$
$$\text{IC:} \quad u(x, 0) = \xi(x), \quad x \in (0, L), \tag{3.12.26}$$
$$\text{BC:} \quad u_x(0, t) = g_1(t), \quad u_x \in (L, t) = g_2(t), \quad t > 0.$$

(c) The one-dimensional heat equation with initial condition (ID) and the *mixed boundary conditions* (BC) is as follows:

$$u_t(x, t) = ku_{xx}(x, t), \quad x \in (0, L), \quad t > 0,$$
$$\text{IC:} \quad u(x, 0) = \eta(x), \quad x \in (0, L),$$
$$\text{BC:} \quad \alpha_1 u(0, t) + \beta_1 u_x(0, t) = h_1(t),$$
$$\alpha_2 u(L, t) + \beta_2 u_x(L, t) = h_2(t), \quad t > 0,$$

(3.12.27)

for some constants α_1, α_1, β_1, and β_2.

We note that when the BCs in all three types of PDEs in (3.12.25), (3.12.26), and (3.12.27) are reduced to $u(0, t) = u(L, t) = 0$ (laterally insulated bars) or $u_x(0, t) = u_x(L, t) = 0$ (end-insulated bars), they all can be solved by using the Fourier series. Given in the succeeding sections are the solutions to the previous examples:

Solution of (3.12.25)
In (3.12.25), let $f_1(t) = f_2(t) = 0$. Then, the solutions are of the form:

$$u(x, t) = \sum_{n=1}^{\infty} b_n e^{-k(n\pi/L)^2 t} \sin \frac{n\pi x}{L},$$

(3.12.28)

where $\{b_n\}_{n=1}^{\infty}$ can be determined from the Fourier sine series of the initial function φ, that is,

$$\varphi(t) = \sum_{n=1}^{\infty} b_n \sin \frac{n\pi x}{L},$$

(3.12.29)

$$b_n = \frac{2}{L} \int_0^L \varphi(x) \sin \frac{n\pi x}{L} dx, \quad n \geq 1.$$

(3.12.30)

Solution of (3.12.26)
In (3.12.26), let $g_1(t) = g_2(t) = 0$. Then, the solutions are of the form:

$$u(x, t) = \frac{a_0}{2} + \sum_{n=1}^{\infty} a_n e^{-k(n\pi/L)^2 t} \cos \frac{n\pi x}{L},$$

(3.12.31)

where $\{a_n\}_{n=0}^{\infty}$ can be determined from the Fourier cosine series of the initial function ξ, that is,

$$\xi(t) = \frac{a_0}{2} + \sum_{n=1}^{\infty} a_n \cos \frac{n\pi x}{L},$$

(3.12.32)

$$a_n = \frac{2}{L} \int_0^L \xi(x) \cos \frac{n\pi x}{L} dx, \quad n \geq 0.$$

(3.12.33)

Example 3.12.7

Let us solve the following first-dimensional heat equation with initial and *Dirichlet-type boundary conditions*:

$$u_t(x,t) = 2u_{xx}(x,t), \quad x \in (0,8), \quad t > 0,$$

$$\text{IC:} \quad u(x,0) = \varphi(x) = \begin{cases} x & \text{if } 0 < x < 4, \\ 8-x & \text{if } 4 < x < 8, \end{cases}$$

$$\text{BC:} \quad u(0,t) = u(8,t) = 0, \quad t > 0.$$

Solution

Assuming $k = 2$ in (3.12.28), then it follows from (3.12.28), (3.12.29), and (3.12.30) that:

$$b_n = \frac{2}{8}\left[\int_0^4 x\sin\frac{n\pi x}{8}\,dx + \int_4^8 (8-x)\sin\frac{n\pi x}{8}\,dx\right]$$

$$= \begin{cases} 0, & \text{if } n = 2k, \quad k = 1,2,\ldots, \\ (-1)^k \dfrac{32}{(2k-1)^2 \pi^2}, & \text{if } n = 2k-1, \quad k = 1,2,\ldots, \end{cases}$$

and

$$u(x,t) = \sum_{n=1}^{\infty} b_n e^{-2(n\pi/8)^2 t}\sin\frac{n\pi x}{8}$$

$$= \frac{32}{\pi^2}\sum_{n=1}^{\infty}\frac{(-1)^k}{(2k-1)^2} e^{-\frac{(2k-1)^2\pi^2 t}{32}}\sin\frac{(2k-1)\pi x}{8}.$$

Example 3.12.8

We now want to solve the following heat equation with initial value and *Neumann-type boundary conditions*:

$$u_t(x,t) = \frac{3}{2}u_{xx}(x,t), \quad x \in (0,12), \quad t > 0,$$

$$\text{IC:} \quad u(x,0) = \xi(x) = \begin{cases} 2x & \text{if } 0 < x < 4, \\ 8 & \text{if } 4 < x < 12, \end{cases}$$

$$\text{BC:} \quad u_x(0,t) = u_x(12,t) = 0, \quad t > 0.$$

Solution

By using (3.12.32) and (3.12.33), we have:

$$a_0 = \frac{2}{12}\left[\int_0^4 2x\,dx + \int_4^{12} 8\,dx\right] = \frac{40}{3},$$

$$a_n = \frac{2}{12}\left[\int_0^4 2x\cos\frac{n\pi x}{12}\,dx + \int_4^{12} 8\cos\frac{n\pi x}{12}\,dx\right]$$

$$= -\frac{48\left(1-\cos\dfrac{n\pi}{3}\right)}{n^2\pi^2}, \quad n \geq 1,$$

and

$$u(x,t) = \frac{a_0}{2} + \sum_{n=1}^{\infty} a_n e^{-3/2(n\pi/12)^2 t}\cos\frac{n\pi x}{12}$$

$$= \frac{20}{3} - \frac{48}{\pi^2}\sum_{n=1}^{\infty}\frac{1-\cos\dfrac{n\pi}{3}}{n^2}e^{-n^2\pi^2 t/96}\cos\frac{n\pi x}{12}.$$

EXERCISES

3.1. For each of the following ordinary differential equations, determine its order and whether or not it is linear:

a. $3x^2 y'' - xy' + 2y = \sin x$.

b. $(2 - y^2)(dy/dx^2) - x(dy/dx) + y = e^x$.

c. $y''' + (\cos x)y'' + 3y' - (\cos^2 x)y = x^4$.

d. $y'' - (y')^2 + 2y = 0$.

3.2. For each of the following differential equations, verify that the given function is its solution on some interval:

a. $y(x) = \cosh x, \; y'' - y = 0$.

b. $y(x) = \cos x(\ln\cos x) + x\sin x, \; y'' + y = \sec x, \; 0 < x < \pi/2$.

c. $y(x) = \tan((x^3/3) + 5)x, \; y' = x^2(1 + y^2)$.

d. $y(x) = (x^2/3) - (1/x), \; xy' + y = x^2$.

3.3. Solve the following initial value problems by the separation of variables method:

a. $(xy^2 + x)dx - (x^2 y - y)dy = 0, \; y(2) = 5$.

b. $2y' + \sin(x + y) = \sin(x - y), \; y(0) = \pi$.

c. $xy\,dx + \sqrt{1-x^2}\,dy = 0, \quad y(0) = 3e$.

d. $\sqrt{1-y^2}\,dx + y\sqrt{1-x^2}\,dy = 0, \quad y(0) = 1$.

3.4. Solve the following linear differential equations:

a. $y' - 2y = x^2 e^{2x}$.

b. $y' + y\tan x = \cos x$.

c. $y' + (1/x)y = 2\ln x + 1$.

d. $(x + y^2)\,dy = y\,dx$.

3.5. Solve the following initial value problems:

    a. $xy' + 2y = \sin x$, $y(\pi) = 1/\pi$.

    b. $(xy' - 1) \ln x = 2y = \sin x$, $y(e) = 0$.

3.6. Show that the function $g'(y)$, given in (3.5.17), is independent of x.

3.7. Verify that the function $f(x, y)$, as defined by (3.5.15), satisfies the conditions in (3.5.13).

3.8. Determine whether or not each of the following equations is exact. If so, find the solution.

    a. $(2xy^2 + 2y) + (2x^2y + 2x)y' = 0$, $y(1) = 3$.

    b. $((y/x) + 6x)dx + (\ln x - 2)dy = 0$, $y(e) = 1$.

    c. $2x \cos^2 y \, dx + (2y - x^2 \sin^2 y) \, dy = 0$, $y(1) = 0$.

    d. $(x \ln y - xy) \, dx + (y \ln x - xy) \, dy = 0$, $y(1) = 1$.

3.9. Find the integrating factor for each of the following differential equations and solve them:

    a. $2xy \, dx + (y^2 - 3x^2) \, dy = 0$, $y(2) = -3$.

    b. $e^x \, dx + (e^x \cot y + 2y \csc y) \, dy = 0$.

3.10. Show that after substitution (3.6.11), the differential equation (3.6.10) will be separable.

3.11. Show that the following differential equations are homogeneous and then solve them:

    a. $xy' - y = xe^{y/x}$.

    b. $3xyy' = x^2 - y^2$, $y(1) = 1$.

    c. $(2x - y)y' = 4y - 3x$, $y(0) = 2$.

    d. $xy' - y = x^2 + xy/x - y$, $y(1) = 3$.

3.12. Solve the following Bernoulli differential equations:

    a. $y' + y = xe^x \sqrt{y}$, $y(0) = 4$.

    b. $y' - (\tan x)y = (\cos x)y^4$, $y(0) = 3$.

    c. $y' = xy^3 - y$, $y(0) = 1$.

3.13. Solve the following initial value problems:

    a. $y' = (x + y + 1)^2$, $y(0) = \sqrt{3} - 1$.

    b. $y' - 2 = \sqrt{y - 2x + 3}$, $y(1) = 3$.

3.14. Suppose an extra-large cup of hot tea at 90°C is handed to a customer at a coffee shop. Two minutes later, the temperature of the tea becomes 80°C. The temperature of the coffee shop is kept at 25°C. After how many minutes is the tea's temperature will be at 60°C, which is the temperature preferred by the customer?

3.15. Suppose a pack of frozen food at 32°F is taken out from the freezer by a customer and is sitting in the shopping cart in a grocery store at 70°F.

Five minutes later, the temperature of the pack becomes 40°F. When will the pack's temperature become 60°F?

3.16. Find the general solution of the following differential equations. When the initial condition is provided, find the particular solution as well.

a. $y'' + 2y' - 3y = 0$.

b. $4y'' - 4y' + y$, $y(0) = -1$, $y'(0) = -2$.

c. $y'' + 6y' + 13y$, $y(0) = 3$, $y'(0) = -1$.

3.17. Use the method of undetermined coefficients to find the general solution of the following nonhomogeneous differential equations. When the initial condition is provided, find the particular solution as well.

a. $y'' - 2y' + y = 2e^x$.

b. $y'' + y = x \sin 2x$, $y(0) = 5/9$, $y'(0) = 2$.

c. $y'' - y = (4x - 6)e^{-x}$.

d. $y'' - 2y' + y = 5 \cos 2x + \sin 2x$, $y(0) = -1$, $y'(0) = 2$.

3.18. Use the variation of parameter method to find the general solution of the following differential equations:

a. $y'' + 9y = \tan 3x$.

b. $y'' + y = \sec x$.

c. $y'' + 6y' + 9y = (e^{-3x}/x^3)$, $y(1) = 0$, $y'(1) = 0$.

3.19. Solve the following Cauchy–Euler differential equations:

a. $x^2 y'' + xy' + 4y = 0$, $x > 0$.

b. $x^2 y'' - xy' + y = x$, $x > 0$, $y(1) = 1$, $y'(1) = 4$.

3.20. For logistic differential equation (3.1.20), show that when $\alpha \neq 1$, similar to the case when $\alpha = 1$, we can find the solution of (3.1.20) with an initial value to be (3.3.21).

3.21. Solve the following systems of differential equations by elimination and Laplace transform methods:

a. $\begin{cases} y + x' = 0 \\ y' - 2x - 2y = 0 \end{cases}$ $x(0) = y(0) = 2$.

b. $\begin{cases} x' = y - z, \\ y' = y + x, \\ z' = x + z \end{cases}$ $x(0) = 2$, $y(0) = 3$, $z(0) = 4$.

3.22. Use the Laplace transform to solve the following initial value problems with discontinuous input:

a. $y'' + y = f(x) = \begin{cases} 2, & 0 \leq x < \pi \\ 0, & x \leq \pi, \end{cases}$ $y(0) = y'(0) = 0$.

b. $y'' - 3y' = f(x) = \begin{cases} 2, & 0 \leq x < 2 \\ 4, & x \geq 2, \end{cases}$ $y(0) = 1$, $y'(0) = -1$.

c. $y'' + 4y = f(x) = \begin{cases} \sin t, & 0 \le x < \pi \\ \cos, & x \ge \pi, \end{cases}$ $y(0) = y'(0) = 0.$

d. $y'' + 4y' + 4y = f(x) = \begin{cases} e^x, & 0 \le x < 2, \\ e^x - 1, & x \ge 2, \end{cases}$ $y(0) = 1, \; y'(0) = 0.$

3.23. Solve the following initial value problems using Laplace transform:

a. $y'' + 2y' + 5y = 0, \; y(0) = 2, \; y'(0) = -2.$

b. $y'' - 2y' + 3y = 0, \; y(0) = 3, \; y'(0) = 1.$

c. $y'' + 2y' = \cos(3x), \; y(0) = y'(0) = 2.$

d. $y'' - 6y' + 15y = 2\sin(3t), \; y(0) = -1, \; y'(0) = 2.$

e. $y^{(4)} - y = 0, \; y(0) = 1, \; y'(0) = y''(0) = y'''(0) = 0.$

f. $y'' - 3y' + 2y = \begin{cases} 6e^{-t}, & \text{if } t \in (0, 2), \\ 0, & \text{if } t \in (2, \infty), \end{cases}$ $y(0) = 0, \; y'(0) = 1.$

3.24. Solve the following initial value problems with impulse inputs:

a. $y'' + y' = \delta(t-1) + \delta(t-2), \; y(0) = 1, \; y'(0) = -2.$

b. $y'' + 9y = e^{-t} + \delta(t-1), \; y(0) = 0, \; y'(0) = 3.$

c. $y'' + 16y = \cos t + \delta(t - \pi), \; y(0) = 0, \; y'(0) = 1.$

d. $y'' + 3y' + 2y = 1 + \delta(t-1), \; y(0) = -1, \; y'(0) = 1.$

3.25. Suppose a mechanical system has a mass of 100 kg, a damping of 200 kg/s, and a spring constant k newtons per meter. The mass is initially at equilibrium position with initial velocity of 20 m/s. For each case of (a) $k = 0$ and (b) $k = 100$, determine the displacement.

3.26. Suppose a damped mechanical system has parameters $m = 4$ kg, $b = 11$ kg/m, and $k = 9$ N/m. At time $t = 0$, the weight is at the equilibrium position with initial velocity of 90 m/s. Determine the motion of the weight as a function of time.

3.27. Find the charge Q and current $I = dQ/dt$ in the given RLC circuit if at $t = 0$ the charge on the capacitor and current in the circuit are zero:

a. $(1/10)Q'' + 4Q + 200Q = 0.$

b. $(1/10)Q'' + 4Q + 200Q = \cos 4t.$

c. $(1/20)Q'' + 6Q + 100Q = 2\sin 3t.$

d. $(1/5)Q'' + 6Q + 45Q = \sin 3t + \cos 3t.$

e. $LQ'' + RQ' + (1/C)Q = E(t).$

3.28. Find the recurrence relation and two linearly independent solutions in power series of $x, -x_0$ for the given differential equations:

a. $y'' + y = 0, \; x_0 = 0.$

b. $y'' - y = 0, \; x_0 = 0.$

c. $y'' - xy' - y = 0, \; x_0 = 0.$

3.29. Solve the following first-order partial differential equations:
  a. $y(\partial z/\partial x) + x(\partial z/\partial y) = x - y$.
  b. $xy(\partial z/\partial x) - x^2(\partial z/\partial y) = yz$.

3.30. Solve the heat equation with Dirichlet-type boundary conditions:

$$u_t = 2u_{xx}, \quad 0 < x < 3; t > 0,$$
$$u(0, t) = u(3, t) = 0,$$
$$u(x, 0) = 3\sin 6\pi x - 7\sin 8\pi x + 4\sin 12\pi x.$$

3.31. Consider the following heat equation:

$$u_t = u_{xx}, \quad 0 < x < 4; t > 0,$$
$$u(0, t) = u(4, t) = 0,$$
$$u(x, 0) = 3\sin 5x - 7\sin 7x.$$

Solve the given boundary conditions problem with:
a. Dirichlet type
b. Neumann type

3.32. Solve the heat equation with Neumann-type boundary conditions:

$$u_t = u_{xx}, \quad 0 < x < 6; t > 0,$$
$$u_x(0, t) = u_x(6, t) = 0,$$
$$u(x, 0) = x + 1.$$

CHAPTER FOUR

Difference Equations

We are going to reiterate what we said at the introduction of Chapter 3 and add to it that the theories of differential and difference equations form two extreme representations of real-world problems. For example, we saw a simple population growth model when represented as a differential equation. For the corresponding discrete analog, we refer to May (1975), which tells it all. The abstract contains the following statements:

> For biological populations with nonoverlapping generations, population growth takes place in discrete time steps and is described by difference equations. Some of the simplest such nonlinear difference equations can exhibit a remarkable spectrum of dynamical behavior, from stable equilibrium points, to stable cyclic oscillations between two population points, to stable cycles with four points, then eight, 16, etc., points, through to a chaotic regime in which (depending on the initial population value) cycles of *any* period, or even totally aperiodic but bounded population fluctuations, can occur. This rich dynamical structure is overlooked in conventional linearized stability analyses; its existence in the simplest and fully deterministic nonlinear ("density dependent") difference equations is a fact of considerable mathematical and ecological interest.

The field of difference equations has wide applications. The modern development of calculus of differences began with a memoir by Poincaré published in 1885. The theory of difference equations, the methods used, and their wide application have progressed to such an extent that they occupy a central position in applicable analysis. In fact, in the last 12 years, hundreds of research articles and several monographs have been published and many international conferences and numerous special sessions have been convened. However, it seems that only minimal progress has been made in the development of a systematic theory of nonlinear difference equations.

To see how a difference equation develops, suppose we want to study the behavior changes of a subject under certain experimental conditions. For instance, we may think of a subject undergoing a sequence of events, starting

Difference and Differential Equations with Applications in Queueing Theory, First Edition.
Aliakbar Montazer Haghighi and Dimitar P. Mishev.

with the perception of a stimulus, followed by the performance of a response, and ending with the occurrence of an environmental event. So let us imagine an experiment in which a subject is repeatedly exposed to this sequence of events (stimulus–response–environmental event). We consider dividing the experiment into stages, each stage being a trial during which the subject is run through the sequence. The subject's level of performance is then a function of the trial number, denoted by n. Now we let P_n denote the probability of the response (during the specified time interval following the stimulus) in the nth trial. The number P_0 denotes the initial value describing the disposition of the subject toward the response when the subject is first introduced to the experiment proper. Hence, the function p will be defined with the domain containing n-values $0, 1, 2, \ldots$. Assuming P to be a probability implies that $0 \le P_n \le 1$, $n = 0, 1, \ldots$, which includes the extremes of no response and certain response with the values 0 and 1, respectively.

The calculus of difference and differential–difference equations has increasingly attracted the interest of mathematicians, scientists, engineers, and other professionals. This is perhaps due to the availability of high-speed computers and the numerous applications of difference equations to engineering, sciences (such as physics, chemistry, biology, probability, and statistics), economics, and psychology. The mathematical theory is of interest in itself, especially in view of the analogy of the theory to differential equations. In regard to the latter, as we have already alluded to, one main difference between difference equations and differential equations is that difference equations are equations that involve discrete changes of an unknown function, while differential equations involve instantaneous rates of changes of an unknown function.

Hence, the theory and solutions of difference equations in many ways are parallel to the theory and solutions of differential equations. In fact, the derivative of a function of one variable defined as a limit of a difference quotient is the main reason for many analogies between the differential calculus and the calculus of finite differences. In other words, as we may recall from differential calculus, the derivative of a given function $f(x)$ is defined as:

$$f'(x) = \lim_{h \to 0} \frac{f(x+h) - f(x)}{h} = \lim_{h \to 0} \frac{\Delta[f(x)]}{h},$$

if the limit exists, where Δ denotes the *difference operator*, that is, $\Delta f(x) = f(x+h) - f(x)$. For higher-order differences, we may continue the Δ operator.

Example 4.1.
As an example of the similarities, we consider $\Delta^2 f(x) = \Delta[\Delta f(x)]$, and so on. Now let us denote by $Df(x)$ the derivative of $f(x)$, that is, when D applies to a function, it results in the derivative of the function. Thus, we may denote higher derivatives of $f(x)$ by $D[Df(x)]$ or $D^2 f(x)$ and so on. Hence, similarities should now be clear. To generalize the idea, let $y = f(x)$, then will we have the following sample table:

Differential Calculus	Difference Calculus
$Dy = \lim_{h \to 0} \dfrac{\Delta[f(x)]}{h}$	$\Delta y = f(x+h) - f(x)$
$D^n = D[D^{(n-1)}f(x)], n = 1, 2, \ldots$	$\Delta^n y = \Delta[\Delta^{n-1}f(x)]$
$D(cy) = cDy$, where c is a constant	$\Delta(cy) = c\Delta y$, where c is a constant

Example 4.2.
As derivatives calculate the rate of change, difference operators are similar. For instance, the average speed in a time interval h, from x to $x + h$, can be calculated as the ratio of distances of differences of y and x; that is, let $h = \Delta x$ and $y(x)$ the displacement along a straight line. Then, the speed will be $\Delta y(x)/\Delta x$. The instantaneous speed at time instant x is the limit (as $\Delta x \to 0$) of the average speed. Hence,

$$\lim_{\Delta x \to 0} \frac{\Delta y(x)}{\Delta x} = Dy(x).$$

Example 4.3.
Let $y(x)$ represent the total cost of producing x units of a production in a factory. The average rate, then, of the change of cost with respect to the number of units produced within an interval of production number variation from x to $x + \Delta x$ is $\Delta y(x)/\Delta x$. The corresponding instantaneous rate of change will be $Dy(x)$, which is referred to as the marginal cost at output x.

A difference equation surfaces when the system under consideration depends upon one or more discrete variables. Generally, a difference equation is an equation that contains sequence differences. Thus, a difference equation is solved by finding a sequence that satisfies the equation. The sequence so found is called a solution of the equation. As it can be seen, we will be using the term "sequence" in relation to the discussion of difference equation. Because of this, we start by formally defining the basic terms needed.

4.1. BASIC TERMS

We have seen the definition of a function and its applications in the previous chapters. An immediate use of this concept is the following definition. A *sequence* is a function whose domain is the set of nonnegative integers. In other words, a sequence is a function, say f, from either $\{0, 1, 2, \ldots\}$ or $\{0, 1, 2, \ldots, n\}$, where n is some natural number, to a set, say S, of objects such as $\{s, s, \ldots\}$ or $\{s, s, \ldots, s\}$; the result could be an infinite sequence $\{s_0, s_1, s_2, \ldots\} = \{s_i, i \in N \cup \{0\}\}$ or a finite sequence $\{s_0, s_1, s_2, \ldots, s_n\}$, where n is some natural number. Hence, a sequence is a list of numbers such as $s_0, s_1, s_2, \ldots, s_n$ or $s_0, s_1, s_2, \ldots$.

We note that sequences play important roles in mathematics, engineering, science, and other areas of application of mathematics. We also note that the domain of a sequence could be the set of nonnegative integers or the set of natural numbers, although earlier we chose the former.

Example 4.1.1.
Let $S = \{3, 2, 4, 3, 5, 6, 1\}$. The finite sequence f, then, is the function such that $f: 1 \to 3, 2 \to 2, 3 \to 4, 4 \to 3, 5 \to 5, 6 \to 6, 7 \to 1$. For an infinite sequence, let $S = \{3, 2, 4, 3, \ldots\}$. The infinite sequence f, then, is the function such that $f: 1 \to 3, 2 \to 2, 3 \to 4, 4 \to 3, \ldots$.
We now start the subject of the chapter.

Definition 4.1.1.
Given a function F and a set S, for k in S and some positive integer n, a *difference equation* over a given set S is an equation of the form:

$$F(k; y_k, y_{k+1}, \ldots, y_{k+n}) = 0. \tag{4.1.1}$$

The *order* of a difference equation is the difference between the highest and the lowest indices that appear in the equation.

Notes:

1. A difference equation is an equation that involves differences. Thus, $\Delta y(x) + 4y(x) = 0$ is a difference equation. However, just by looking at the equation, one cannot tell if it is defined on the set of real numbers, positive integers, or some other set of numbers. Hence, the set on which the equation is to be defined on must be explicitly stated, otherwise it may be assumed. Thus, for instances, we may assume S is the set of positive integers. Hence, we use the symbol k or n instead of x to denote the number in the domain of the functions related by difference equations. We also write y_k or y_n for $y(x)$ to indicate the value of y at k or n, respectively. Thus, equation $\Delta y(x) + 4y(x) = 0$ may be written as $\Delta y_k + 4y_k = 0$. The range of values of the integer k, of course, must be stated. We further note that the choices of k or n are due to discreteness property of the difference equation.

2. Unless otherwise stated, all difference operators are taken with a difference interval of length 1.

3. The equation (4.1.1) is a difference equation of order n, because $k + n - k = n$.

4. The shifting indices has no bearing in the order of the difference equation. Hence, the following equation:

$$F(k+r; y_{k+r}, y_{k+r+1}, \ldots, y_{k+n+r}) = 0 \tag{4.1.2}$$

still is of order n, which is equivalent to (4.1.1).

Definition 4.1.2.
By a *solution* of the difference equation, equation (4.1.1), it is meant a sequence
$\{y_k\}$ or $\{y_0, y_1, y_2, \ldots\}$, which satisfies the difference equation (4.1.1) for all values
of k in S. A *general solution* of (4.1.1) is one, which involves exactly n arbitrary
constants. A *particular solution* is a solution that is obtained from the general
solution by assigning values to the arbitrary constants.

Example 4.1.2.
The sequence $\{y_k\} = \{2^k\}$, $k = 0, 1, 2, \ldots$, is a solution of the difference equation
$y_{k+1} - 2y_k = 0$. This is because $y_{k+1} - 2y_k = 2^{k+1} - 2 \cdot 2^k = 0$, $k = 0, 1, 2, \ldots$.

Example 4.1.3.
The following sequence

$$\{y_k\} = \left\{1 - \frac{2}{k}\right\}, k = 1, 2, \ldots$$

is a solution of the difference equation $(k + 1)y_{k+1} + ky_k = 2k - 3$, $k = 1, 2, \ldots$
This is because:

$$y_{k+1} = 1 - \frac{2}{k+1}.$$

Hence, we should have:

$$(k+1)\left(1 - \frac{2}{k+1}\right) + k\left(1 - \frac{2}{k}\right) = 2k - 3.$$

Simplifying the left-hand side of this equation, we obtain $(k + 1) - 2 + k - 2 = 2k - 3$. This is the same as the right hand-side of the given equation.

Example 4.1.4.
Let us show that the function y given by:

$$y_n = 2^n(3 + 4n), n = 0, 1, \ldots \tag{4.1.3}$$

is a solution of:

$$y_{n+2} - 4y_{n+1} + 4y_n = 0, \quad n = 1, 1, \ldots. \tag{4.1.4}$$

Answer
For (4.1.3) to be a solution of (4.1.4), (4.1.3) should satisfy (4.1.4), that is,
substituting (4.1.3) into (4.1.4), we should show that the result is 0. Doing that,
we have:

$$2^{n+2}[3+4(n+2)]-4\cdot 2^{n+1}[3+4(n+1)]+4\cdot 2^{n}[3+4(n)]=0.$$

Thus, it is true that (4.1.3) is a solution of (4.1.4).

Definition 4.1.3.
A difference equation of the form:

$$f_0(k)y_{k+n}+f_1(k)y_{k+n-1}+\cdots+f_{n-1}(k)y_{k+1}+f_n(k)y_k = f(k), \qquad (4.1.5)$$

where each of f and f_i, $i=0,1,2,\ldots,n$, is a function of k defined for all values of $k\in S$, is called *linear* over the set S. If a difference equation is not linear, it is called *nonlinear*. The linear difference equation (4.1.5) is called of *order n*, if both $f_0(k)$ and $f_n(k)$ are different from zero at each point of S. In other words, the order of a difference equation, such as (4.1.5), is the difference between the highest and the lowest indices that appear in the equation (4.1.5) if and only if:

$$f_0(k)f_n(k)\neq 0, \forall k \in S. \qquad (4.1.6)$$

If $f(k)$ on the right-hand side of (4.1.5) is zero, then (4.1.5) is called a *homogeneous linear difference equation of order n*; otherwise, it is called *nonhomogeneous* or *inhomogeneous*.

Example 4.1.5.
The following are examples of linear and nonlinear difference equations over S, where for each equation, S is defined, order is listed, and it is stated if it is linear, nonlinear, homogeneous, or nonhomogeneous:

$7y_{k+1}-8y_k=0$, $S=\{0,1,2,\ldots\}$, order 1, linear, homogeneous.
$y_{k+2}-6y_k=2k$, $S=\{0,1,2\ldots\}$, order 2, linear, nonhomogeneous.
$ky_{k+2}-7y_{k+1}+3y_k=0$, $S=\{1,2\ldots\}$, order 2, linear, homogeneous.
$5y_{k+2}-y_k^3=0$, $S=\{0,1,2\ldots\}$, order 2, nonlinear, homogeneous.
$y_{k+3}-3y_{k+2}+6y_{k+1}-4y_k=-2k+5$, $S=\{0,1,2,\ldots\}$, order 3, linear, nonhomogeneous.

In general, an *ordinary difference equation* is a relation of the form:

$$y_{k+n}=F(k; y_k, y_{k+1},\ldots y_{k+n-1}), \quad k=0,1,2,\ldots, \qquad (4.1.7)$$

where the relation F is a well-defined function for each of its arguments. Hence, if an *initial condition* (or *value*) is given, the other elements of the sequence can be found. Now consider the nth order difference equation defined in (4.1.7). The following theorem guarantees not only the existence of a solution of a difference equation, but when a solution exists, it guarantees it is unique.

Theorem 4.1.1. *Existence and Uniqueness Theorem*
Suppose there are n initial conditions $y_0, y_1, \ldots, y_{n-1}$. Then, corresponding to each given initial condition, equation (4.1.7) has a unique solution.

Proof:
As we saw earlier, once $y_0, y_1, \ldots, y_{n-1}$ are known, the difference equation with $k = 0$ uniquely determines y_n. As soon we have y_n, we obtain the difference equation with $k = 1$, y_{n+1}. Continuing this process, all y_k for all $k \geq n$ will be found.

4.2. LINEAR HOMOGENEOUS DIFFERENCE EQUATIONS WITH CONSTANT COEFFICIENTS

In this section, we will discuss linear difference equations in detail.

Definition 4.2.1.
Generally, an *nth-order linear homogeneous difference equations with constant coefficients* is an equation of the form:

$$y_{k+n} + a_1 y_{k+n-1} + \cdots + a_{n-1} y_{k+1} + a_n y_k = 0, \tag{4.2.1}$$

where $a_1, a_2, \ldots, a_n$ are n given constants with $a_n \neq 0$.

Before giving examples of linear difference equations to solve, we give the following example of how we may generate an nth-order difference equation given the general member of a sequence.

Example 4.2.1.
Suppose the general term, y_k, of the sequence $\{y_k\}$, which is the solution of a difference equation to be generated, is given by:

$$y_k = C3^k, \tag{4.2.2}$$

where C is an arbitrary constant. Having one arbitrary constant implies that the difference equation should be of the first order. Thus:

$$y_{k+1} = C3^{k+1} = 3C3^k. \tag{4.2.3}$$

Therefore, from (4.2.2) and (4.2.3), we have:

$$y_{k+1} = 3y_k. \tag{4.2.4}$$

Now, to solve a homogeneous difference equation, we offer two methods of solution, namely, *recursive* and *characteristic equation*.

First, we shall discuss the *recursive method*.

Example 4.2.2.

Let us solve the first-order linear homogenous equation with constant coefficients:

$$y_{k+1} - 3y_k = 0, \quad k = 0, 1, 2, \dots. \tag{4.2.5}$$

Solution

We rewrite (4.2.5) as:

$$y_{k+1} = 3y_k, \quad k = 0, 1, 2, \dots. \tag{4.2.6}$$

The values of y_{k+1}, $k = 0, 1, 2, \dots$, can be found recursively as follows:

$$y_1 = 3y_0, \quad y_2 = 3y_1 = 3^2 y_0, \quad y_3 = 3y_2 = 3^3 y_0, \dots.$$

Hence, by mathematical induction, we will have a solution in the form $y_k = 3^k y_0$. If the value of y_0 is given, then the solution is completely determined.

Now, we shall discuss the *characteristic equation method*.

Definition 4.2.2.

For the nth-order linear homogeneous difference equation with constant coefficient given by (4.2.1), the nth degree polynomial equation:

$$r^n + a_1 r^{n-1} + \dots + a_{n-1} r + a_n = 0, \tag{4.2.7}$$

is called a *characteristic equation* (or *auxiliary equation*).

Theorem 4.2.1.

Let $r_i, i = 1, 2, \dots, n$ be a root of the characteristic equation (4.2.7).

(a) Then:

$$y_k^{(i)} = r_i^k, \quad i = 1, 2, \dots, n \tag{4.2.8}$$

is a solution of (4.2.1).

(b) If all n roots of the characteristic equation (4.2.7) are district real numbers, then the general solution of (4.2.1) is:

$$y_k = c_1 y_k^{(1)} + c_2 y_k^{(2)} + \dots + c_n y_k^{(n)}, \tag{4.2.9}$$

where $c_i, i = 1, 2, \dots, n$, are n arbitrary constants.

(c) If roots r_i of the characteristic equation (4.2.7) are with multiplicity m_i, $i = 1, 2, \ldots, l$, such that:

$$\sum_{i=1}^{l} m_i = n, \tag{4.2.10}$$

then the general solution for (4.2.1) is:

$$\begin{aligned} y_k = r_1^k \left(c_1^{(1)} + c_2^{(1)} k + \cdots + c_{m_1}^{(1)} k^{m_1 - 1} \right) + r_2^k \left(c_1^{(2)} + c_2^{(2)} k + \cdots + c_{m_2}^{(2)} k^{m_2 - 1} \right) \\ + \cdots + r_{m_l}^k \left(c_1^{(l)} + c_2^{(l)} k + \cdots + c_{m_l}^{(l)} k^{m_l - 1} \right), \end{aligned} \tag{4.2.11}$$

where $c_i^{(j)}$, $i = 1, \ldots, l; j = 1, 2, \ldots, m_i$, are arbitrary constants.

Proof:
See Mickens (1990, p. 124).

Example 4.2.3.
Let us solve the following linear second-order homogenous difference equation with constant coefficients b_1 and $b_2 \neq 0$:

$$y_{k+2} + b_1 y_{k+1} + b_2 y_k = 0. \tag{4.2.12}$$

Solution
Using characteristic equation method, let r_1 and r_2 be the two roots of the characteristic equation:

$$r^2 + b_1 r + b_2 = 0 \tag{4.2.13}$$

of (4.2.12). Then, the solution, Y_k, of (4.2.12), with two arbitrary constants c_1 and c_2, may be of the form in one of the following three cases:

Case 1. If r_1 and r_2 are *two distinct real roots*, then the general solution of (4.2.12) is:

$$Y_k = c_1 r_1^k + c_2 r_2^k. \tag{4.2.14}$$

Case 2. If r_1 and r_2 are *double real roots*, say r, then the general solution of (4.2.12) is:

$$Y_k = (c_1 + c_2 k) r^k. \tag{4.2.15}$$

Case 3. If r_1 and r_2 are *complex conjugate roots*, in their polar form as $\rho(\cos\theta \pm i \sin\theta)$, then the general solution of (4.2.12) is:

$$Y_k = c_1 \rho^k \cos(k\theta + c_2). \tag{4.2.16}$$

Example 4.2.4.
Let us solve the so-called *Fibonacci difference equation*:

$$y_{k+2} = y_{k+1} + y_k. \tag{4.2.17}$$

Solution
The characteristic equation of (4.2.17) is $r^2 - r - 1 = 0$, which has two distinct real roots:

$$r_1 = \frac{1-\sqrt{5}}{2} \quad \text{and} \quad r_1 = \frac{1+\sqrt{5}}{2}.$$

Hence, the general solution of (4.2.17), denoted by Y_k, is:

$$Y_k = a_1 \left(\frac{1-\sqrt{5}}{2}\right)^k + a_2 \left(\frac{1+\sqrt{5}}{2}\right)^k,$$

where a_1 and a_2 are arbitrary constants.

Example 4.2.5.
Let us find the general solution of the linear second-order equation with constant coefficients:

$$9y_{k+2} + 6y_{k+1} + y_k = 0. \tag{4.2.18}$$

Solution
The characteristic equation for this difference equation is $9r^2 + 6r + 1 = 0$, which has a double real root $r = -1/3$. Hence, the general solution of (4.2.18), denoted by Y_k, is:

$$Y_k = (c_1 + c_2 k)\left(-\frac{1}{3}\right)^k,$$

where c_1 and c_2 are arbitrary constants.

Example 4.2.6.
Once again, we want to find the general solution of the linear second-order homogenous difference equation:

$$y_{k+2} + 4y_k = 0. \tag{4.2.19}$$

Solution
The characteristic equation for the difference equation (4.2.19) is $r^2 + 4 = 0$, which has complex conjugate roots $r_1 = 2i$ and $r_1 = -2i$. Applying the polar coordinates, it is known that given a complex number $a + bi$, we have

$\rho = \sqrt{a^2 + b^2}$, $\theta = \tan^{-1}(b/a)$ and, hence, $a + bi = \rho(\cos\theta + i\sin\theta)$, $-\pi < \theta < \pi$. Thus, since $2i = 0 + 2i$ and $-2i = 0 - 2i$, we will have $\rho = 2$ and $\theta = \pi/2$. Therefore, the general solution of (4.2.19), denoted by Y_k, is:

$$Y_k = c_1 2^k \cos\left(\frac{k\pi}{2} + c_2\right),$$

where c_1 and c_2 are arbitrary constants.

In the following example, we will show how we may generate a linear difference equation with constant coefficient. It is an example of a single-server queueing model with a finite buffer. More examples will be given in the next chapter.

Example 4.2.7.
Let us consider the following linear second-order homogenous difference equation with constant coefficients p, $0 < p < 1$, and $q = 1 - p$:

$$P_k = pP_{k+1} + qP_k. \tag{4.2.20}$$

Equation (4.2.20) describes the classic *gambler's ruin problem*. It is a classic problem in the theory of stochastic processes. See Haghighi et al. (2011b, p. 446). The problem is as follows: There are two players A and B gambling against each other. The following assumptions rule the game:

(i)   Each play ends up with a win for one player and a loss for the other.
(ii) The probability of a win is p, $0 < p < 1$, and of a loss is $q = 1 - p$.
(iii) Players A and B each start with an asset of N_A units and N_B units, respectively, with total starting assets as N units (i.e., $N_A + N_B = N$).
(iv) The game stops when player A wins N_B units or loses N_A units. The latter case is referred to as *player A will be ruined*.

The question is how to calculate the probability that player A will be ruined under the given scenario.

Answer
To answer the question, we need to calculate the probability of ruin for the player A. As assumed, the initial asset of player A is N_A; however, that changes during the course of the games. Therefore, let us suppose that player A's asset at any time during the play is k units. We let P_k denote the probability that player A will be ruined when his asset is k units.

Note that if $k = 0$, then A is already ruined and hence the play will stop. Also note that if $k = N$, the play will stop, since that means A has won N_B units. These conditions may be formulated as $P_0 = 1$ and $P_N = 1$, which are called *boundary conditions*.

Thus, equation (4.2.20) describes the difference equation calculating P_k. This is because, suppose A has k units, then after the next game he will have $k + 1$ units if he wins, and the probability of being ruined will be P_{k+1}. However, if he loses the game, then his assets will reduced to $k - 1$ and the probability of being ruined will be P_{k-1}. Since these are the only two possibilities, equation (4.2.20) is developed.

We solve (4.2.20) by characteristic equation method as follows. The characteristic equation of (4.2.20) is:

$$pr^2 - r + q = 0. \tag{4.2.21}$$

If $p \neq q$, then equation (4.2.21) has two distinct real roots $r_1 = 1$ and $r^2 = q/p$. Thus, the general solution for (4.2.20) would be:

$$P_k = C_1 + C_2 \left(\frac{q}{p}\right)^k, \tag{4.2.22}$$

where C_1 and C_2 are two arbitrary constants. We use the boundary conditions to evaluate these constants. Hence, substituting 0 and N for k in (4.2.22) yields:

$$C_1 = \frac{1}{1 - \left(\frac{p}{q}\right)^N} \quad \text{and} \quad C_2 = \frac{\left(\frac{p}{q}\right)^N}{1 - \left(\frac{p}{q}\right)^N}.$$

Therefore:

$$P_k = \frac{\left(1 - \frac{p}{q}\right)^{N-k}}{1 - \left(\frac{p}{q}\right)^N}, \quad p \neq q. \tag{4.2.23}$$

In case that $p = q$, then the characteristic equation (4.2.21) will have a root $r = 1$ with multiplicity 2. Hence, the general solution, in this case, would be:

$$P_k = c_1 + c_2 k, \tag{4.2.24}$$

where c_1 and c_2 are two arbitrary constants. Again, substituting 0 and N for k in (4.2.24) will obtain $c_1 = 1$ and $c_2 = -1/N$. Therefore:

$$P_k = 1 - \frac{k}{N}, \quad p = q. \tag{4.2.25}$$

The following theorem shows relationship between linear differential and difference equations.

Theorem 4.2.2.

For arbitrary constants $c_{i,j+1}$ and c_j, let:

$$y(x) = \sum_{i=1}^{m} \left(\sum_{j=0}^{n_i-1} c_{i,j+1} x^j \right) e^{r_i x} + \sum_{j=(n_1+\cdots+n_m)+1}^{n} c_j e^{r_j x}, \tag{4.2.26}$$

where $n_i \geq 1$, $i = 1, 2, \ldots, m$, with $n_1 + n_2 + \ldots + n_m \leq n$, is the general solution of the nth-order linear differential equation:

$$\frac{d^n y(x)}{dx^n} + a_1 \frac{d^{n-1} y(x)}{dx^{n-1}} + \cdots + a_{n-1} \frac{dy(x)}{dx} + a_n y(x), \tag{4.2.27}$$

where a_i, $i = 1, 2, \ldots, n$, $a_n \neq 0$, are given constants Also, let the characteristic equation:

$$r^n + a_1 r^{n-1} + \cdots + a_n = 0 \tag{4.2.28}$$

has r_j simple roots and r_i roots with multiplicity n_i, $i = 1, 2, \ldots, m$. Additionally, let y_k be the general solution of the difference equation:

$$y_{k+n} + a_1 y_{k+n-1} + \cdots + a_n y_k = 0. \tag{4.2.29}$$

Then:

$$y_k = \left. \frac{d^k y(x)}{dx^k} \right|_{x=0} \tag{4.2.30}$$

and

$$y_k = \sum_{i=1}^{m} \left(c_{i1} + \sum_{l=1}^{n_i-1} \gamma_{i,l} k^l \right) r_i^k + \sum_{j=(n_1+\cdots+n_m)+1}^{n} c_j r_j^k. \tag{4.2.31}$$

Proof

See Mickens (1990, p. 139).

Example 4.2.8.

Consider the second-order differential equation:

$$y'' - 4y' + 2y = 0. \tag{4.2.32}$$

The characteristic equation corresponding to (4.2.32) is:

$$r^2 - 4r + 2 = 0, \tag{4.2.33}$$

with two distinct real roots 5/2 and 3/2. Hence, the general solution of equation (4.2.32) is:

$$y(x) = c_1 e^{\frac{5}{2}x} + c_2 e^{\frac{3}{2}x}, \tag{4.2.34}$$

where c_1 and c_2 are arbitrary constants.

Now the difference equation associated with (4.2.32) is:

$$y_{k+2} - 4y_{k+1} + 2y_k = 0. \tag{4.2.35}$$

Thus, from (4.2.33), the general solution of (4.2.35) is:

$$y_k = A_1 \left(\frac{5}{2}\right)^k + A_2 \left(\frac{3}{2}\right)^k, \tag{4.2.36}$$

where A_1 and A_2 are arbitrary constants. Note that (4.2.36) was obtained from the characteristic equation (4.2.33) that is associated with both (4.2.32) and (4.2.35). That is, we did not use the relationship between the differential and difference equations.

Now let us use (4.2.30) to obtain (4.2.36). Thus, from (4.2.34) we have:

$$\frac{d^k}{dx^k} \left(c_1 e^{\frac{5}{2}x} + c_2 e^{\frac{3}{2}x} \right) = c_1 \left(\frac{5}{2}\right)^k e^{\frac{5}{2}x} + c_2 \left(\frac{3}{2}\right)^k e^{\frac{3}{2}x}. \tag{4.2.37}$$

Setting $x = 0$ in (4.2.37), we obtain (4.2.36), of course with arbitrary constants c_1 and c_2 instead of A_1 and A_2.

4.3. LINEAR NONHOMOGENEOUS DIFFERENCE EQUATIONS WITH CONSTANT COEFFICIENTS

Now we consider the first-order linear nonhomogeneous:

$$y_{k+1} - ay_k = b, \quad k = 0, 1, 2, \ldots, \tag{4.3.1}$$

with initial condition $y_0 = c$, where a, b, and c are constants and $a \neq 0$. We offer two methods of solving equation (4.3.1): (a) by characteristic equation and (b) by recursive method.

4.3.1. Characteristic Equation Method

From Theorem 4.2.1, the characteristic equation of homogeneous part, when $b = 0$,

$$y_{k+1} - ay_k = 0, \quad k = 0, 1, 2, \ldots, \tag{4.3.2}$$

of (4.3.1) is $r - a = 0$, with only one root $r = a$. Hence, the general solution of the (4.3.2) is:

$$Y_k = C(a)^k, \quad k = 0, 1, 2, \ldots, \tag{4.3.3}$$

where C is an arbitrary constant.

To find the particular solution of (4.3.1), we consider different cases for values of a and b, and offer numerical examples.

4.3.1a. Case 1: $a = 1$.

Consider the equation (4.3.1), with $a = 1$ as its initial condition. The general solution (4.3.3) then becomes $Y_k = C$, with $y_0 = c$. Thus, the graph of (4.3.3) is just a horizontal line that passes through $(0, c)$. Hence, the solution sequence, for the homogeneous case, when $a = 1$ converges to c, is $\lim_{k \to \infty}\{y_k\} = c$.

For a particular solution, denoted by $_p y_k$, when $a = 1$, we try a solution of the form:

$$_p y_k = C \cdot k. \tag{4.3.4}$$

Assuming that (4.3.4) is a solution of the homogeneous part, we substitute it in (4.3.2) and obtain:

$$_p y_{k+1} - a_p y_k = C(k+1) - a \cdot C \cdot k = C[1 + (1-a)k]. \tag{4.3.5}$$

Substituting $a = 1$ in (4.3.5), we obtain:

$$_p y_{k+1} = {_p y_k} + C. \tag{4.3.6}$$

Equating (4.3.6) equivalently with (4.3.1) implies that $C = b$. Therefore,

$$_p y_k = b \cdot k \tag{4.3.7}$$

is the desired particular solution. Thus, the general solution of (4.3.1) is:

$$y_k = C + b \cdot k, \quad k = 0, 1, \ldots. \tag{4.3.8}$$

To find the arbitrary constant C, we use the initial condition and, hence, substituting $y_0 = c$ in (4.3.3), we have $y_0 = c = C + 0 \cdot k = C$. Thus, from (4.3.8), the general solution of (4.3.1) is:

$$y_k = c + b \cdot k, \quad k = 0, 1, \ldots. \tag{4.3.9}$$

We note that the graph of the solution sequence (4.3.9) is a line which passes through $(0, c)$. If $b > 0$, then the slope of the line is positive and the solution sequence is increasing, divergent, and $\lim_{k \to \infty}\{y_k\} = \infty$. If $b < 0$, then the slope of the line is negative and the solution sequence is decreasing, divergent, and $\lim_{k\to\infty}\{y_k\} = -\infty$.

Example 4.3.1.
Let us solve the difference equation $y_{k+1} - y_k = 3$ with initial condition $y_0 = 2$.

Solution
Comparing the given equation with (4.3.1), we see that $a = 1$, $b = 3$ (positive), and $c = 2$. Hence, from (4.3.9), the solution is $y_k = 2 + 3k$, $k = 0, 1, 2, \ldots$ The graph of the solution sequence is a line with a positive slope of 3 that passes through $(0,2)$. The solution sequence is increasing, divergent, and $\lim_{k\to\infty}\{y_k\} = \infty$.

Example 4.3.2.
Let us solve the difference equation $y_{k+1} - y_k = -3$ with initial condition $y_0 = 1$.

Solution
In this example, $a = 1$, $b = -3$ (negative), and $c = 1$. Thus, from (4.3.9), the solution is $y_k = 1 - 3k$, $k = 0, 1, 2, \ldots$ The graph of the solution sequence in this case is a line with a negative slope of -3 that passes through $(0, 1)$. The solution sequence is decreasing, divergent, and $\lim_{k\to\infty}\{y_k\} = -\infty$.

4.3.1b. Case 2: $a \neq 1$.
In this case, we try a particular solution of the form:

$$_p y_k = C. \tag{4.3.10}$$

As we did in the previous case, assuming that (4.3.10) is a solution, we substitute it in (4.3.2). Hence, we have:

$$_p y_{k+1} - a\,_p y_k = C - a \cdot C = C(1-a) = b. \tag{4.3.11}$$

From (4.3.11), we have:

$$C = \frac{b}{1-a}. \tag{4.3.12}$$

Thus, a particular solution, in this case, would be:

$$_p y_k = \frac{b}{1-a}, \tag{4.3.13}$$

and hence, the general solution of (4.3.1) in this case is:

$$y_k = Ca^k + \frac{b}{1-a}. \tag{4.3.14}$$

Once again, to find the arbitrary constant C, we use the initial condition $y_0 = c$. Hence, substituting $y_0 = c$ in (4.3.14), we will have $y_0 = c = C + b$, and thus $C = b - c$. Therefore, from (4.3.14), the general solution of (4.3.1) in this case is:

$$y_k = \left(c - \frac{b}{1-a}\right)a^k + \frac{b}{1-a}, \quad k = 0, 1, 2, \ldots. \qquad (4.3.15)$$

4.3.1c. Case 3: $a = -1$.

Let us try a particular solution of the form (4.3.14), again. Substituting $a = -1$ and $_p y_k = C$ in (4.3.1), we obtain $2C = b$ or $C = b/2$. Since $b \neq 0$, $C \neq 0$. Therefore, $_p y_k = b/2$ is the desired particular solution. Hence, the general solution of (4.3.1) in this case is:

$$y_k = C(-1)^k + \frac{b}{2}. \qquad (4.3.16)$$

Using the initial condition, $y_0 = c$, the constant C in (4.3.16) can be found as:

$$C = c - \frac{b}{2}. \qquad (4.3.17)$$

Since $C \neq 0$, $c \neq b/2$. Now, substituting (4.3.17) in (4.3.16), the solution of (4.3.1) is:

$$y_k = \left(c - \frac{b}{2}\right)(-1)^k + \frac{b}{2}, \qquad (4.3.18)$$

where $c \neq b/2$. Therefore, the solution is an alternating sequence that passes through $(0, c)$, and $\lim_{k \to \infty}\{y_k\}$ does not exist. The solution sequence diverges because $\lim_{k \to \infty}|y_k| \neq 0$.

Example 4.3.3.

Let us solve the difference equation $y_{k+1} + y_k = 2$ with initial condition $y_0 = 4$.

Solution

In this example, $a = -1$, $b = 2$, and $c = 4$. Thus, from (4.3.9), the general solution under the given initial condition is $y_k = 3(-1)^3 + 1$, $k = 0, 1, 2, \ldots$, which is an alternating sequence that passes through $(0, 4)$ and $\lim_{k \to \infty}\{y_k\}$ does not exist.

4.3.1d. Case 4: $a > 1$.

As in Case 2, in this case, we try a particular solution of the form $_p y_k = C$. Thus, the general solution of (4.3.1), in this case, would be the same as (4.3.15), but with $c > b/(1 - a)$ or $c < b/(1 - a)$. Therefore, the graph of the solution sequence is an exponential curve that passes through $(0, c)$. In case that $c > b/(1 - a)$, the solution sequence is increasing, divergent, and $\lim_{k\to\infty}\{y_k\} = \infty$. In case that $c < b/(1-a)$, the solution sequence is decreasing, divergent, and $\lim_{k\to\infty}\{y_k\} = -\infty$.

Example 4.3.4.
Let us solve the difference equation $y_{k+1} - 4y_k = -3$ with initial condition $y_0 = 5$.

Solution
From (4.3.15), the solution is $y_k = 4(4)^k + 1$, $k = 0, 1, 2, \ldots$ The graph of the solution sequence is an exponential curve that passes through $(0, 5)$, is increasing, divergent, and $\lim_{k\to\infty}\{y_k\} = \infty$.

Example 4.3.5.
Let us solve the difference equation $y_{k+1} - 7y_k = -5$ with initial condition $y_0 = -3$.

Solution
Again from (4.3.15), the solution is:

$$y_k = -\frac{13}{6}(7)^k + \frac{5}{6}, \quad k = 0, 1, 2, \ldots .$$

The graph of the solution sequence is an exponential curve that passes through $(0, 7)$ is decreasing, divergent, and $\lim_{k\to\infty}\{y_k\} = -\infty$.

4.3.1e. Case 5: $0 < a < 1$.
As in Case 4, in this case, the general solution of (4.3.1) would be the same as (4.3.15), but with $c > b/(1 - a)$ or $c < b/(1 - a)$. Therefore, the graph of the solution sequence is an exponential curve steadily getting closer to the line $y = b/(1 - a)$, which passes through $(0, c)$. In case that $c > b/(1 - a)$, the solution sequence is decreasing, converging to $b/(1 - a)$, and $\lim_{k\to\infty}\{y_k\} = b/(1 - a)$. In case that $c < b/(1 - a)$, the solution sequence is increasing, that is, divergent.

Example 4.3.6.
Let us solve the difference equation

$$y_{k+1} - \frac{1}{3}y_k = 1,$$

with initial condition $y_0 = 2$.

Solution
From (4.3.26), the solution is:

$$y_k = \frac{1}{2}\left(\frac{1}{3}\right)^k + \frac{3}{2}, \quad k = 0, 1, 2, \ldots .$$

The graph of the solution sequence is an exponential curve that passes through $(0, 2)$, is increasing, divergent, and $\lim_{k \to \infty} \{y_k\} = 3/2$.

Example 4.3.7.
Let us solve the difference equation

$$y_{k+1} - \frac{2}{5} y_k = -2,$$

with initial condition $y_0 = -4$.

Solution
Again from (4.3.15), the solution is:

$$y_k = -\frac{2}{3}\left(\frac{2}{5}\right)^k - \frac{10}{3}, \quad k = 0, 1, 2, \ldots.$$

The graph of the solution sequence is an exponential curve that passes through $(0, -4)$, is increasing, divergent, and $\lim_{k \to \infty} \{y_k\} = -10/3$.

4.3.1f. Case 6: $-1 < a < 0$.
As in Case 5, in this case, the general solution of (4.3.1) is the same as (4.3.26). The graph of the solution sequence is steadily getting closer to the line $y = b/(1 - a)$, which passes through $(0, c)$, converges to $b/(1 - a)$, and $\lim_{k \to \infty} \{y_k\} = b/(1 - a)$.

Example 4.3.8.
Let us solve the difference equation

$$y_{k+1} + \frac{1}{2} y_k = 2,$$

with initial condition $y_0 = 2$.

Solution
From (4.3.15), the solution is:

$$y_k = \frac{2}{3}\left(-\frac{1}{2}\right)^k + \frac{4}{3}, \quad k = 0, 1, 2, \ldots.$$

The graph of the solution sequence is an exponential curve that passes through $(0, 2)$, is convergent, and $\lim_{k \to \infty} \{y_k\} = 4/3$.

4.3.1g. Case 7: $a < -1$.
As in previous cases, in this case, the general solution of (4.3.1) is the same as (4.3.15). The graph of the solution is an alternating sequence that is alternating

farther away from zero, passes through $\{0, c\}$, diverges, and $\lim_{k \to \infty}\{y_k\}$ does not exist.

Example 4.3.9.
Let us solve the difference equation $y_{k+1} + 2y_k = 4$ with initial condition $y_0 = -1$.

Solution
From (4.3.15), the solution is:

$$y_k = -\frac{7}{3}(-2)^k + \frac{4}{3}, \quad k = 0, 1, 2, \dots.$$

The graph of the solution sequence is an exponential curve that passes through $(0, -1)$, is increasing, divergent, and $\lim_{k \to \infty}\{y_k\}$ does not exist.

4.3.1h. Case 8: $a \neq 1, c = b/(1 - a)$.
In this case, the general solution is $y_k = b/(1 - a)$. Therefore, the graph of the solution sequence is a horizontal line that passes through $(0, c)$. The solution sequence is constant, converges to $b/(1 - a)$, and $\lim_{k \to \infty}\{y_k\} = b/(1 - a)$.

Example 4.3.10.
Let us solve the difference equation

$$y_{k+1} - 3y_k = -\frac{1}{2},$$

with initial condition $y_0 = 1/4$.

Solution
The solution is:

$$y_k = \frac{1}{4}, \quad k = 0, 1, 2, \dots.$$

The graph of the solution sequence is an exponential curve that passes through $\{0, 1/4\}$, converges, and $\lim_{k \to \infty}\{y_k\} = 1/4$.

4.3.2. Recursive Method

In this method, we rewrite (4.3.1) as:

$$y_{k+1} = ay_k + b, \quad k = 0, 1, 2, \dots. \tag{4.3.19}$$

For $k = 0$, we have the initial value $y_0 = c$. Substituting this value in (4.3.19), we will have:

$$y_1 = ay_0 + b = ac + b. \tag{4.3.20}$$

For $k = 1$, from (4.3.19) and (4.3.20), we obtain:

$$y_2 = ay_1 + b = a(ac + b) + b = a^2 c + b(1+a). \tag{4.3.21}$$

For $k = 2$, from (4.3.19) and (4.3.21), we obtain:

$$y_3 = ay_2 + b = a\left[a^2 c + b(1+a) + b\right] + b = a^3 c + b(1+a+a^2). \tag{4.3.22}$$

Equations (4.3.20) through (4.3.22) show a pattern that we can conjecture the general term and prove it by mathematical induction. Thus, we choose the solution for the initial problem (4.3.1) as:

$$y_k = a^k c + b(1 + a + a^2 + \cdots + a^{k-1}). \tag{4.3.23}$$

To follow the induction principles, first, note that the quantity in parentheses on the right-hand side of (4.3.23) is the sum of the first k terms of a geometric progression with ratio a. Thus, we have:

$$y_k = \begin{cases} a^k c + b\left(\dfrac{1-a^k}{1-a}\right), & a \neq 1, \quad k = 0,1,2,\ldots, \\ a + b \cdot k, & a = 1. \end{cases} \tag{4.3.24}$$

Now for y_{k+1} as the last part of the induction, from (4.3.24), when $a \neq 1$, we have:

$$\begin{aligned} y_{k+1} &= a^{k+1} c + b\left(\frac{1-a^{k+1}}{1-a}\right) \\ &= a\left[a^k c + b\left(\frac{1-a^{k+1}}{1-a}\right)\right] + b, \quad k = 0,1,\ldots \\ &= a^{k+1} c + b\left[1 + \frac{a(1-a^k)}{1-a}\right], \quad k = 0,1,\ldots \\ &= a^{k+1} c + b\left[\frac{1-a+a(1-a^k)}{1-a}\right], \quad k = 0,1,\ldots \\ &= a^{k+1} c + b\left[\frac{1-a^{k+1}}{1-a}\right], \quad k = 0,1,\ldots. \end{aligned}$$

We leave it as an exercise to show that the induction could be completed when $a = 1$. Thus, we have proved the following theorem:

Theorem 4.3.1. *Existence and Uniqueness Theorem*
The linear first-order initial-valued differential equation given by (4.3.1) has a unique solution given by (4.3.24).

Proof:
We have already given the proof of existence and showed that (4.3.24) is the solution of (4.3.1). Now we need to prove the uniqueness of the solution. In other words, suppose that $y_k(x)$, $k = 0, 1, 2, \ldots$, is a solution, then there is a constant, say C, such that:

$$y_k = \begin{cases} Ca^k + b\left(\dfrac{1-a^k}{1-a}\right), & a \neq 1, \quad k = 0, 1, 2, \ldots, \\ C + b \cdot k, & a = 1, \end{cases}$$

(4.3.25)

which will provide only one solution for each value of k. To see the uniqueness, all we have to do is equate C with y_0. Of course, for $k = 0$, the value of y_0 is given as $y_0 = c$. Thus, (4.3.24) is a special case of (4.3.25). In other words, we can conclude that *if $y_k(x)$, $k = 0, 1, 2, \ldots$, is a solution of (4.3.1), then there is a constant, say C, such that (4.3.25) is a unique solution for each value of k, when the initial value is given.*

Example 4.3.11.
Let us solve the initial value first-order linear differential equation:

$$y_{k+1} - y_k = 4, \quad k = 0, 1, 2, \ldots, \quad y_0 = 3.$$

(4.3.26)

Solution
Using the characteristic equation method, since $a = 1$ and $b = 4$, from (4.3.8) and (4.3.26), we will have:

$$y_k = 3 + 4k, \quad k = 0, 1, 2, \ldots.$$

But that is exactly what we can get from the second equation of (4.3.24).

Example 4.3.12.
In this example, let us solve the initial-value, first-order linear differential equation:

$$2y_{k+1} - y_k = 4, \quad k = 0, 1, 2, \ldots, \quad y_0 = 3.$$

(4.3.27)

Solution
Rewriting (4.3.27) in the form (4.3.1), we will have:

$$y_{k+1} - \frac{1}{2}y_k = 2, \quad k = 0, 1, 2, \ldots, \quad y_0 = 3.$$

(4.3.28)

Hence, in this case, $a = 1/2$, $b = 2$, and $c = 3$. Thus, using characteristic equation method, from (4.3.15) we have:

$$y_k = \left(3 - \frac{2}{1-\frac{1}{2}}\right)\left(\frac{1}{2}\right)^k + \frac{2}{1-\frac{1}{2}} = 4 - \left(\frac{1}{2}\right)^k, \quad k = 0, 1, 2, \ldots,$$

which is the same as we have from the first equation of (4.3.24), as:

$$y_k = 3\left(\frac{1}{2}\right)^k + 2\left[\frac{1-\left(\frac{1}{2}\right)^k}{1-\left(\frac{1}{2}\right)}\right] = 3\left(\frac{1}{2}\right)^k + 4\left[1 - \left(\frac{1}{2}\right)^k\right], \quad k = 0, 1, 2, \ldots.$$

We have already discussed the behavior of the solution sequence of (4.3.1) when solving (4.3.1) by characteristic equation. Now having proved the existence and uniqueness of the solution, we can discuss the behaviors of the solution for the different cases. We start with the case $a = 1$.

Theorem 4.3.2
If the sequence $\{y_k, k = 0, 1, \ldots\}$ is the solution of (4.3.1) when $a = 1$, which is the linear first-order nonhomogeneous:

$$y_{k+1} = y_k + b, \quad k = 0, 1, 2, \ldots, y_0 = c, \qquad (4.3.29)$$

then, (a) if $b = 0$, $\{y_k\}$ is a constant sequence, (b) if $b > 0$, the solution sequence will diverge to $+\infty$, and (c) if $b < 0$, it will diverge to $-\infty$.

Proof:
The solution for the initial value problem (4.3.29) is given by the second equation in (4.3.24) as $y_k = c + b \cdot k$. Hence, when $b = 0, y_k = c$, for all $k = 0, 1, 2, \ldots$, the solution sequence will be a constant sequence. This proves part (a). Now to prove part (b), that is, to prove that for $b > 0$, the solution sequence divergence, we have to show that for a given positive integer, say M, there exist a corresponding integer N such that that $y_p > M$ for all $k \geq N$. We do this by contradiction.

Now, if $c \geq M$, then since $b > 0$, $y_k > M$, $\forall k \geq 0$. Let us choose $N = 1$, and we have to be sure that $y_k > M$, $\forall k \geq 0$. If $M > c$, then from $y_k = c + b \cdot k > M$, we have $bk > M - c$ or $k > M - c/b$. Hence, if we choose $N > M - c/b$, y_k will be greater than M for all $k \geq N$. A similar argument can be done for the case $b < 0$, and that completes the proof of the theorem.

All examples so far have been first-order nonhomogeneous. The following is a *second-order* example.

Example 4.3.13.
Consider the initial value problem:

$$y_{k+2} - 5y_{k+1} - 6y_k = 3^k, \quad k = 0, 1, 2, \ldots, \quad y_0 = -2, y_1 = 3. \qquad (4.3.30)$$

Solution

To solve (4.3.30), we want to use the Z-transform we studied in Chapter 2 (Section 2.3). Recall that the Z-transform of the sequence $\{f_n\}$, denoted by $Z\{f_n\} \equiv F(z)$ was defined as $F(z) \equiv \sum_{n=0}^{\infty} f_n z^{-n}$ or $F(z) \equiv \sum_{n=-\infty}^{\infty} f_n z^{-n}$. We choose the first definition based on the information given in (4.3.30). Hence, we denote by $F(z)$ the Z-transform of $\{y_k\}$, that is,

$$F(z) \equiv Z\{y_k\} = \sum_{n=0}^{\infty} y_k z^{-k}. \tag{4.3.31}$$

Note that from (4.3.31), we have:

$$Z\{y_{k+1}\} = \sum_{n=0}^{\infty} y_{k+1} z^{-k} = z \cdot \sum_{n=0}^{\infty} y_{k+1} z^{-(k+1)} = z[F(z) - y_0], \tag{4.3.32}$$

$$Z\{y_{k+2}\} = \sum_{n=0}^{\infty} y_{k+2} z^{-k} = z^2 \cdot \sum_{n=0}^{\infty} y_{k+2} z^{-(k+2)} = z^2 [F(z) - y_0 - y_1 z^{-1}], \tag{4.3.33}$$

and for a as a constant,

$$Z(a^k) = \sum_{n=0}^{\infty} a^k z^{-k} = \sum_{n=0}^{\infty} \left(\frac{a}{z}\right)^k = \frac{1}{1 - \dfrac{a}{z}} = \frac{z}{z-a}. \tag{4.3.34}$$

Now, applying (4.3.31) on (4.3.30), we will have:

$$Z\{y_{k+2}\} - 5Z\{y_{k+1}\} - 6Zy_k = Z\{3^k\}. \tag{4.3.35}$$

From (4.3.35), using (4.3.30), (4.3.31), (4.3.32), (4.3.33), and (4.3.34), we will have the following:

$$z^2 \left[F(z) + 2 - \frac{3}{z}\right] - 5z[F(z) + 2] - 6F(z) = \frac{z}{z-3}. \tag{4.3.36}$$

Calculating $F(z)$ from (4.3.36), we will have:

$$F(z) = \frac{z}{(z-3)(z^2 - 5z - 6)} + \frac{-2z^2 + 13z}{z^2 - 5z - 6},$$

or

$$\frac{F(z)}{z} = \frac{1}{(z-3)(z^2 - 5z - 6)} + \frac{-2z + 13}{z^2 - 5z - 6}. \tag{4.3.37}$$

Applying the partial fraction method on the right-hand side of (4.3.37), and solving for $F(z)$, we will have:

$$F(z) = -\frac{1}{12} \cdot \frac{z}{z-3} + \frac{4}{21} \cdot \frac{1}{z-6} - \frac{59}{28} \cdot \frac{1}{z+1}.$$ (4.3.38)

Now, inverting (4.3.38), we will have our answer as:

$$y_k = -\frac{1}{12} \cdot 3^k + \frac{4}{21} \cdot 6^k - \frac{59}{28} \cdot (-1)^k, \quad k = 2, 3, \dots.$$ (4.3.39)

4.3.9. Solving Differential Equations by Difference Equations.
Differential equations may be solved numerically using difference equations. The following example is a numerical solution of differential equation by the *Euler method*. The *Euler method* approximates the differential equation:

$$\frac{dy}{dx} = f(x, y), \quad y(x_0) = y_0$$ (4.3.40)

by the difference equation

$$y_{n+1} = y_n + \Delta(x) y_n',$$ (4.3.41)

where $y_n = y(x_n)$, $y_n' = y'(x_n) = f(x_n, y_n)$, and $\Delta(x) = x_{n+1} - x_n$.

Example 4.3.14.
Let us apply the Euler method to solve the differential equation:

$$y' + 3y = 2 - e^{-5x}, \quad y(0) = 2.$$ (4.3.42)

Solution
The differential equation (4.3.42) is a nonhomogeneous linear differential equation with constant coefficient. Hence, the characteristic equation of its homogeneous part is $r + 3 = 0$, with the general solution of homogenous part, denoted by y_H, as:

$$y_H(x) = Ce^{-3x},$$ (4.3.43)

where C is an arbitrary constant. For a particular solution, denoted by y_P, we will try:

$$y_P(x) = A + Be^{-5x},$$ (4.3.44)

where A and B are constants to be found. Substituting (4.3.44) in (4.3.42), we will have $A = 2/3$ and $B = 1/2$, giving the particular solution as:

$$y_P(x) = \frac{2}{3} + \frac{1}{2}e^{-5x},$$ (4.3.45)

and the general solution to (4.3.42) as:

$$y(x) = Ce^{-3x} + \frac{2}{3} + \frac{1}{2}e^{-5x}. \tag{4.3.46}$$

Applying the initial value given in (4.3.42) on (4.3.46), we will have $C = 5/6$. Thus, the exact solution to the given differential equation in (4.3.42) is:

$$y(x) = \frac{2}{3} + \frac{5}{6}e^{-3x} + \frac{1}{2}e^{-5x}. \tag{4.3.47}$$

From (4.3.47), we see that the exact value of the solution when $x = 3$ is $y(3) = 0.6668$. We now use the Euler method with a step size of $\Delta(x) = 0.1$ to find the approximate values of the solution of (4.3.42) at points $x = 0.1, 0.2, \ldots, 3$. We are hoping to see the iterations will converge to the exact value 0.6668 as they do, and the limit is 0.6667. The 0.0001th error is due to approximation and it is expected. To do so, we rewrite (4.3.42) as:

$$y' = 2 - e^{-5x} - 3y = f(x, y), \tag{4.3.48}$$

with $x_0 = 0$ and $y_0 = y(0) = 2$. The following program using MATLAB will perform the iterations and values for each step (we skipped some intermediate rows to save space):

```
x_n = 0:.1:3; Delta = 0.1; x_0 = 0; y_0 = 2;
for i = 0:length(x_n)
    if i == 0
        f_xy(i+1) = 2 - exp(-5*i) - 3*y_0;
        y_n(i+1) = y_0+ Delta *f_xy(i+1);
    else
        f_xy(i+1) = 2 - exp(-5*x_n(i)) - 3*y_n(i);
        y_n(i+1) = y_n(i)+ Delta *f_xy(i+1);
    end
end
```

x_n	y_n
0	1.5000
0.1000	1.1500
0.2000	0.9443
0.3000	0.8243
0.6000	0.6921
0.8000	0.6726
2.6000	0.6666
2.9000	0.6667
3.0000	0.6667
Answer	0.6667

4.4. SYSTEM OF LINEAR DIFFERENCE EQUATIONS

Up to this point, we have been dealing with difference equations with one variable. However, we have many cases in which the difference equation has more than one variable. Here in this book, we only deal with those of two variables. Thus, before we start addressing systems of difference equation, since there are many examples that are involved with equations with two variables, we give the following definition:

Definition 4.4.1.
As defined in Chapter 3, consider $z = f(x, y)$ as a function with two variables x and y. We write differences of z with respect to x and y. That is, once we regard y as a constant, we denote the difference as:

$$\Delta_x f(x, y) = f(x+h, y) - f(x, y). \tag{4.4.1}$$

Similarly, we regard x as a constant and denote the difference as:

$$\Delta_y f(x, y) = f(x, y+k) - f(x, y). \tag{4.4.2}$$

Relations (4.4.1) and (4.4.2) are called *partial differences* of the function $z = f(x, y)$. An equation that contains partial differences is referred to as a *partial difference equation*.

Note that we do not choose to go into the detailed definition of partial difference equations due to their involvement with different operators that we will not be using in this text. For a detailed discussion, we refer the reader to Jordan (1979).

We offer two methods for solving a system of linear difference equations: recursive and generating function. Generating functions, as defined in Chapter 2, play an extremely useful role in solving difference equations.

First, we shall discuss *recursive function*.

Example 4.4.1.
Let us consider the following finite system of $N+1$ linear difference equations with $N+1$ unknowns $P_0, P_1, \ldots, P_N$, where N is a finite number:

$$\begin{cases} aP_0 = bP_1, \\ (a+b)P_1 = aP_0 + bP_2, \\ \vdots \\ (a+b)P_{N-1} = aP_{N-2} + bP_N, \\ \sum_{n=0}^{N} P_n = 1. \end{cases} \tag{4.4.3}$$

Solution

Starting with the first equation of (4.4.3), it is easy to see that, recursively, the solution to the system (4.4.3) is:

$$
\begin{cases}
P_1 = \left(\dfrac{a}{b}\right) P_0, \\[2ex]
P_2 = \left(\dfrac{a}{b}\right)^2 P_0, \\[2ex]
\vdots \\[2ex]
P_N = \left(\dfrac{a}{b}\right)^N P_0.
\end{cases}
\tag{4.4.4}
$$

From the last equation of (4.4.3), the normalizing equation, we have:

$$
P_0 = \frac{1}{1 + \left(\dfrac{a}{b}\right) + \left(\dfrac{a}{b}\right)^2 + \cdots + \left(\dfrac{a}{b}\right)^N}.
\tag{4.4.5}
$$

If $a = b$, then the denominator of (4.4.5) will sum to $N + 1$. However, if $a \neq b$, it is a *geometric progression*. It is well known that for constants c and r, the geometric progression $\sum_{k=0}^{n} c r^k$, for $|r| < 1$, equals $c(1 - r^{n+1})/1 - r$. Hence:

$$
P_0 =
\begin{cases}
\dfrac{1}{N+1}, & \text{if } a = b \\[3ex]
\dfrac{1 - \dfrac{a}{b}}{1 - \left(\dfrac{a}{b}\right)^{N+1}}, & \text{if } a \neq b.
\end{cases}
\tag{4.4.6}
$$

Therefore, from (4.4.4) and (4.4.6), we will have:

$$
P_n =
\begin{cases}
\dfrac{1}{N+1}, & \text{if } a = b, \quad n = 0, 1, 2, \ldots, \\[3ex]
\dfrac{1 - \dfrac{a}{b}}{1 - \left(\dfrac{a}{b}\right)^{N+1}} \left(\dfrac{a}{b}\right)^n, & \text{if } a \neq 1, \quad n = 0, 1, 2, \ldots.
\end{cases}
\tag{4.4.7}
$$

4.4.1. Generating Functions Method

Example 4.4.2.

Let X have a binomial distribution with parameters n and p. That is, as discussed in Chapter 1, we repeat a Bernoulli trial n times, $n = 0, 1, \ldots$, with probability of success in each trial as p (and, thus, of failure as $q = 1 - p$). Let us denote by $P_{k,n}$ the probability of exactly k successes, that is,

$$P_{k,n} = P(X = k), \quad k = 0, 1, 2, \ldots, n, \tag{4.4.8}$$

and

$$\sum_{k=0}^{n} P_{k,n} = 1. \tag{4.4.9}$$

Here, the idea is to find $P_{k,n}$ by developing a difference equation and then solve it by generating a function method.

Solution
The event that $(k + 1)$ successes occur in $(n + 1)$ trials is possible only in the following two mutually exclusive ways:

1. at the end of the nth trial, $(k + 1)$ successes have already been achieved and at the $(n + 1)$st trial, a failure will occur, or
2. at the end of the nth trial, k successes have occurred and at the $(n + 1)$ st trial, another success will occur.

The events being independent, the probability of the event in case (1) is the product of the probability of $(k + 1)$ successes in the first n trial and the probability of a failure at the $(n + 1)$st trial. Hence, the probability of the event in that case is $qP_{k+1,n}$. Similarly, the probability of the event in case (2) is $pP_{k,n}$. Therefore:

$$P_{k+1,n+1} = qP_{k+1,n} + pP_{k,n}, \quad k = 0, 1, 2, \ldots, n, \quad n = 0, 1, 2, \ldots. \tag{4.4.10}$$

We note that an equation such as (4.4.10) is referred to as the *partial difference equation* for the distribution function $P_{k,n}$.

Now the event of no successes in $(n + 1)$ trials is possible only if no success has been observed at the end of the nth trial and then a failure at the $(n + 1)$ th trial. Thus, as we did for the other case, we have:

$$P_{0,n+1} = qP_{0,n}, \quad n = 0, 1, 2, \ldots. \tag{4.4.11}$$

Assuming that the 0th trial is the starting point, the number of successes in that trial is zero. This will give us the initial conditions as:

$$P_{0,0} = 1 \quad \text{and} \quad P_{k,0} = 0, \quad k > 0. \tag{4.4.12}$$

Of course, we have the given normalizing equation (4.4.9).

Having developed a set of difference equations (4.4.10) and (4.4.11) with initial conditions (4.4.12) and normalizing equation (4.4.9), we want to apply the generating function to solve it for $P_{k,n}$. Additionally, we want to find the mean and variance of the random variable X.

Therefore, to move on, we first list all the information as follows:

$$
\begin{cases}
P_{0,n+1} = qP_{0,n}, & n = 0, 1, 2, \ldots, \\
P_{k+1,n+1} = qP_{k+1,n} + pP_{k,n}, & k = 0, 1, 2, \ldots, n, \quad n = 0, 1, 2, \ldots, \\
\displaystyle\sum_{k=0}^{n} P_{k,n} = 1, \\
P_{0,0} = 1, \\
P_{k,0} = 0. & k > 0.
\end{cases}
\tag{4.4.13}
$$

Now we define the generating function $G_n(z)$, on k, for the sequence $\{P_{k,n}\}$ as follows:

$$
G_n(z) = \sum_{k=0}^{\infty} P_{k,n} z^k, \quad n = 0, 1, 2, \ldots.
\tag{4.4.14}
$$

Of course, there is one such generating function for each value of n. To apply (4.4.14) on the second equation of (4.4.13), we multiply each term of it by z^{k+1}, sum over k, and obtain:

$$
\sum_{k=0}^{\infty} P_{k+1,n+1} z^{k+1} = q \sum_{k=0}^{\infty} P_{k+1,n} z^{k+1} + p \sum_{k=0}^{\infty} P_{k,n} z^{k+1}, n = 0, 1, 2, \ldots.
\tag{4.4.15}
$$

Using (4.4.14), (4.4.15) can be rewritten as:

$$
G_{n+1}(z) = (q + pz)G_n(z) + (P_{0,n+1} - qP_{0,n}), n = 0, 1, 2, \ldots.
\tag{4.4.16}
$$

Using the initial conditions, that is, the last two equations of (4.4.13), equation (4.4.16) may further be rewritten as:

$$
G_{n+1}(z) = (q + pz)G_n(z), n = 0, 1, 2, \ldots,
\tag{4.4.17}
$$

which is a first-order difference equation for the generating function $G_n(z)$. To solve (4.4.17), letting $Y^n \equiv G_n(z)$, substituting in (4.4.17), we obtain:

$$
Y^{n+1} = (q + pz)Y^n \quad \text{or} \quad Y^n(Y_n - (q + pz)) = 0.
\tag{4.4.18}
$$

It is clear that $Y^n \neq 0$; otherwise, there will not be any equation. Thus, dividing both sides of equation (4.4.18) by Y^n, we obtain $Y = q + pz$, and raising both sides to power n, we will have $Y^n = (q + pz)^n$ or

$$G_n(z) = (q + pz)^n, \quad n = 0, 1, 2, \ldots. \tag{4.4.19}$$

Using the binomial theorem, (4.4.19) will be:

$$G_n(z) = (q + pz)^n = \sum_{k=0}^{n} \binom{n}{k} (pz)^k q^{n-k}$$

or

$$G_n(z) = (q + pz)^n = \sum_{k=0}^{n} \left[\binom{n}{k} p^k q^{n-k} \right] z^k. \tag{4.4.20}$$

The coefficients of (4.4.20) are the distribution function we were looking for, that is,

$$P_{k,n} = P(X = k) = \binom{n}{k} p^k q^{n-k}, \quad k = 0, 1, 2, \ldots n. \tag{4.4.21}$$

Note that the distribution function we found in (4.4.21) is the binomial probability distribution with parameters n and p we saw in Chapter 1, and (4.2.19) is the probability generating function for the binomial distribution we found in Chapter 2.

To find the mean, variance, and all moments of the distribution, we take derivatives of the distribution and evaluate them at $z = 1$. Hence, for the binomial distribution (4.4.21), we have:

$$G_n'(z) = n(q + pz)^{n-1} p,$$

where $p + q = 1$ implies that:

$$E(X) = G_n'(1) = n(q + p)^{n-1} p = np.$$

Moreover:

$$G_n''(z) = n(n-1)(q + pz)^{n-2} p^2.$$

Thus:

$$E[X(X-1)] = n(n-1)(q + p)^{n-2} p^2 = n(n-1)p^2.$$

In contrast, $[E(X)]^2 = (np)^2$. Hence:

$$\begin{aligned} Var(X) &= E(X^2) - [E(X)]^2 = E[X(X-1)] + E(X) - [E(X)]^2 \\ &= n(n-1)p^2 + np - (np)^2 = np - np^2 = np(1-p) = npq. \end{aligned}$$

Example 4.4.3.

Suppose from a real-life situation such as the one in Example 4.4.2, we were able to develop the following system of difference equations, in which u_k is defined to be independent of the number of trials and the values of $u_0, u_1, \ldots$ are assumed to be known values. We are to use the generating function method to solve the *system of partial difference equations* for $P_{k,n}$:

$$\begin{cases} P_{0,n+1} = (1-u_0)P_{0,n}, & n = 0, 1, 2, \ldots, \\ P_{k+1,n+1} = (1-u_{k+1})P_{k+1,n} + u_k P_{k,n}, & k = 0, 1, 2, \ldots \quad n = 0, 1, 2, \ldots, \\ P_{0,0} = 1, & \\ P_{k,0} = 1, & k = 1, 2, \ldots. \end{cases} \tag{4.4.22}$$

Solution

Let k be fixed. We define the generating function of $P_{k,n}$ as follows:

$$G_k(z) = \sum_{n=0}^{\infty} P_{k,n} z^n, \quad k = 1, 2, \ldots. \tag{4.4.23}$$

We now apply (4.4.23) on the first equation of (4.4.22) and obtain:

$$\sum_{n=0}^{\infty} P_{0,n+1} z^{n+1} = (1-u_0) \sum_{n=0}^{\infty} P_{0,n} z^{n+1},$$

$$\sum_{n=1}^{\infty} P_{0,n} z^n = (1-u_0) z \sum_{n=0}^{\infty} P_{0,n} z^n,$$

$$G_0(z) - P_{0,0} = (1-u_0) z G_0(z),$$

and hence,

$$[1-(1-u_0)z]G_0(z) = P_{0,0}. \tag{4.4.24}$$

Applying the initial conditions, that is, the last two equations of (4.4.22) and solving equation (4.4.24), we will have:

$$G_0(z) = \frac{P_{0,0}}{1-(1-u_0)z} = \frac{1}{1-(1-u_0)z}. \tag{4.4.25}$$

Similarly, we apply the generating function on the second equation of (4.4.22) and performing some algebra. Thus:

$$\sum_{n=0}^{\infty} P_{k+1,n+1} z^{n+1} = (1-u_{k+1}) \sum_{n=0}^{\infty} P_{k+1,n} z^{n+1} + u_k \sum_{n=0}^{\infty} P_{k,n} z^{n+1},$$

$$\sum_{n=1}^{\infty} P_{k+1,n} z^n = (1-u_{k+1}) z \sum_{n=0}^{\infty} P_{k+1,n} z^n + u_k z \sum_{n=0}^{\infty} P_{k,n} z^n,$$

$$G_{k+1}(z) - P_{k+1,0} = (1-u_{k+1})zG_{k+1}(z) + u_k z G_k(z), \quad k = 0, 1, \ldots,$$

$$[1-(1-u_{k+1})z]G_{k+1}(z) = u_k z G_k(z), \quad k = 0, 1, 2, \ldots.$$

Thus:

$$G_{k+1}(z) = \frac{u_k z}{[1-(1-u_{k+1})z]} G_k(z), \quad k = 0, 1, 2, \ldots. \tag{4.4.26}$$

Equation (4.4.26) is a homogeneous first-order difference equation, which can be solved recursively to get:

$$G_k(z) = G_0(z) \prod_{i=0}^{k-1} \frac{z u_i}{1 - z(1-u_{i+1})}, \quad k = 0, 1, 2, \ldots. \tag{4.4.27}$$

Substituting (4.4.25) into (4.4.27), we obtain:

$$G_k(z) = \frac{1}{1-z(1-u_0)} \prod_{i=0}^{k-1} \frac{z u_i}{1 - z(1-u_{i+1})}, \quad k = 0, 1, 2, \ldots. \tag{4.4.28}$$

To find the probabilities $P_{k,n}$, we need to invert the generating functions found in (4.4.28). To do this, note that:

$$\frac{1}{1-z(1-u_0)} = \sum_{n=0}^{\infty} [(1-u_0)z]^n.$$

Hence, from the first equation of (4.4.22), we have:

$$P_{0,n} = (1-u_0)^n, \quad n = 0, 1, \ldots \tag{4.4.29}$$

Now, for $k = 1$ from (4.4.28) we have:

$$G_1(z) = \frac{u_0 z}{[1-z(1-u_0)][1-z(1-u_1)]}. \tag{4.4.30}$$

Using the method of partial fractions, (4.4.30) can be written as:

$$G_1(z) = \frac{u_0}{u_1 - u_0} \left[\frac{1}{1-z(1-u_0)} - \frac{1}{1-z(1-u_1)} \right], u_1 \neq u_0. \tag{4.4.31}$$

As before, inverting generating functions in (4.4.31) when $u_1 \neq u_0$, that is, writing (4.4.31) in power series and choosing coefficients, yields:

$$P_{1,n} = \frac{u_0}{u_1 - u_0} \left[(1-u_0)^n - (1-u_1)^n \right], \quad n = 0, 1, 2, \ldots \tag{4.4.32}$$

The rest of the probabilities can be found similarly and we leave it as an exercise.

Example 4.4.4.
Using the generating function method, we want to solve the following system of first-order difference equations, in which a and b are positive constants, and $\sum_{n=0}^{\infty} P_n = 1$:

$$bP_{n+1} = (a+b)P_n - aP_{n-1}, n = 1, 2, \ldots,$$ (4.4.33)

$$bP_1 = aP_0.$$ (4.4.34)

Solution
We note that the system considered in this example originates from a single-server queueing model $M/M/1$ or a birth-and-death model, a variation of which we will discuss in the next chapter. We also note that the system under consideration can be solved iteratively. However, we want to show how the generating function method works. Hence, let us define the generating function $G(z)$ for the sequence $\{P_n, n = 0, 1, 2, \ldots,\}$ as:

$$G(z) = \sum_{n=0}^{\infty} P_n z^n.$$ (4.4.35)

As we did before, multiplying both sides of the equation of (4.4.35), summing over n, applying (4.4.35), and performing some algebra, we will have:

$$(a+b)\sum_{n=1}^{\infty} P_n z^n = a\sum_{n=1}^{\infty} P_{n-1} z^n + b\sum_{n=1}^{\infty} P_{n+1} z^n$$

or

$$(a+b)\left[\sum_{n=0}^{\infty} P_n z^n - P_0\right] = az\sum_{n=0}^{\infty} P_n z^n + \frac{b}{z}\sum_{n=2}^{\infty} P_n z^n$$

or

$$(a+b)[G(z) - P_0] = azG(z) + \frac{b}{z}[G(z) - P_0 - P_1 z]$$

or

$$\left(a+b-az-\frac{b}{z}\right)G(z) = (a+b)P_0 - \frac{b}{z}P_0 - bP_1$$

or

$$\frac{(a+b)z - az^2 - b}{z}G(z) + aP_0 = (a+b)P_0 + bP_1 - \frac{b}{z}P_0 - bP_1$$

or

$$[(a+b)z-az^2-b]G(z)=b(z-1)P_0.$$

Thus:

$$G(z)=\frac{b(z-1)}{(a+b)z-az^2-b}P_0. \tag{4.4.36}$$

Now replacing z by 1 in (4.4.36), we will have:

$$G(1)=1=\lim_{z\to 1}\frac{b(z-1)}{(a+b)z-az^2-b}P_0=\frac{b}{b-a}P_0$$

or

$$P_0=\frac{b-a}{b}=1-\frac{a}{b}. \tag{4.4.37}$$

Substituting (4.4.37) in (4.4.36) yields:

$$G(z)=\frac{(b-a)(z-1)}{(a+b)z-az^2-b}=\frac{(b-a)(z-1)}{(z-1)(-az+b)}=\frac{b-a}{b-az}. \tag{4.4.38}$$

Using the Maclaurin expansion of $G(z)$, that is,

$$\sum_{n=0}^{\infty}\frac{G^{(n)}(0)}{n!}z^n$$

on (4.4.38), we will have:

$$G(z)=\left(1-\frac{a}{b}\right)\left[1+\frac{a}{b}z+\left(\frac{a}{b}\right)^2z^2+\cdots\right]. \tag{4.4.39}$$

Thus:

$$P_0=1-\frac{a}{b},\quad P_1=\left(1-\frac{a}{b}\right)\frac{a}{b},\quad P_2=\left(1-\frac{a}{b}\right)\left(\frac{a}{b}\right)^2,\dots.$$

Therefore, the solution we are looking for is:

$$P_n=\left(1-\frac{a}{b}\right)\left(\frac{a}{b}\right)^n,\quad n=0,1,2,\dots. \tag{4.4.40}$$

4.5. DIFFERENTIAL–DIFFERENCE EQUATIONS

In addition to differences in an equation, there may also be derivatives or integrals. In such cases, we refer to the equation as a *differential–difference equation* or as a *integral–difference equation*, respectively. We only consider deferential-difference equations in this book. These types of equations may be solved by various special techniques such as generating functions and Laplace transforms.

   We note that the recursive method does not work in such cases due to the presence of the differential part.

Example 4.5.1.
Let us solve the following differential–difference equation

$$y'_{n+1}(x) = y_n(x), \quad n = 0, 1, 2, \ldots \tag{4.5.1}$$

by generating function method, given the following initial conditions:

$$y_n(0) = \delta_{n,0} = \begin{cases} 1, & \text{if } n = 0, \\ 0, & \text{if } n \neq 0, \end{cases} \quad n = 0, 1, 2, \ldots. \tag{4.5.2}$$

Solution
We define the generating function $G(z)$ for the sequence $\{y_n(x), n = 0, 1, 2, \ldots\}$ as:

$$G(x, z) = \sum_{n=0}^{\infty} y_n(x) z^n. \tag{4.5.3}$$

Applying (4.5.3) on (4.5.1), we obtain:

$$\sum_{n=0}^{\infty} y'_{n+1}(x) z^n = \sum_{n=0}^{\infty} y_n(x) z^n$$

or

$$\frac{1}{z} \sum_{n=1}^{\infty} y'_n(x) z^n = \sum_{n=0}^{\infty} y_n(x) z^n. \tag{4.5.4}$$

Equation (4.5.4) can be written in terms of generating function $G(x, z)$ as:

$$\frac{1}{z} \left[\frac{\partial}{\partial x} G(x, z) - y'_0(x) \right] = G(x, z). \tag{4.5.5}$$

Since from (5.4.2), $y'_0(x) = 0$, equation (4.5.5) becomes:

$$\frac{\partial G(x, z)}{\partial x} - z G(x, z) = 0. \tag{4.5.6}$$

Equation (4.5.6) is a linear first-order differential equation with respect to x and, therefore, the general solution of it is:

$$G(x, z) = C e^{xz}, \tag{4.5.7}$$

where C is an arbitrary constant. To find the constant C from (4.5.2), (4.5.3), and (4.5.7), we have:

$$G(0, z) = C = \sum_{n=0}^{\infty} y_n(0) z^n = 1.$$

Thus, $C = 1$ and the solution of the equation (4.5.1) would be:

$$G(x, z) = e^{xz} = \sum_{n=0}^{\infty} \frac{(xz)^n}{n!} = \sum_{n=0}^{\infty} \frac{(x)^n}{n!} z^n. \tag{4.5.8}$$

Since the coefficient of z^n in the expansion (4.5.8) is $y_n(x)$, we find the solution as:

$$y_{n(x)} = \frac{x^n}{n!}, \quad n = 0, 1, 2, \dots. \tag{4.5.9}$$

We will have more examples of differential difference equations in Chapter 5 in the discussion of transient cases of different types of queues.

Example 4.5.2.
We will now consider a system of differential–difference equation. This system is generated for a pure birth process in the theory of stochastic processes called the *Poisson process*. Consider the system:

$$\begin{cases} P_0'(t) = -\lambda P_0(t), \\ P_n'(t) = -\lambda P_n(t) + \lambda P_{n-1}(t), \quad n = 1, 2, \dots, \end{cases} \tag{4.5.10}$$

with initial condition $P_0(0) = \delta_{i0}$, where δ_{i0} is the *Kronecker delta function* defined as:

$$\delta_i 0 = \begin{cases} 1, & \text{if } i = 0, \\ 0, & \text{if } i \neq 0, \end{cases} \tag{4.5.11}$$

and normalizing equation $\sum_{n=0}^{\infty} P_n = 1$. We want to find $P_n(t), n = 0, 1, 2, \dots$

Solution

To solve the system (4.5.10), we will first use the Laplace transform and recursive method and then the generating function method.

Recursive Method

Let us denote by $P_0^*(s)$ and $P_n^*(s), n = 1, 2, \ldots$ the Laplace transforms of $P_0(t)$ and $P_0(t), n = 1, 2, \ldots$, respectively. Then, from (4.5.11), we have:

$$\mathcal{L}[P_0'(t)] = sP_0^*(s) - P_0(0) = sP_0^*(s) - 1 \tag{4.5.12}$$

and

$$\mathcal{L}[P_n'(t)] = sP_n^*(s) - P_n(0) = sP_n^*(s). \tag{4.5.13}$$

Applying (4.5.12) and (4.5.13) on (5.4.10), we will have:

$$\begin{cases} sP_0^*(s) - 1 = -\lambda P_0^*(s), \\ sP_n^*(s) = -\lambda P_n^*(s) + \lambda P_{n-1}^*(s), \quad n = 1, 2, \ldots \end{cases} \tag{4.5.14}$$

or

$$\begin{cases} (s + \lambda)P_0^*(s) = 1, \\ (s + \lambda)P_n^*(s) = \lambda + P_{n-1}^*(s), \quad n = 1, 2, \ldots. \end{cases} \tag{4.5.15}$$

Recursively, we will have the following:

$$P_0^*(s) = \frac{1}{(s + \lambda)},$$

$$P_1^*(s) = \frac{\lambda}{(s + \lambda)} P_0^*(s) = \frac{\lambda}{(s + \lambda)} \frac{\lambda}{(s + \lambda)} P_0^*(s) = \frac{\lambda}{(s + \lambda)^2},$$

$$\vdots$$

$$P_n^*(s) = \frac{\lambda^n}{(s + \lambda)^{n+1}}. \tag{4.5.16}$$

Now inverting (4.5.16), we will have:

$$P_n(t) = \mathcal{L}^{-1}[P_n^*(s)] = \mathcal{L}^{-1}\left[\frac{\lambda^n}{(s + \lambda)^{n+1}}\right] = \lambda^n \frac{t^n e^{-\lambda t}}{n!}, \quad n = 1, 2, \ldots \tag{4.5.17}$$

and

$$P_0(t) = \mathcal{L}^{-1}[P_0^*(s)] = \mathcal{L}^{-1}\left[\frac{1}{(s + \lambda)}\right] = e^{-\lambda t}. \tag{4.5.18}$$

Combining (4.5.17) and (4.5.18), we will have:

$$P_n(t) = \frac{(\lambda t)^n}{n!} e^{-\lambda t}, \quad n = 0, 1, 2, \ldots, \tag{4.5.19}$$

which is the Poisson probability distribution function.

Generating Function Method
We now want to solve the system (4.5.10) using the probability generating function. Therefore, we define:

$$G(z, t) = \sum_{n=0}^{\infty} P_n(t) z^n, \quad |z| < 1. \tag{4.5.20}$$

Applying (4.5.11) on (4.5.20), we will have:

$$G(z, 0) = z^i. \tag{4.5.21}$$

Carrying out differentiation of (4.5.20) term by term, within the region of convergence, we obtain:

$$\frac{\partial}{\partial t} G(z, t) = \sum_{n=0}^{\infty} \frac{\partial}{\partial t} [P_n(t)] z^n = P_0'(t) + \sum_{n=0}^{\infty} P_n'(t) z^n.$$

Now, multiplying the first equation of system (4.5.10) by z^n and summing over $n = 1, 2, \ldots$, we will have:

$$\frac{\partial}{\partial t} G(z, t) - P_0'(t) = -\lambda [G(z, t) - P_0(t) - z G(z, t)]. \tag{4.5.22}$$

Using the second equation of system (4.5.10), we will have:

$$\frac{\partial}{\partial t} G(z, t) = G(z, t) [\lambda(z - 1)]. \tag{4.5.23}$$

Solving the system consisting of (4.5.22) and (4.5.23), we leave it as an exercise to show that we will have:

$$G(z, t) = C e^{\lambda(z-1)t}$$

or

$$G(z, t) = z^i e^{\lambda(z-1)t}. \tag{4.5.24}$$

Thus, $P_n(t)$ are the coefficients of the expansion of (4.5.24). Therefore:

$$P_n(t) = \text{coefficient of } z^n \text{ in } G(z,t)$$

$$= \begin{cases} \dfrac{(\lambda t)^{n-1}}{(n-1)!} e^{-\lambda t}, & n = i, i+1, i+2, \dots \\ 0, & n = 0, 1, 2, \dots, i-1. \end{cases} \tag{4.5.25}$$

Since a Poisson process is a Markov chain, say $\{X(t), t \in (0, \infty)\}$, as we will see in the next chapter, with stationary transition probabilities, we will have:

$$P\{X(t+\Delta t) - X(\Delta t) = n \mid X(\Delta t) = i\} = P\{X(t+\Delta t) = i+n \mid X(\Delta t) = i\}$$

$$= \frac{(\lambda t)^k}{k!} e^{-\lambda t}, \quad i, k = 0, 1, 2, \dots; \quad t, \Delta t > 0,$$

$$\tag{4.5.26}$$

which is the same as (4.5.19).

Example 4.5.3.
Consider the following set of differential–difference equations (that belongs to the busy period of a system of parallel queue considered in the next chapter):

$$\begin{cases} P_n'(t) = -(\lambda + n\mu) P_n(t) + (n+1)\mu P_{n+1}(t), & k = n < m-1, \\ P_n'(t) = -(\lambda + n\mu) P_n(t) + \lambda P_{n-1}(t) + (n+1)\mu P_{n+1}(t), & k < n < m-1, \\ P_n'(t) = -(\lambda + m\mu) P_n(t) + m\mu P_{n+1}(t), & k = n = m, \\ P_n'(t) = -(\lambda + m\mu) P_n(t) + \lambda P_{n-1}(t) + m\mu P_{n+1}(t), & n > m, \end{cases} \tag{4.5.27}$$

where k is a fixed number less than or equal to m, subject to $P_k(0) = \delta_{i,k}$. The system (4.5.27) can be written in a reduced form:

$$\begin{cases} P_n'(t) = -(\lambda + n\mu) P_n(t) + \lambda P_{n-1}(t) + (n+1)\mu P_{n+1}(t), & k \le n < m-1, \\ P_n'(t) = -(\lambda + m\mu) P_n(t) + \lambda P_{n-1}(t) + m\mu P_{n+1}(t), & n \ge m, \end{cases} \tag{4.5.28}$$

where the term $\lambda P_{n-1}(t)$ will not appear in the first equation of (4.5.28) when $k = n$ or in the second equation of (4.5.28) when $k = n = m$. In such a case, the first equation becomes redundant.

We want to solve the system (4.5.27), or equivalently (4.5.28) under the conditions stated earlier. Let $P_n^*(s)$ denote the Laplace transform of $P_n(t)$. Then, applying Laplace transform on (4.5.28), we will have:

$$\begin{cases} (\lambda + s + n\mu) P_n^*(s) - 1 = \lambda P_{n-1}^*(s) + (n+1)\mu P_{n+1}^*(s), & k \le n < m-1, \\ (\lambda + s + m\mu) P_n^*(s) = \lambda P_{n-1}^*(s) + m\mu P_{n+1}^*(s), & n \ge m. \end{cases} \tag{4.5.29}$$

We note that -1 on the left-hand side of the first equation of (4.5.29) will appear only when $k = n$. Moreover, when $k = n = m$, then the first equation of

(4.5.29) becomes redundant and the second equation of (4.5.29) will have −1 on its left.

We now define the probability generating function of the Laplace transform as follows:

$$G(z, s) = \sum_{n=k}^{\infty} P_n^*(s) z^n, \quad |z| < 1. \tag{4.5.30}$$

Multiplying both equations of (4.5.29) by the appropriate powers of z and adding over n (the powers), we will have:

$$(\lambda + s + m\mu) G(z, s) - \sum_{n=k}^{m-1} \mu(m-n) p_n^*(s) z^n - z^k$$

$$= \lambda z G(z, s) + \frac{m\mu}{z} G(z, s) + \sum_{n=k}^{m-1} \mu(n+1-m) P_{n+1}^*(s) z^n. \tag{4.5.31}$$

Solving for $G(z, s)$ from (4.5.31), we have:

$$G(z, s) = \frac{z^{k+1} + \mu(z-1) \sum_{n=k}^{m-1} (m-n) P_n^*(s) z^n - k\mu P_k^*(s) z^k}{-\lambda z^2 + (\lambda + s + m\mu) z - m\mu}. \tag{4.5.32}$$

The denominator of (4.5.32) has two roots, denoted by z_1 and z_2, as:

$$z_1 = \frac{\lambda + s + m\mu + \sqrt{(\lambda + s + m\mu)^2 - 4\lambda m\mu}}{2\lambda} \tag{4.5.33}$$

and

$$z_2 = \frac{\lambda + s + m\mu - \sqrt{(\lambda + s + m\mu)^2 - 4\lambda m\mu}}{2\lambda}, \tag{4.5.34}$$

with $|z_2| < 1$. From (4.5.30), we see that the probability generating function defined exists within the unit circle. Hence, since denominator of (4.5.32) is zero at z_2 within the unit circle, the numerator of (4.5.32) should also vanish at that value. Thus, substituting z_2 in the numerator of (4.5.32), we will have:

$$z_2^{k+1} + \mu(z_2 - 1) \sum_{n=k}^{m-1} (m-n) P_n^*(s) z_2^n - k\mu P_k^*(s) z_2^k = 0$$

or

$$(1 - z_2) \sum_{n=k}^{m-1} (m-n) P_n^*(s) z_2^n + k P_k^*(s) z_2^k = \frac{z_2^{k+1}}{\mu}. \tag{4.5.35}$$

The system (4.5.35) consists of $(m - k)$ unknowns $P_n^*(s)$, $k \le n \le m - 1$. We leave it as an exercise to find these $P_n^*(s)$'s from the first equation of (4.5.29) and (4.5.35). Having found the unknowns, one will be invert them and find the $P_n(t)$. We also leave it as an exercise for the cases $k = m$ or $k = m - 1$ that the system (4.5.35) is sufficient to determine $P_n^*(s)$. We finally note that when $k = m$, the sum in (4.5.35) will not appear.

4.6. NONLINEAR DIFFERENCE EQUATIONS

Although we do not use nonlinear cases in this book, we give the following examples to show how a nonlinear equation may be solved and be modeled.

Example 4.6.1.
We want to solve the following nonlinear difference equation:

$$y_{k+1} - y_k + k(y_{k+1})(y_k) = 0, \quad y_1 = 2. \tag{4.6.1}$$

Solution
There are different techniques to solve difference equations, and here in this example we use substitution and recursion as follows: Divide both sides of (4.6.1) by $y_k + {}_1 y_k$ and obtain:

$$\frac{1}{y_{k+1}} - \frac{1}{y_k} = k. \tag{4.6.2}$$

Now, let

$$u_k = \frac{1}{y_k}. \tag{4.6.3}$$

Then, using (4.6.3) and (4.6.1), equation (4.6.2) can be rewritten as:

$$u_{k+1} - u_k = k, \quad u_1 = \frac{1}{2}. \tag{4.6.4}$$

From (4.6.4), going downward, we have $u_k = u_{k-1} + k - 1, u_{k-1} = u_{k-2} + k - 2, \ldots$ Hence, we can recursively, for $k, k - 1, k - 2, \ldots, 1$, have the following:

$$u_{k+1} = u_k + k = u_{k-1} + (k-1) + k = u_{k-2} + (k-2) + (k-1) + k = \cdots$$

or

$$u_{k+1} = u_1 + 1 + 2 + \cdots + k = \frac{1}{2} + \frac{k(k+1)}{2} = \frac{k^2 + k + 1}{2}. \tag{4.6.5}$$

Thus, from (4.6.3) and (4.6.5), we have:

$$y_{k+1} = \frac{2}{k^2 + k + 1}.$$

Therefore, the solution of (4.6.1) is:

$$y_k = \frac{2}{(k-1)^2 + k - 1 + 1} = \frac{2}{k^2 - k + 1}, \quad k = 1, 2, \ldots.$$

Example 4.6.2.
This example is a simple nonlinear difference equation extracted from the contents of a paper by Robert M. May from the Department of Biology at Princeton University, New Jersey, USA. It is titled "Biological Populations Obeying Difference Equations: Stable Points, Stable Cycles, and Chaos."

As we have already mentioned, population growth is usually modeled as a continuous process and described by differential equations. In some biological situations (such as man), this is how it is done and generations overlap. However, in some other biological situations (such as 13-year periodical cicadas), population growth takes place at discrete intervals of time and generations are completely nonoverlapping. In such cases, the appropriate mathematical model is in terms of nonlinear difference equations. For a single species, the simplest such differential equations, with no time delays, lead to very simple dynamics: a familiar example is the logistic, mentioned in Chapter 3, (3.1.20):

$$\frac{dN}{dt} = rN\left(1 - \frac{N}{k}\right), \quad \forall r > 0, \tag{4.6.6}$$

with a globally stable equilibrium point at $N = k$. However, the corresponding simplest difference equations, with their built-in time lag in the operation of regulatory mechanisms, can have a complicated dynamical structure.

Specifically, we now discuss some models studied by May (1975).

Model 4.6.1.
Some researchers such as Macfadyen (1963) and Cook (1965) considered

$$N_{t+1} = N_t e^r \left(1 - \frac{N_t}{k}\right) \tag{4.6.7}$$

to be the difference equation analog of the logistic differential equation (3.1.20), with r and k as the usual growth rate and carrying capacity, respectively.

For a single species, the difference equations arising in population biology are usually discussed as having either a stable equilibrium point or unstable, growing oscillations. Some of the most elementary of these nonlinear difference equations show a spectrum of dynamical behavior, which, as r increases, goes from a stable equilibrium point to stable cyclic oscillations between two population points to stable cycles with four points, then eight points, and so on, through to a regime that can only be described as "chaotic" (a term used by Li & Yorke, 1975). For any given value of r, in this chaotic regime there are cycles of period $2, 3, 4, 5, \ldots, n, \ldots$, where n is a positive integer, along with an uncountable number of initial points for which the system does not eventually settle into any finite cycle; whether the system converges on a cycle, and, if so, which cycle, depends on the initial population point (and some of the cycles may be attained only from infinitely unlikely initial points).

May (1975) illustrated this range of behavior through a graph (fig. 1 in May, 1975). He considered the spectrum of dynamical behavior of the population density, N_t/k, as a function of time t, as described by the difference equation (4.6.7) for various values of r. Specifically, the following is found on figure 1 in May (1975):

   (a) $r = 1.8$, stable equilibrium point
   (b) $r = 2.3$, stable two-point cycle
   (c) $r = 2.6$, stable four-point cycle

Figure parts (d), (e), and (f) were in the chaotic regime, where the detailed character of the solution depends on the initial population value, with

   (d) $r = 3.3$, $(N_0/k = 0.075)$
   (e) $r = 3.3$, $(N_0/k = 1.5)$
   (f) $r = 5.0$, $(N_0/k = 0.02)$

May (1975) not only illustrated the stability character of equation (4.6.7) as a function of increasing r in a graphic form as explained earlier, he also set it out in a table format as in Table 4.6.1.

TABLE 4.6.1. Dynamics of a Population Described by the Difference Equation (4.6.7)

Dynamical Behavior	Value of Growth Rate (r)	Illustration
Globally stable equilibrium point	$2 > r > 0$	Fig. 1a
Globally stable two-point cycle	$2.526 > r > 2$	Fig. 1b
Globally stable four-point cycle	$2.656 > r > 2.526$	Fig. 1c
Stable cycle, period 8, giving way in turn to cycles of period 16, 32, and so on as r increases	$2.692 > r > 2.656$	
Chaos (cycles of arbitrary period or aperiodic behavior, depending on initial condition)	$r > 2.692$	Fig. 1d–f

Model 4.6.2.
Other authors such as Maynard Smith (1968), Krebs (1972), May (1973a), and Levine et al. (1977) have considered:

$$N_{t+1} = N_t \left[1 + r \left(1 - \frac{N_t}{k} \right) \right] \qquad (4.6.8a)$$

or, equivalently, by letting:

$$x = \left(\frac{r}{1+r} \right) \left(\frac{N}{k} \right),$$
$$x_{t+1} = (1+r) x_t (1 - x_t), \qquad (4.6.8b)$$

becomes the analog of the logistic differential equation, probably the simplest nonlinear difference equation one could write. However, it seems that equation (4.6.8) is less satisfactory than (4.6.7) because of its unbiological feature that the population can become negative if at any point N_t exceeds $k(1 + r)/r$. Thus, stability properties here refer to stability within some specific neighborhood, unlike equation (4.6.7), where, for example, the stable equilibrium point at $N = k$ is globally stable (for all $N > 0$) for $2 > r > 0$. With this proviso, the stability behavior of equation (4.6.8) is strikingly similar to that of equation (4.6.7). May (1975) illustrates this in a table format (see Table 4.6.2).

Some other simple difference equations, which are mainly culled from the entomological literature and which exhibit the phenomenon, are as follows:

Model 4.6.3.
The following equation:

$$N_{t+1} = \frac{\lambda N_t}{(1 + a N_t)^b}, \qquad (4.6.9)$$

has been used by Hassell (1975) to provide a two-parameter fit to a wide range of field and laboratory data on single-species population growth; for relatively small values of b or of λ, there is a globally stable point; the conjunction of

TABLE 4.6.2. Dynamics of a Population Described by the Difference Equation (4.6.8)

Dynamical Behavior	Value of Growth Rate, r
Stable equilibrium point	$2 > r > 0$
Stable two-point cycle	$2.449 > r > 2$
Stable four-point cycle	$2.544 > r > 2.449$
Stable cycle, period 8, then 16, 32, and so on	$2.570 > r > 2.544$
Chaos	$r > 2.570$

moderate values of b and λ produces stable cycles; and relatively large values of b and λ leads to chaos.

Model 4.6.4.
The density dependent form:

$$N_{t+1} = \left[\lambda_1 + \frac{\lambda_2}{1 + e^{A(N_t - B)}} \right] N_t \qquad (4.6.10)$$

discussed by Pennycuik et al. (1968), Usher (1972), and (in a limiting step-function form) by Williamson (1974), can also exhibit all three regimes as the two parameters A and B are varied.

Model 4.6.5.
Similarly, the class of models

$$N_{t+1} = \frac{\lambda N_t}{1 + (aN_t)^b}, \qquad (4.6.11)$$

the possible stable points of which have been discussed by Maynard Smith (1974), can show all three types of behavior as b and λ vary.

Model 4.6.6.
The density dependent equation:

$$N_{t+1} = \begin{cases} \lambda N_t, & \text{if } N_t < B \\ \dfrac{\lambda N_t}{\left(\dfrac{N_t}{B} \right)^b}, & \text{if } N_t > B, \end{cases} \qquad (4.6.12)$$

is discussed by Varley et al. (1973). In this model, as a consequence of the pathological discontinuity, the stable point regime $(0 < b < 2)$ gives way directly to the chaotic regime $(b > 2)$, with no intervening regime of stable cycles.

Model 4.6.7.
In all the previous examples, the parameters can take values such that the curve relating N_{t+1} to N_t has a hump. On the contrary, the form

$$N_{t+1} = \frac{\lambda N_t}{1 + aN_t}, \qquad (4.6.13)$$

which is sometimes called the *logistic difference equation* (Skellam, 1952; Leslie, 1957; Utida, 1967; Pielou, 1969), gives a monotonic curve relating

N_{t+1} and N_t, and consequently it always leads to a globally stable equilibrium point.

For population biology in general, and for temperate-zone insects in particular, the implication is that even if the natural world was 100% predictable, the dynamics of populations with "density-dependent" regulation could nonetheless, in some circumstances, be indistinguishable from chaos, if the intrinsic growth rate r is large enough.

Example 4.6.3.
Robert May (1976) studied the following logistic biological map model:

$$x_{n+1} = rx_n(1-x_n), \qquad (4.6.14)$$

where x_n, $0 < x_n < 1$, represents the ratio of the existing population to the maximum possible population at time n. x_0 represents the initial ratio at time 0. r is a positive number representing the combined rate for reproduction and starvation; it is sometimes called the *driving parameter*.

The way equation (4.6.14) works is that a fixed value for r and an initial population x_0 are given. Then all x_n, $n = 1, 2, \ldots$, will be calculated recursively. The model calculates reproduction as r increases and n goes to infinity. For low value of r, when n goes to infinity, $\{x_n\}$ converges to a single number. However, as r becomes almost 3, when n goes to infinity, $\{x_n\}$ no longer converges and it oscillates between two values.

When the growth rate starts to decrease, starvation starts. When r is between 0 and 1, the population eventually vanishes. When r is between 1 and 2, regardless of the initial population, the population eventually approaches a single value $(r - 1)/r$. When r is between 2 and 3, the population fluctuates around $(r - 1)/r$ and eventually approaches to it. As r further increases, the sequence will oscillate between four values. This pattern continues to 8, 16, and then "chaos." When r becomes 3.57, then $\{x_n\}$ neither converges nor oscillates, but its value becomes random. However, at a particular value of r, the sequence starts to oscillate again.

EXERCISES

4.1. Find the general solution of each of the following homogeneous difference equations:

    a. $y_{n+2} - 4y_n = 0$
    b. $y_{n+2} + 4y_{n+1} + y_n = 0$
    c. $4y_{n+2} - 6y_{n+1} + 3y_n = 0$
    d. $y_{n+2} + y_{n+1} + y_n = 0$

4.2. Find particular solutions for Exercise 4.1 satisfying the initial conditions $y_0 = 0$ and $y_1 = 1$.

4.3. Find the general solution of each of the following nonhomogeneous difference equations:

a. $y_{n+2} - 5y_{n+1} + 6y_n = 5$

b. $y_{n+2} - 3y_{n+1} + 2y_n = 1$

c. $y_{n+2} - y_{n+1} - 2y_n = n^2$

d. $y_{n+2} - 2y_{n+1} + y_n = 3 + 4n$

e. $y_{n+2} - 2y_{n+1} + y_n = 2^n(n + 1)$

f. $8_{n+2} - 6y_{n+1} + y_n = 5\sin(n\pi/2)$

4.4. Find the solution of the first-order difference equations with the given initial condition and write out the first five values of y_k in sequence form:

a. $y_{n+1} = y_n + 2$, $\quad y_0 = 3$

b. $y_{n+1} + y_n - 3 = 0$, $\quad y_0 = 2$

c. $y_{n+1} = 3y_n - 1$, $\quad y_0 = 1$

d. $y_{n+1} + 3y_n = 0$, $\quad y_0 = 5$

e. $3y_{n+1} = 2y_n + 4$, $\quad y_0 = -4$

f. $2y_{n+1} - y_n = 4$, $\quad y_0 = 3$

4.5. Find the generating function $G(z)$ of the sequence $\{a_n\}$:

a. $a_n = 1 - 3n$

b. $a_n = (2/5)(1/5)^n$

c. $a_n = (n + 2)(n + 3)$

d. $a_n = n(n + 1)/2$

4.6. Find the sequence $\{a_n\}$ having the generating function $G(z)$:

a. $G(z) = 2/(1 - z) + z/(1 - 2z)$

b. $G(z) = 2/(1 - z)^2 + z/(1 - z)$

c. $G(z) = z^2/(1 - z)$

4.7. Solve the partial difference equation using the generating function method:

$$2Y_{k,n} = Y_{k-1,n} + Y_{k,n-1}, \quad Y_{k,0} = 0, \quad Y_{0,0} = 0, \quad Y_{0,n} = 1, \quad n = 1, 2, \ldots.$$

4.8. Prove Theorem 4.3.1 for the case $a = 1$.

4.9. Redo Examples 4.3.1 through 4.3.10 using the recursive method, Theorem 4.3.1.

4.10. Complete the solution of Example 4.3.10 to show that the induction could be completed when $a = 1$.

4.11. Complete the solution of Example 4.4.3 by finding other probabilities similar to what was done to obtain $P_{1,n}$ in (4.4.32).

4.12. Consider the system of the following partial difference equation:

$$\begin{cases} P_{1,1} = \rho P_{0,0}, & \\ P_{i,0} = \rho P_{i-1,0} & 1 \le i \le n, \\ P_{i,0} = \theta P_{i-1,0} & n \le i, \\ P_{i,1} = (1+\rho) P_{i-1,1} - \rho P_{i-2,1} & 2 \le i \le n, \\ P_{i,1} = (1+\rho) P_{i-1,1} - \rho P_{i-2,1} - \omega P_{i-1,0} & n+1 \le i, \end{cases}$$

where $\rho = \lambda/\mu (<1)$, $\theta = \lambda/\lambda + \gamma$, $\omega = \gamma/\mu$, λ, μ, θ, ω, and γ are constants. Assume that $P_i = P_{i,0} + P_{i,1}$. Show that that if $\rho \ne \theta$ and $\alpha = 1 - \theta/\theta + n(1 - \theta)$, then:

$$P_i = \alpha(1 - \rho^{i+1}), \quad 0 \le i \le n-1,$$

$$P_i = \frac{\alpha\left[(1 - \rho^{n-1})\rho^{i-n+2} + (1-\rho)(\rho^{i-n+2} - \theta^{i-n+2})\right]}{p - \theta}, \quad n \le i.$$

4.13. Consider the system of differential–difference equations:

$$\begin{cases} F_0'(t) = -\lambda F_0(t) \\ F_n'(t) = \lambda F_{k-1}(t) - (\lambda + j\mu) F_n(t), \quad n = 1, 2, \dots, \end{cases}$$

subject to the initial condition $F_n(0) = p_n$, where λ, μ, and j are constants. Show that the unique solution to this system is $F_n(t) = p_n e^{-\lambda}$.

4.14. Solve systems (4.5.23) and (4.5.24) and show the solution is (4.5.25).

4.15. a. Find the unknowns $P_n^*(s)$ in (4.5.35). Having found the unknowns, invert them and find $P_n(t)$.

b. For the cases $k = m$ or $k = m - 1$, show that the system (4.5.35) is sufficient to determine $P_n^*(s)$.

CHAPTER FIVE

Queueing Theory

In this chapter we utilize what we have discussed in the previous chapters, to the extent necessary, and consider their applications in queueing theory. We start with some definitions as our basic chapter vocabulary.

5.1. INTRODUCTION

A *stochastic process* is a sequence of random variables, say $\{X(t), t \in T\}$, where t is an epoch (a time point) belonging to a set T (denumerable or nondenumerable). If T is countable (or denumerable or finite), then the process is called a *discrete-time* process, while if it is continuous, the process is called *continuous-time* stochastic process. When the index set, T, is nonnegative, the discrete-time process is denoted by $\{X_n, n = 0, 1, \ldots\}$, while for the continuous-time process it is denoted by $\{X(t), t \geq 0\}$. The set of all possible values of random variable $X(t)$, denoted by S, is called the *state space* of the process. The state space may be finite, countable, or noncountable. The starting value of the process is called the *initial state* and is denoted by $X(0)$ for the continuous case and X_0 for the discrete case. The index set T is called the *parameter space*. Thus, a stochastic process could have any one of the following four categories: (1) discrete state space and discrete parameter space, (2) continuous state space and discrete parameter, (3) continuous state space and discrete parameter space, and (4) continuous state space and continuous parameter space.

As an example, consider the arrival of customers at a store. The number of customers that arrived during the interval of time $[0, t]$ is a stochastic process in which the state space is $S = \{0, 1, 2, \ldots\}$ and the parameter space is $T = [0, \infty)$.

A *point process* is a strictly increasing sequence of real numbers, say $\{t_1, t_2, \ldots, t_n, \ldots, t_1 \geq 0\}$, with $t_1 < t_2 < \ldots$, which does not have a finite limit point, that is, $\lim_{n \to \infty} t_n = +\infty$. The numbers t_n, $n = 1, 2, \ldots$, are called *event times* or *epochs*. The t_ns do not necessarily have to be points of time; they could be

Difference and Differential Equations with Applications in Queueing Theory, First Edition.
Aliakbar Montazer Haghighi and Dimitar P. Mishev.
© 2013 John Wiley & Sons, Inc. Published 2013 by John Wiley & Sons, Inc.

points such as the location of potholes in a road or arrivals in a waiting line. In such cases, since the number of potholes, for instance, is finite in the real world, we need to consider finite samples from the point process. In case the points are different in type, for instance, the type of customers arriving to a store, the process is called a *marked point process*.

Let $N(t), t > 0$ denote the number of events that occurred in a time interval $(0, t]$ at epochs t_n. The sequence $\{N(t), t \geq 0\}$ is called the *counting process* belonging to the point process $\{t_1, t_2, \ldots\}$. A counting process is called a *random point process* when the epochs t_n are random variables, $t_1 < t_2, \ldots$ and $P\{\lim_{n \to \infty} t_n = +\infty\} = 1$ For $n = 1, 2, \ldots, \tau_n = t_n - t_{n-1}$, the interevents, with $t_0 = 0$, are called *renewal periods*. If $\{\tau_1, \tau_2, \ldots\}$ is a sequence of independent identically distributed (iid) random variables, then it is called a *renewal process*. If $N(t), t > 0$, represents the number of events occurring in $(0, t]$, then $\{N(t), t \geq 0\}$ is called the *renewal counting process*.

Consider a stochastic process $\{X(t), t \in T\}$ and a set of discrete points $\{t_0, t_1, t_2, \ldots\}$. If distribution properties of the process for $t_i, i = 0, 1, 2, \ldots, n$, are the same, we say that the process is *regenerating* itself or is a *recurrent point process*. Since the renewal periods are iid, a renewal process is also a recurrent point process.

Let $\tau_i, i = 1, 2, \ldots$, be independent. Further assume that $F_1(t) \equiv P\{\tau_1 \leq t\}$ and $\tau_i, i = 2, 3, \ldots$, are identically distributed as $F(t) \equiv P\{\tau_i \leq t\}, i = 2, 3, \ldots$, $F(t) \neq F_1(t)$. The sequence $\{\tau_1, \tau_2, \ldots\}$ is then called a *delayed renewal process*. The random time point, t_n, at which the nth renewal occurs is given by $t_n = \sum_{i=1}^{n} \tau_i, n = 1, 2, \ldots$. The random point process $\{t_1, t_2, \ldots\}$ is called the *time points of renewal process*.

Definition 5.1.1
Consider epochs $\theta_n, n = 1, 2, \ldots$, as random variables. We denote the number of events occurring in the interval $(0, t]$ by the random variable $X(t)$, which $X(t) = \max\{n, \theta_n \leq t\}$. A continuous-time stochastic process $\{X(t), t \geq 0\}$ with state space $\{0, 1, 2, \ldots\}$ is then called the *random counting process* belonging to the random point process $\{\theta_1, \theta_2, \ldots, \theta_n, \ldots\}$. The following are three properties of a random counting process:

a. $X(0) = 0$.
b. $X(\theta_1) \leq X(\theta_2)$, for $\theta_1 \leq \theta_2$.
c. For any θ_1 and θ_2 such that $0 \leq \theta_1 \leq \theta_2$, denote the number of events occurred in $(\theta_1, \theta_2]$ by $X(\theta_1, \theta_2)$. Then, assume $X(t) = X(0, t), X(\theta_1, \theta_2)$ is called the *increment* and $X(\theta_1, \theta_2) = X(\theta_2) - X(\theta_1)$.

5.2. MARKOV CHAIN AND MARKOV PROCESS

Recall that two events are independent means that the outcome of an experiment at one time is not affected by the outcome of another time such as tossing a coin two times. In contrast, suppose we were to choose integers randomly between 1 and 10 without replacement. Such events are dependent since choosing without replacement increases the chances for the next numbers to be chosen. This example leads us to the following concept.

Definition 5.2.1

A discrete-time stochastic process $\{X_n, n = 0, 1, \ldots\}$ with discrete state space S is called a *discrete-time Markov chain with the state space S* if for a given sequence of values from the state space (i.e., $x_i \in S$) the following property holds:

$$P\{X_n = x_n \mid X_0 = x_0, X_1 = x_1, \ldots, X_{n-1} = x_{n-1}\} = P\{X_n = x_n \mid X_{n-1} = x_{n-1}\}.$$
(5.2.1)

If the state space, S, is finite, the chain is called a *finite-state Markov chain* or a *finite Markov chain*. Equation (5.2.1) is the *Markov property* for the Markov chain. The right-hand side of (5.2.1), that is, $P\{X_n = x_n \mid X_{n-1} = x_{n-1}\}$, is called a *one-step transition probability* from state x_{n-1} to state x_n. In general, the conditional probability $P\{X_n = j \mid X_{n-1} = i\}, i, j \in S$, is called the *transition probability* from state i to state j, denoted by p_{ij}, that is,

$$p_{ij} = \{X_n = j \mid X_{n-1} = i\}, \quad i, j \in S.$$
(5.2.2)

where $i = j$, p_{ii} is the probability that the Markov chain stays in state i for another step. The transition probabilities p_{ij} are nonnegative and sum to 1 along each row, that is,

$$0 \le p_{ij} \le 1, \forall i, j \in S, \text{ and } \sum_{j \in S} p_{ij} = 1, \forall_i \in S.$$
(5.2.3)

The one-step transition probabilities may be summarized in a matrix form as follows:

$$\mathbf{P} = \begin{pmatrix} p_{00} & p_{01} & p_{02} & \cdots & \vdots & \cdots \\ p_{10} & p_{11} & p_{12} & \cdots & \vdots & \cdots \\ \vdots & \vdots & \vdots & \cdots & \vdots & \cdots \\ p_{i0} & p_{i1} & p_{i2} & \cdots & \vdots & \cdots \\ \vdots & \vdots & \vdots & \cdots & p_{ij} & \cdots \\ \vdots & \vdots & \vdots & \cdots & \vdots & \ddots \end{pmatrix}.$$
(5.2.4)

The matrix $\mathbf{P}$ has two properties $0 \le p_{ij} \le 1$ and $\Sigma_{j \in S} \, p_{ij} = 1, \forall i \in S$ and it is called a *stochastic matrix*. If columns also sum to 1, then it is called a *doubly stochastic matrix*. When the one-step transition probabilities are independent of time, the Markov chain is called *homogeneous*.

What the Markov property (5.2.1) says is that the process or chain "forgets" the history of states except its current state and that the future state depends only on the current one. Hence, (5.2.1) is sometimes called the *forgetfulness* property of a Markov chain. Equation (5.2.1) shows some type of dependency among random variables X_n. For example, a fair coin "forgets" how many times it has visited the "tails" state and the "heads" state at the current toss. That is, $P\{$"heads" on this toss|"tail" on the previous 10 tosses$\}$ is still 1/2. We should caution the reader that p_{ij} in some books and other literatures is for the transition from y to x rather than from x to y that we defined. See, for example, Karlin and Taylor (1975) and Kemeny and Snell (1960).

Example 5.2.1

As a particular discrete-time counting process, we consider the following Bernoulli process. Suppose we divide a continuous-time interval into equal-length discrete disjoint subintervals such that at most one event (success or failure) can occur in a subinterval. A counting process is then called a *Bernoulli counting process* if (1) the number of successes (or failures) that can occur in each subinterval is at most 1; (2) all subintervals are equally likely to have an event with probability p; and (3) the occurrence of events in nonoverlapping subintervals are independent of each other. It can be proved that the total number of events (say, successes), denoted by X_n, in a Bernoulli counting process at the end of the nth subinterval, $n = 1, 2, \ldots$, with initial state $X_0 = 0$, has the following binomial distribution (refer to Section 1.2 of Chapter 1):

$$P\{X_n = k\} = \binom{n}{k} p^k (1-p)^{n-k} \quad k = 0, 1, \ldots, n.$$

This is because since X_n is a sum of the outcomes of n identical and independent Bernoulli trials. Thus, the one-step transition probabilities of a state x are $P\{x \rightarrow x+1\} = p$ and $P\{x \rightarrow x\} = 1 - p$. Note that the Bernoulli counting process, as defined, is time homogeneous since the probability of the occurrence of the event (say, success) does not change with time. However, if instead of equal-length interval we would have taken time segment intervals, then the process could have been nonhomogenous.

For a Bernoulli counting process, we may be interested in the probability distribution of number in the number of intervals between two successive occurrences of events of interest, say U. That is, if the event of interest, say success, occurred on the nth trial after the start of counting, that is, $U = n$, we should have no successes in the prior $n - 1$ trials. In other words, we should have $n - 1$ failures in a row before a success occurs. Thus, $P\{U = u\} = (1 - p)^{u-1} p$, $u = 1, 2, \ldots$ In case we want the probability of more than u intervals for a success to occur, then we need to sum the probabilities as follows:

$$P\{U > u\} = \sum_{i=u+1}^{\infty} (1-p)^{i-1} p$$

$$= p \times \left[\sum_{k=0}^{\infty} (1-p)^k - \sum_{k=0}^{u-1} (1-p)^k \right]$$

$$= p \times \left[\frac{1}{1-(1-p)} - \frac{1-(1-p)^u}{1-(1-p)} \right]$$

$$= (1-p)^u.$$

For example, if the probability of a success is 10%, then the probability that it would take more than five time intervals to have a success will be $(1 - 0.1)^5 = 0.9^5 = 59\%$.

Now suppose the following are known: the unit of time to be u and the length of an interval is taken as x units of time (could be a fraction of a unit). Hence, there are $n = u/x$ intervals in a unit of time. For instance, if the length of interval is 1 minute, and a unit of time is 1 hour, then there would be $60/1 = 60$ intervals in an hour. Suppose also that the expected number of success in a unit of time (1 hour) (or the *rate of success*) is given as μ. From Chapter 1, we know that the expected value for a Binomial distribution with parameters n and p is np. Hence, $np = \mu$ or $p = \mu/n = (x/u)\mu$. In other words, the probability of success is proportional to the length of the interval with the constant of proportionality being the rate of success. The rate of success, μ, may be estimated by the ratio of the number of successes in τ units of time (a unit being u) and τ, that is, $\hat{\mu} = [(\text{number of success in } \tau \text{ units of time})/\tau]$. For instance, if there are 150 successes in 10 hours, then the success rate is estimated at $150/10 = 15$ per hour. For more examples and details, the reader is referred to Higgins and Keller-McNulty (1995).

Definition 5.2.2
For the general case, the *n-steps transition probability* from state x to state y in n steps, denoted by $p_{xy}^{(n)}$, $x, y \in S$, where S is the state space, are the conditional probabilities:

$$p_{xy}^{(n)} = P\{X_{m+n} = y \mid X_m = x\}, \quad x, y \in S, \tag{5.2.5}$$

for every $m \geq 0$ for which $P(X_m = x) > 0$, where $p_{xy}^{(1)} = p_{xy}$ and by convention $p_{xy}^{(0)} = 1$, when $x = y$ and 0, otherwise.

Definition 5.2.3
A stochastic process is called a *Markov counting process* if it has the following properties: (1) it has the Markov property (2) all possible states are natural numbers, and (3) for each state x, the only possible transitions are $x \to x$, $x \to x + 1, x \to x + 2, \ldots$.

Based on Definition 5.2.2, to reach y from state x in $m + n$ steps, we will have:

$$p_{xy}^{(m+n)} = \sum_{i=0}^{n} p_{xi}^{(n)} p_{iy}^{(m)}, \quad \forall m, n, x, y \geq 0. \tag{5.2.6}$$

Relation (5.2.6) is an example of a *Chapman–Kolmogorov equation*. Equation (5.2.6) in its matrix form is:

$$\mathbf{P}^{m+n} = \mathbf{P}^m \mathbf{P}^n. \tag{5.2.7}$$

Relation (5.2.7) may be obtained directly through the iterative method as follows: $\mathbf{P}^n = \mathbf{P}^{n-1}\mathbf{P}$. The matrix $\mathbf{P}^n$ is called the *nth-step transition matrix*. In other words, the n-step transition probabilities $p_{xy}^{(n)}$ are the (x, y)th elements of the nth power of the one-step transition matrix (see Neuts 1973).

We can find the probabilities when the initial state is a random variable. To find the probability that the Markov chain $\{X_n, n \geq 0\}$ with state space S is in state x after one step, we add up probabilities of all possible ways, starting from the initial state x to state y. Thus, we will have:

$$P\{X_1 = y\} \sum_x P\{X_0 = x\} p_{xy}, \quad y \in S. \tag{5.2.8}$$

Now if we let π_n (if there would be a possibility of confusion, we write $\pi_n(y)$ to indicate where the state is at time n) be a row vector whose elements consist of probabilities $P\{X_n = y\}, y \in S$, then, from (5.2.8) we will have:

$$\pi_1 = \pi_0 \mathbf{P}, \quad \text{or} \quad \pi_1(y) = \pi_0(x)\mathbf{P} \tag{5.2.9}$$

and for transition after n steps, we will have:

$$\pi_n = \pi_0 \mathbf{P}^n, \quad \text{or} \quad \pi_n(y) = \pi_0(x)\mathbf{P}^n, \tag{5.2.10}$$

that is,

$$P\{X_n = y\} = \sum_x P\{X_0 = x\} p_{xy}^{(n)} = \sum_x \pi_0(x) p_{xy}^{(n)}, \quad y \in S. \tag{5.2.11}$$

Example 5.2.2

We perform a sequence of Bernoulli trials in which the outcomes are "success" or "failure," denoted by 1 and 0, respectively. We assume that trials are dependent, that is, the outcome of a trial depends upon the outcome of the previous trial (i.e., Markov dependence). Let the conditional probability of "success" on the $(n + 1)$st trial, given that the nth trial's outcome was also a "success" be α, that is, $p_{11} = \alpha$. Let also the conditional probability of a "success" on the

$(n + 1)$st trial, given that the nth trial's outcome was a "failure" be β, that is, $p_{01} = \beta$. Similarly, we assume $p_{00} = 1 - \beta$ and $p_{10} = 1 - \alpha$. α and β are assumed to be independent of trials before the nth trial.

A specific case of this example could be the case when a signal as 1 or 0 is sent through a series of three relay stations after it leaves the source. Errors could cause a wrong signal sent from one place to the next. Hence, we can write a 2 by 2 transition matrix $\mathbf{P}$, in which the rows represent the signal transmitted and the columns represent the signals received. Now if we let $X_n = 1$, when the outcome of the nth trial is a "success" and $X_n = 0$, when it is a "failure," based upon the assumptions, the sequence of random variables $\{X_n, n \geq 0\}$ form a two-state Markov chain on the state space $\{0, 1\}$. In this case, the one-step transition probability matrix will be:

$$\mathbf{P} = \begin{pmatrix} 1-\beta & \beta \\ 1-\alpha & \alpha \end{pmatrix}. \tag{5.2.12}$$

For instance, if $\alpha = 0.8$ and $\beta = 0.3$, that is, if a signal was sent as 1 and was received as 1, it has a probability of $p_{11} = 0.8$, while the probability that a 0 signal was sent and a 1 was received is $p_{01} = 0.7$, and so on. Therefore:

$$\mathbf{P} = \begin{matrix} & \begin{matrix} 0 & & 1 \end{matrix} \\ \begin{matrix} 0 \\ 1 \end{matrix} & \begin{pmatrix} p_{00} & p_{01} \\ p_{10} & p_{11} \end{pmatrix} \end{matrix} = \begin{pmatrix} .7 & .3 \\ .2 & .8 \end{pmatrix} \quad \text{and} \quad \mathbf{P}^3 = \begin{pmatrix} .475 & .525 \\ .350 & .650 \end{pmatrix}.$$

Thus, if we start by sending a 0, the third station will receive 0 with probability of 0.475, while if we start with a 1 signal, we will have a 1 at the third station with a probability of 65%.

Suppose the pattern observed is that 75% of the time a 1 signal is initially transmitted and 25% of the time a 0 signal is initially transmitted, that is, the initial signal randomly transmitted 1's with the probability $\alpha_1 = 0.75$ and 0's with the probability $\alpha_0 = 1 - \alpha_1 = 0.25$. The third station will then receive a 1 with probability $(0.75)(0.65) = 0.4875 = 49\%$. The probability of a correct signal being received at the third station is:

$$P\{X_0 = 0\} p_{00}^{(3)} + P\{X_0 = 1\} p_{11}^{(3)} = (.25)(.475) + (.75)(.650) = 6063 = 60\%.$$

We end this section with an example of a stochastic process that is not Markovian.

Example 5.2.3. *Polya's Urn Model*
A ball is drawn from an urn containing b black and r red balls with replacement. Observing the color of the ball drawn, c balls of that color and d balls of the opposite color is added after a draw. Hence, for the second draw, there are $b + r + c + d$ balls to draw from, where c and d are arbitrary integers. The procedure continues. In case c or d or both are negative, the process will stop as soon as there are no balls left in the urn. Note that if $c = -1$ and $d = 0$, the

procedure becomes drawing without replacement that will terminate after $b + r$ draws.

Now we define the random variable $X_n = 1$ or 0, according to whether the ball drawn on the nth drawing was black or red, respectively. The sequence $\{X_n, n = 1, 2, \ldots\}$ is not a Markov chain. Here is a counterexample. Let us consider the conditional probability that the third draw resulted a black given that the second draw was also a black, that is, $P\{X_3 = 1|X_2 = 1\}$. Note that the sample space of size n in this case contains a sequence of total n b and r's. The probability of a black or a red at the first drawing, assuming an equally likely chance, is $b/(b + r)$ or $r/(b + r)$, respectively. Under the condition specified, if the first drawing resulted in red, the probability that the second drawing results in black would be $(b + c)/(b + r + c + d)$, that is,

$$P\{X_2 = 1|X_1 = 1\} = \frac{b+c}{b+r+c+d}.$$

Using a similar argument, we can find the conditional probability of having the result of the third drawing as red, given we had a red at the second drawing. For absolute probabilities, the probability of a black at first drawing and another black at the second drawing would be:

$$P\{X_1 = 1, X_2 = 1\} = \frac{b}{b+r} \cdot \frac{b+c}{b+r+c+d}.$$

Still for the third black, we will have the joint probability as:

$$P\{X_1 = 1, X_2 = 1, X_3 = 1\} = \frac{b}{b+r} \cdot \frac{b+c}{b+r+c+d} \cdot \frac{b+2c}{b+r+2c+2d}.$$

It can be shown by mathematical induction that the probability of a black ball drawn at any trial is $b/(b + r)$. It can also be proved by induction that for any $m < n$, the probabilities that the mth and the nth drawings result in a pair of (black, black) or (black, red) are:

$$P\{X_m = 1, X_n = 1\} = \frac{b(b+c)}{(b+r)(b+r+c)}$$

and

$$P\{X_m = 1, X_n = 0\} = \frac{br}{(b+r)(b+r+c)},$$

respectively. Thus,

$$P\{X_3 = 1|X_2 = 1\} = \frac{b+c}{b+r+c},$$

while

$$P\{X_3 = 1 \,|\, X_2 = 1, X_1 = 1\} = \frac{b + 2c}{b + r + 2c}.$$

For more examples and detail, see Feller (1968).

Definition 5.2.4

A state x for which $p_{xx} = 1$ (so that $p_{xy} = 0$ for $y \neq x$, that is, P(leaving an absorbing state) $= 0$) is called an *absorbing state* (or an absorbing barrier). If we let f_x be the probability that starting from state x the Markov chain will ever reenter state x, then state x is called a *recurrent* state if $f_x = 1$ and called a *transient* state if $f_x < 1$, that is, the state is a nonabsorbing state. If every state of a Markov chain can be reached from every other state in some positive step, the chain is called *irreducible*; otherwise, it is called *reducible*. The chain is called *periodic with period d* if $p_{xx}^n = 0$, when n is not divisible by n and d is the largest integer having such property. If $d = 1$, the state is called *aperiodic*. Moreover, if a state is recurrent and the expected number of steps for the first return to itself is finite, it is called a *positive recurrent state*. If all elements of the n-step transition probability matrix is nonzero for some $n > 0$, the chain is called *regular*. Furthermore, a state is called *ergotic* if it is positive recurrent and aperiodic. Finally, a chain is called an *ergotic Markov chain* if all its states are ergotic.

Example 5.2.4

The finite Markov chain with state space $S = \{0, 1, 2\}$ and transition probability matrix

$$\mathbf{P} = \begin{array}{c} \\ 0 \\ 1 \\ 2 \end{array}\begin{array}{c} \begin{array}{ccc} 0 & 1 & 2 \end{array} \\ \left(\begin{array}{ccc} 0 & \frac{1}{3} & \frac{2}{3} \\ \frac{1}{2} & 0 & \frac{1}{2} \\ \frac{3}{4} & \frac{1}{4} & 0 \end{array} \right) \end{array}$$

is irreducible. Of course, $\mathbf{P}$ is a stochastic matrix since the sum of each row is 1. With 0s on the main diagonal, it seems as though none of the states 0, 1, or 2 is accessible. Although this is true for a one-step transition, with additional steps each may be accessed. Here is how. Note that we are using the arrow ($\rightarrow$) denoting transition from one state to another.

Note that it is *sufficient, but not necessary* for a Markov chain to be irreducible if $p_{xy}^{(n)} > 0$ for all x, y for some $n \geq 1$. For instance, the Markov chain with two stats 0 and 1 and the transition matrix:

$$\mathbf{P} = \begin{pmatrix} 0 & 1 \\ 1 & 0 \end{pmatrix}$$

TABLE 5.2.1. Transitions among States of a Markov Chain

Transitions	Probability
$0 \to 0$	0
$0 \to 1$	$\left(\dfrac{1}{3}\right)$
$0 \to 2$	$\left(\dfrac{2}{3}\right)$
$1 \to 0$	$\left(\dfrac{1}{2}\right)$
$1 \to 1$	0
$1 \to 2$	$\left(\dfrac{1}{2}\right)$
$2 \to 0$	$\left(\dfrac{3}{4}\right)$
$2 \to 1$	$\left(\dfrac{1}{4}\right)$
$2 \to 2$	0

is irreducible, but we see zero elements in

$$\mathbf{P}^{2n} = \begin{pmatrix} 1 & 0 \\ 0 & 1 \end{pmatrix} \quad \text{and} \quad \mathbf{P}^{2n+1} = \begin{pmatrix} 0 & 1 \\ 1 & 0 \end{pmatrix},$$

for $n \geq 0$.

From matrix $\mathbf{P}$, transitions among states are listed in Table 5.2.1.

The following theorem supports the earlier statement. For the proof, refer to Neuts (1973, p. 362).

Now using the product rule for probabilities, Table 5.2.2 shows some examples of how to access states in more than one step.

Theorem 5.2.1

If a Markov chain is irreducible and aperiodic (i.e., $d(x) = 1$, $\forall x$), then $p_{xy}^{(n)} > 0$ for all x, y for all n sufficiently large.

The following is an important theorem for a Markov chain.

Theorem 5.2.2

If a Markov chain with state space S is irreducible and ergotic, then $\lim_{n \to \infty} p_{xy}^n$ exists for every x and y in S and is independent of x.

As an immediate result from Theorem 5.2.2, let $\pi_y \equiv \lim_{n \to \infty} p_{xy}^{(n)}$. Then the limiting probability that the chain will be in state y at time n is the long-term proportion of time that the chain will stay in the state y and is determined by the system of equations:

**TABLE 5.2.2. Accessing States of a
Markov Chain in More than One Step**

Transitions	Probability
$0 \rightarrow 1 \rightarrow 0$	$\left(\dfrac{1}{3}\right) \times \left(\dfrac{1}{2}\right) = \left(\dfrac{1}{6}\right)$
$0 \rightarrow 2 \rightarrow 0$	$\left(\dfrac{2}{3}\right) \times \left(\dfrac{3}{4}\right) = \left(\dfrac{1}{2}\right)$
$1 \rightarrow 0 \rightarrow 1$	$\left(\dfrac{1}{2}\right) \times \left(\dfrac{1}{3}\right) = \left(\dfrac{1}{6}\right)$
$1 \rightarrow 2 \rightarrow 1$	$\left(\dfrac{1}{2}\right) \times \left(\dfrac{3}{4}\right) = \left(\dfrac{3}{8}\right)$
$2 \rightarrow 0 \rightarrow 2$	$\left(\dfrac{3}{4}\right) \times \left(\dfrac{2}{3}\right) = \left(\dfrac{1}{2}\right)$
$2 \rightarrow 1 \rightarrow 2$	$\left(\dfrac{1}{4}\right) \times \left(\dfrac{1}{2}\right) = \left(\dfrac{1}{8}\right)$

$$\pi_y = \sum_{i=0}^{\infty} \pi_i p_{iy}, \quad y \in S, \text{ with } \sum_{y \in S} \pi_y = 1. \tag{5.2.13}$$

This is because of the Chapman–Kolmogorov equations, where:

$$p_{xy}^{(n+1)} = \sum_{i=0}^{\infty} p_{xi}^{(n)} p_{iy}, \quad x, y \in S. \tag{5.2.14}$$

Now as the Theorem 5.2.2 suggests, taking the limit of both sides of (5.2.14), we will have (5.2.13).

Definition 5.2.5
Let $\pi = (\pi_0, \pi_1, \ldots)$ be a row vector, equation (5.2.13) can be rewritten as:

$$\pi = \pi \mathbf{P} \tag{5.2.15}$$

and for the Markov chain $\{X_n, n \geq 0\}$,

$$\lim_{n \to \infty} P\{X_n = y\} = \pi_y, \quad y \in S. \tag{5.2.16}$$

The set $\{\pi_y\}$ found in (5.2.16) that is the unique solution of the system of equation (5.2.13) when the number of steps becomes large is called the *stationary or steady-state distribution of the Markov chain*.

If the Markov chain has a finite transition probability matrix with nonzero elements at some point in time, say k, the chain eventually attains its stationary case. In addition, for a regular Markov chain with a transition probability matrix $\mathbf{P}$, there exists a matrix Π with all identical rows and nonzero elements such that:

$$\lim_{k \to \infty} \mathbf{P}^k = \Pi, \tag{5.2.17}$$

where each row of Π is equal to π and each element of π is a probability with the sum of elements equal to 1. The vector π is called the *steady-state* or *stationary probability vector* for the chain (see Bhat, 2008).

Example 5.2.5
Consider the Markov chain defined in Example 5.2.4. We showed that the chain was ergodic. Now by raising $\mathbf{P}$ to different powers, we will observe that for $k \geq 31$, all rows of the power matrices are equal, and of course all elements are nonzero, as, for instance:

$$\mathbf{P}^{32} = \begin{array}{c} \\ 0 \\ 1 \\ 2 \end{array} \begin{array}{ccc} 0 & 1 & 2 \\ \left(\begin{array}{ccc} 0.3962 & 0.2264 & 0.3774 \\ 0.3962 & 0.2264 & 0.3774 \\ 0.3962 & 0.2264 & 0.3774 \end{array} \right) \end{array}.$$

Hence, the steady-state distribution of the chain is $\pi = (0.3962, 0.2264, 0.3774)$. In other words, regardless of in what state the chain started, (in long run) after about 31 steps, (steady-state) probabilities to be in states 0, 1, and 2 are 0.3962, 0.2264 and 0.3774, respectively.

Definition 5.2.6
A stochastic process $\{X(t), t \in T\}$, where T is the space parameter, with discrete-valued random variable in continuous-time $X(t)$ is said to have *independent increments* if for any $n \geq 1$, and arbitrary sequence $\{t_0, t_1, \ldots, t_{n+1}\}$ with $t_0 < t_1 < \ldots < t_{n+1}$, $X(t_1) - X(t_0), X(t_2) - X(t_1), \ldots, X(t_n) - X(t_{n-1})$ are independent. Such a process is called *stationary* if $X(t_2 + \Delta t) - X(t_1 + \Delta t)$ has the same distribution as $X(t_2) - X(t_1)$, $\forall t_1, t_2, (t_1 + \Delta t), (t_2 + \Delta t) \in T, \Delta t > 0$.

Definition 5.2.7
Consider a discrete-valued random variable in continuous-time $X(t)$. If the stochastic process $\{X(t), t \geq 0\}$ for any $n \geq 1$ and arbitrary sequence $\{t_0, t_1, \ldots, t_{n+1}\}$ with $t_0 < t_1 < \ldots < t_{n+1}$ and $\{x_0, x_1, \ldots, x_{n+1}\}, x_i \in \{0, \pm1, \pm2, \ldots\}$, the *Markov property* will become as follows:

$$\begin{aligned} P\{X(t_n) = x_n \,|\, X(0) = x_0, X(t_1) = x_1, \ldots, X(t_{n-1}) = x_{n-1}\} \\ = P\{X(t_n) = x_n \,|\, X(t_{n-1}) = x_{n-1}\}. \end{aligned} \tag{5.2.18}$$

The stochastic process $\{X(t), t \geq 0\}$ that satisfies (5.2.18) is called a *discrete-space Markov process*.

Definition 5.2.8
Let the stochastic process $\{X(t), t \geq 0\}$ be continuous in time and state, that is, $X(t)$ in continuous-valued random variable and state space is one of the forms

$(-\infty, \infty)$, $[0, \infty)$, or $[0, N]$, where N is a finite number. Then, the $\{X(t), t \geq 0\}$ is called a *continuous-time Markov process* if the *Markov property*

$$
\begin{aligned}
P\{X(t) \leq x \,|\, X(t_1) = x_1, X(t_2) = x_2, \ldots, X(t_n) = x_n\} \\
= P\{X(t) \leq x \,|\, X(t_n) = x_n\}.
\end{aligned} \tag{5.2.19}
$$

holds for $t_1 < t_2 < \ldots < t_n < t$. Note that the probability in (5.2.19) is given in terms of the cumulative distribution. Hence, the *transition probabilities* of the Markov process in this case is given by (5.2.19) and denoted by $F(x_n, t_n; x, t)$.

Definition 5.2.9
For a Markov process, the *transition probability* from x to y in a time interval of length t will be denoted by $p_{xy}(t)$. Based on (5.2.18), this means that:

$$
p_{xy}(t) = P\{X(s+t) = y \,|\, X(s) = x\}. \tag{5.2.20}
$$

The probability that a Markov process is in state x at time t, $P\{X(t) = x\}$, denoted by $p_x(t)$, that is, $p_x(t) \equiv P\{X(t) = x\}$, is called the *absolute state probability at time t*. The set of probabilities $\{p_x(t), x \in \mathbb{Z}\}$, where $\mathbb{Z}$ is the set of integers, is called the *absolute distribution* of the Markov process at time t, while $\{p_x(0), x \in \mathbb{Z}\}$ is called the *initial probability distribution* of the Markov process.

Note that according to the law of total probability, the absolute probability distribution of the Markov chain at time t satisfies the system of linear difference equations:

$$
p_y(t) = \sum_{x \in \mathbb{Z}} p_x(0) p_{xy}(t), \quad y \in \mathbb{Z}. \tag{5.2.21}
$$

For a Markov process, the Markov property (5.2.18) can be described as follows: let Y be a continuous random variable, then:

$$
P\{Y > s+t \,|\, Y > t\} = P\{Y > s\}, \quad \forall s, t \geq 0. \tag{5.2.22}
$$

It can be shown that the only continuous probability distribution having the forgetfulness property and conversely is the exponential distribution function. For discrete probability mass functions, however, only the geometric mass function has this property.

Definition 5.2.10
Similar to the ones we had for discrete-time Markov chain, equation (5.2.7), for a Markov process with state space S, the equations

$$
p_{xy}(t+\tau) = \sum_{r} p_{xr}(t) p_{ry}(\tau), \quad x, y \in S, \quad t, \tau \geq 0 \tag{5.2.23}
$$

or in matrix form

$$\mathbf{P}(t+\tau) = \mathbf{P}(t)\mathbf{P}(\tau), \quad \forall t, \tau \in [0, \infty) \tag{5.2.24}$$

are called the *Chapman–Kolmogorov equations* for the Markov process, where the elements of $\mathbf{P}(t)$ are $p_{xy}(t)$. In other words, the passage from state x to state y during time $t + \tau$ should occur through some intermediate state r at time t.

Definition 5.2.11
A Markov process, whose transition probabilities $p_{xy}(t)$ do not depend on time t, is called *stationary* or *time-homogeneous* or simply *homogeneous*. In other words, no matter what the current time t is, if the process is in state x, the probability of transition to a state y at time $t + \Delta t$ is $p_{xy}(\Delta t)$. Thus, for any times t and t_1 we have:

$$\begin{aligned} p_{xy}(\Delta t) &= P\{X(t+\Delta t)\,|\,X(t) = x\} \\ &= P\{X(t_1 + \Delta t) = y\,|\,X(t_1) = x\}. \end{aligned} \tag{5.2.25}$$

The probability $\{\pi_y = P_y(0), y \in S\}$ is called *stationary initial probability distribution* if:

$$\pi_y = P_y(t), \quad \forall t \ge 0, y \in S. \tag{5.2.26}$$

It can be shown that for a Markov process that is positive, the stationary distribution $\{\pi_y\}$ exists and satisfies the following:

$$\lim_{t\to\infty} p_{xy}(t) = \lim_{t\to\infty} P\{X(t) = y\} = \pi_y, \quad y \in S, \text{ with } \sum_{y\in S} \pi_y = 1. \tag{5.2.27}$$

Definition 5.2.12
For a continuous-time Markov process with continuous state space, the *transition probability density function (tpdf)* is defined as the density function for a transition from state x at time s to state y at time t, $s < t$. We denote the tpdf by $f(x, s; y, t)$. The $f(x, s; y, t)$ is called *time homogeneous* or *homogeneous* if:

$$f(x, s+\Delta t; y, t+\Delta t) = f(x, s; y, t), \quad s < t \quad \Delta t > 0. \tag{5.2.28}$$

In other words, transitions only depend on the length of time between states, $s - t$, that is, the process moves from state x to state y in time interval of $s - t$. The *Chapman–Kolmogorov equations* for the tpdf becomes:

$$F(x, s; y, t) \equiv \int_{-\infty}^{\infty} f(x, s; z, u) \cdot f(z, u; y, t) dz, \quad s < u < t. \tag{5.2.29}$$

In case that the initial density is concentrated at a point, say x_0, that is, $P\{X(t_0) = x_0\} = 1$, the probability density function (pdf) will be a *Dirac delta function*, that is, $\delta(x - x_0)$, where $\delta(x - x_0) = 0$ if $x \ne x_0$ and $\int_{-\infty}^{\infty} \delta(x - x_0) dx = 1$.

In this case, the pdf of $X(t)$ may be defined as $f(x_0, t_0; x, t)$ and we may denote it simply as $f_x(t)$. Thus, with this notation, the *Chapman–Kolmogorov equations* (5.2.29) becomes:

$$F_x(t) \equiv \int_{-\infty}^{\infty} f(z, u; x, t) \cdot f_x(t) dz, \quad t_0 < u < t. \tag{5.2.30}$$

Note that we may calculate probabilities as we regularly do to find the distribution function by integrating the pdf. For instance:

$$P\{X(t) \in [a, b]\} = P\{X(t) \in (a, b)\} = \int_a^b f_x(t) dx. \tag{5.2.31}$$

We conclude this section by summarizing the transition probabilities and corresponding Chapman–Kolmogorov equations for different types of Markov processes (or chains).

(i) Discrete state, discrete space (parameter):

$$p_{x,y}^{(m,n)} \equiv P\{X_n = y \mid X_m = x\}, \quad m < n, \tag{5.2.32}$$

$$p_{xy}^{(m,n)} = \sum_{i \in S} p_{xi}^{(m,k)} p_{iy}^{(k,n)}, \quad m < k < n. \tag{5.2.33}$$

(ii) Discrete state, continuous space (parameter):

$$p_{x,y}(s, t) \equiv P\{X(t) = y \mid X(s) = x\}, \quad s < t, \tag{5.2.34}$$

$$p_{xy}(s, t) = \sum_{i \in S} p_{xi}(s, u) p_{iy}(u, t), \quad s < u < t. \tag{5.2.35}$$

(iii) Continuous state, discrete space (parameter):

$$F(x_m, m; x, n) \equiv P\{X_n \le x \mid X_m = x_m\}, \quad m < n, \tag{5.2.36}$$

$$F(x_m, m; x, n) \equiv \int_{y \in S} d_y F(x_m, m; y, k) \cdot F(y, k; x, n), \quad m < k < n. \tag{5.2.37}$$

(iv) Continuous state, continuous space (parameter):

$$F(x_n, t_n; x, t) \equiv P\{X(t) \le x \mid X(t_n) = x_n\}, \quad t_n < t, \tag{5.2.38}$$

$$F(x_s, s; x, t) \equiv \int_{y \in s} d_y F(x_s, s; y, u) \cdot F(y, u; x, t), \quad s < u < t. \tag{5.2.39}$$

5.3. BIRTH AND DEATH (B-D) PROCESS

We begin this section with an example for a discrete-state space and continuous parameter space Markov process that illustrates the idea of the birth and death process.

Example 5.3.1. *A Failure–Repair Model with 1 Machine and 1 Repairman, FR(1,1)*

We consider a well-known example of a two-state machine. Suppose in a manufacturing setting there are a fixed number of identical machines subject to occasional breakdowns. A machine is either in up (working) or in down (nonworking) position at any time. It is up until it fails (turning to the down condition). A repairman will repair the failed machine and put it back to the working condition. Let us assume that an "up" machine has a constant average rate of μ failures per period, while the repairman repairs "down" machines at an average rate of λ per period. We assume that the successive up times are iid random variables as well as successive down times, and that they are independent of each other. If we define

$$X(t) = \begin{cases} 1, & \text{if the machine is up at time } t, \\ 0, & \text{if the machine is down at time } t, \end{cases} \tag{5.3.1}$$

then we leave it as an exercise to show that, the process $\{X(t), t \geq 0\}$ is a continuous-time Markov process, with a discrete state pace $S = \{0, 1\}$. In this example, we may think of failures of the machine as deaths and the repaired machines as births.

Definition 5.3.1

Consider a continuous-time Markov process $\{X(t), t \geq 0\}$ with discrete state space $\{0, 1, 2, \ldots\}$ and the transition probability matrix as follows:

$$\mathbf{P} = \begin{pmatrix} r_0 & \lambda_1 & 0 & 0 & 0 & \cdots & \cdots \\ \mu_1 & r_1 & \lambda_2 & 0 & 0 & \cdots & \cdots \\ 0 & \mu_2 & r_2 & \lambda_3 & 0 & 0 & \cdots \\ 0 & 0 & \mu_3 & r_3 & \lambda_4 & 0 & \cdots \\ \vdots & \vdots & \vdots & \cdots & \cdots & \cdots & \cdots \\ \vdots & \vdots & \vdots & \cdots & \vdots & \ddots & \cdots \end{pmatrix}. \tag{5.3.2}$$

The transition probabilities in matrix (5.3.2) may be written as:

$$\begin{cases} p_{x,x} = r_x, & \text{if } x = 0, 1, 2, \ldots, \\ p_{x,x+1} = \lambda_x, & \text{if } x = 0, 1, 2, \ldots, \\ p_{x,x-1} = \mu x, & \text{if } x = 0, 1, 2, \ldots, \\ p_{x,y} = 0, & \text{if } x = 0, 1, 2, \ldots, \quad y \neq x, y \neq x \pm 1, \end{cases}$$

with $p_x = \lambda_x + \mu_x$ and $\lambda_x + r_x + \mu_x = 1$. Then, the process $\{X(t), t \geq 0\}$ is called:

(1) a *pure birth process*, if $\mu_x = 0$, for $x = 1, 2, \ldots$
(2) a *pure death process*, if $\lambda_x = 0$, for $x = 1, 2, \ldots$

(3) a *birth–death process*, sometimes denoted by *B-D process*, if $\lambda_x > 0$, and $\mu_x > 0$

Note that no B-D process allows transitions of more than one step, that is, transitions such as $x \to (x + 2), (x - 3) \leftarrow (x - 1)$ are not allowed. Intuitively, we may think of a birth–death process as a function of continuous time, which changes state in a random fashion after some random amount of waiting time in the current state.

Recall that in Definition 5.2.3, we defined a Markov process with state space as the set of natural numbers with the property that it only accepts one-step transition forward. It was called the counting process. Based on that, we now define a new process.

Definition 5.3.2
A continuous-time process $\{X(t), t \geq 0\}$ with discrete space process is called a *pure birth process* with transition rate λ_x, $\lambda_x > 0$, provided that the following hold:

(i) The process begins in the smallest state, say m, that is,

$$P\{X(0) = m\} = 1. \tag{5.3.3}$$

(ii) The probability that the count increases by 1 (one birth) in a small time interval $[t, t + \Delta t]$ is approximately proportional to the length, Δt, of the time interval. The transition rate, $p_{x,x+1} = \lambda_x$, is assumed to be a function of the state x and may vary from state to state, that is,

$$P\{X(t+\Delta t) = x+1 \mid X(t) = x\} = p_{x,x+1}(\Delta t) = \lambda_x \Delta t + o(\Delta t). \tag{5.3.4}$$

(iii) If Δt is small, the probability of two or more arrivals (the count increases by more than 1) in the time interval $[t, t + \Delta t]$ is negligible. Equivalently, this may be stated as: $p_{x,y} = 0$ for $y > x + 1$ (since $p_{x,y}(\Delta t) = p_{x,y}(\Delta t) + o(\Delta t) = o(\Delta t)$, when $p_{x,y} = 0$), that is,

$$P\{X(t+\Delta t) = x+1 \mid X(t) = x\} = o(\Delta t). \tag{5.3.5}$$

(iv) The count remains the same or increases over time; that is, the count will never decrease, that is,

$$P\{X(t+\Delta t) < x \mid X(t) = x\} = 0. \tag{5.3.6}$$

As a particular case of Definition 5.3.2, if in the pure birth process $\{X(t), t \geq 0\}$, the transition rate is constant, say λ, (independent of a state) and the initial state is 0, that is, $P\{X(0) = 0\} = 1$, then the process is called *Poisson process* with parameter λ, $\lambda > 0$. We leave it as an exercise to prove that in a Poisson process with transition rate λt, the probability mass function $p_x(t)$ is Poisson with parameter λt, that is,

$$p_x(t) = \frac{(\lambda t)^x}{x!} e^{-\lambda t}, \quad x = 0, 1, 2, \dots. \tag{5.3.7}$$

5.4. INTRODUCTION TO QUEUEING THEORY

Queueing theory has played as one of the most dominated theories of stochastic models. Some of the classical examples of queues can be found in cases associated with the transportation industry, machine interference, system design, facility utilization, inventory fluctuation, finance, and conveyor theory.

Historically, the first textbook on this subject, *Queues, Inventories, and Maintenance*, was authored in 1958 by Morse. Not much later, Saaty wrote his famous book *Elements of Queueing Theory with Application* in 1961. Today, over 40 books have been written on queues, over 1000 papers have been published, and several journals publish papers on queues. One of the most recent books on queue is *Queueing Models in Industry and Business*, by Haghighi and Mishev, originally published in 2008 and now in its second edition (2013).

Arrivals to a queue may be referred to as *items, units, jobs, customers,* or *tasks.* The notion of *customer* may have many interpretations: a computer program in line awaiting execution, a broken device waiting in line for repair, a task waiting in a bank teller's line, a suit waiting to be dry cleaned, thousands of people waiting to receive organ implants to save their lives, goods waiting to be conveyed from a distribution center to retail stores of a company, planes waiting for their turns to take off in a busy airport, and so on. Arrivals may come from without, as with tasks entering a retail store, or from within, as with broken machines lining up for repair and reuse in a shop with a fixed number of machines. A queueing system that allows arrivals from without, that is, the source is infinite, will be called an *open* queue; otherwise, the queue is termed a *closed* queue. The number of tasks in the system is referred to as *queue size.* The *state of a queue* at any time is the number of customers in the system at the time. The input may arrive deterministically or probabilistically according to a probabilistic law (distribution). The probabilistic law may be *Poisson, Erlang, Palm, Recurrent, Yule,* and so on. A task may arrive at the system, but does not attend due to variety of reasons. In this case, we say that the task *balked.* In contrast, a task may attend the system, wait a while, and leave the system before receiving service. In this case, we say the task is *reneging.*

A queueing system may be modeled as a birth–death process. Arrivals to the system may want to be served as in a B-D process. In that case, they may have to wait for service. The place where arrivals await for service is called a *buffer* or *waiting room*. The buffer's capacity may be finite (a finite number of waiting spaces) or infinite (where no customers are turned away). We refer to the location of a server as a *counter* or *service station*. A counter may contain one or more *servers* (such as a repairman, grocery checkout clerks, or bank tellers).

A queueing system with one server is called a *single server* and with more than one server is called a *multichannel* or *multiserver*. If no arrival has to wait before starting its service, then the system is called an *infinite-server queue*. In case of a multiserver queue, a task may switch between the server lines. In this case, we say that the task *jockeys*. We assume *service times* as random variables following some given probabilistic law such as constant, negative exponential, and geometric. The duration of time in which one or more servers is busy is called a *busy period*.

A queueing system may have more than one service station. These stations may be set parallel. In that case, each station may have its own buffer or a common buffer for all. In the first case, an arrival may chose a waiting line; in the latter case, however, all arrivals must line up before the same buffer. Each station may have one or several servers. Such a system is called a *parallel queueing system*. In contrast, the stations may be set in series with the same types of servers. In this case, an output of a station may have to join the next station. The system as such is called a *tandem queueing system*.

The service of tasks will be according to a certain principle. The principle sometimes is referred to as the *queue discipline*. The discipline could be service in order of arrival, that is, *first come, first served* (*FCFS*) or *first in, first out* (*FIFO*), *random service, priority service, last come, first served* (*LCFS*), *batch* (*bulk*) *service*, and so on. After being served, served tasks or *outputs* may *exit* from the system or move on to other activities such as *feedback* the service stations, going for *splitting*.

Suppose tasks (such as calls to a telephone exchange) arrive according to some probabilistic law. If an arriving task (call) finds the system empty (a free line) at its arrival time, its service will start (the call will be connected) and it will take as long as necessary (the conversation will last as long as necessary). That is, the service times (holding times) are random variables. However, if an arriving task (call) finds all servers (lines), of which there are a fixed number, busy, then it will be sent either to a waiting facility (buffer), in which case we will speak of a *waiting system*, or it will be refused, in which case we will speak of a *loss system*. Of course, the system may be a combination of these two kinds, in which case we will call it a *combined waiting and loss system*.

The notation $A/B/C/D$ (introduced by Kendall in 1953) generically describes the system. These letters will determine the arrival type and the service type (whether deterministic or probabilistic, Markovian or Erlangian, or even general). They also give a clue to the number of servers in parallel and the capacity of the system (i.e., waiting room plus the service station capacity). For instance, $M/M/1/K$ means that the arrival is of Poisson distribution (Markovian), service time distribution is Markovian with a single server, and the capacity of the waiting room is $K - 1$. In case that the system has unlimited capacity, it is customary to dispense with the last symbol, K, that is, we will have $M/M/1$. Thus, $M/M/2$ means Poisson arrival and exponential (Markovian) service time with two parallel servers and unlimited waiting room capacity.

Notice that this symbolism does not completely describe the system. For instance, the type of service discipline is unclear. Also, the notation assumes that each arrival waits until served. However, for the case of balking and reneging, in which arrivals may not join the queue or may decide to leave before the service starts, the symbols do not sufficiently describe the system. In more complicated systems, this symbolism becomes even more difficult to understand.

5.5. SINGLE-SERVER MARKOVIAN QUEUE, *M/M/*1

As we will see later, an $M/M/1$ queueing system is a birth and death process, in which arrivals are births and departures are deaths. The difference between the birth and death will determine the population seize at any time. Poisson arrivals and exponential service times enable us to use Markovian properties in the queueing models and, as such, they make the analysis easy and produce usable results.

An *M/M/1 queueing system* is a single-server queueing system with Markovian arrival and Markovian service time processes with infinite source of arrivals (open system) that holds the following properties:

(i) The interarrival times of tasks are exponentially distributed (i.e., arrival of tasks follow a Poisson distribution, why?) with mean arrival rate (birth rate) of λ tasks per unit time, that is, on the average, there are $1/\lambda$ time units between arrivals. Thus, if we let the sequence of random variables $\{X(t)\}$ represent the possible number of tasks arriving in the queue during $(0, t]$, then the probability of k tasks arriving within that interval would be:

$$P\{X(t) = k\} = \frac{(\lambda t)^k}{k!} e^{-\lambda t}, \quad k = 0, 1, 2, \dots. \tag{5.5.1}$$

The arrivals being Poisson with parameter λ implies that the interarrival times may be represented by a random variable whose distribution is exponential with parameter λ. Thus, the pdf for the interarrival times will be:

$$f(x) = \begin{cases} \lambda e^{-\lambda x}, & \text{if } x \geq 0, \\ 0 & \text{if } x < 0. \end{cases} \tag{5.5.2}$$

Hence, the expected interarrival time is $1/\lambda$ or $1/$(arrival rate).

(ii) The service times are identically distributed random variables with exponential distribution function with a mean service rate (death rate) of μ services per unit time. μ is the average service rate of tasks or departure out of the service facility. That is, on average, the service time is $1/\mu$ time units per task. Hence, if we denote the service time distribution by $H(x)$, we will have:

$$H(x) = \begin{cases} 1 - e^{-\mu x}, & \text{if} \quad x \geq 0, \\ 0, & \text{if} \quad x < 0. \end{cases} \tag{5.5.3}$$

From (5.5.2), we can see that pdf for service times, denoted by $h(x)$, is:

$$h(x) = \begin{cases} \mu e^{-\mu x}, & \text{if} \quad x \geq 0, \\ 0, & \text{if} \quad x < 0. \end{cases} \tag{5.5.4}$$

Thus, the expected service time is $1/\mu$ or $1/(\text{service rate})$. We note that the ratio of the arrival rate to the service rate plays an important role in the theory of queues. It gives the performance intensity.

(iii) There is only one service facility with one server.

(iv) There is no limit on the system's capacity. The buffer (or the waiting room) has no size limit and no task will be turned away. The system with this fact is sometimes denoted as $M/M/1/\infty$.

(v) The queue discipline is first in, first out (FIFO), that is, tasks are serviced in the same order in which they arrive.

We now go into detail to find the average queue length, that is, the expected number of tasks in the system (possibly one in the service and others in the waiting line). We do this in two parts, time dependent: *transient case* and *stationary or steady-state case*.

Let us define by $\xi(t)$ the state of the system, that is, the number of tasks in the system at time t. We may refer to $\xi(t)$ as the queue size. Then the state space for the $M/M/1$ queueing process in this case will be $\{0, 1, 2, \ldots\}$. Under the given conditions, $\{\xi(t)\}$ is a Markov process, that is, the future stochastic behavior of the system is uniquely determined by the present state and does not depend on the past state of the process. Assuming that initially the system starts with i tasks in it, that is, $P\{\xi(0) = i\} = 1$, the transition probabilities will be defined as follows:

$$p_{in}(t) = P\{\xi(t) = n \,|\, \xi(0) = i\}, \quad i, n \geq 0. \tag{5.5.5}$$

In birth–death terminology, if the population size is n then the infinitesimal transition rates of birth and death are $\lambda_n = \lambda, n \geq 0$ and $\mu_n = \mu, n \geq 1$, respectively. The assumption of rates to be constant (not state dependent) is to simplify our development. The process $\{\xi(t)\}$ is homogeneous since it has the property

$$P\{\xi(t)=n\,|\,\xi(s)=k\}=P_{kn}(t-s), \quad 0\le s<t. \tag{5.5.6}$$

In other words, the transition probability depends only on the difference $t-s$. The transition probabilities $p_{in}(t)$ defined by (5.5.5) satisfy the Chapman–Kolmogorov equation (5.2.23).

Now, based on the properties of the Poisson process and exponential distribution, we will have the following probabilities for transitions during a small interval $(t,t+\Delta t]$:

For arrivals $(n\ge 0)$:

$$P\{\text{exactly one arrival in } (t,t+\Delta t]\}=\lambda\Delta t+o(\Delta t), \tag{5.5.7a}$$

$$P\{\text{no arrival in } (t,t+\Delta t]\}=1-\lambda\Delta t+o(\Delta t), \tag{5.5.7b}$$

$$P\{\text{more than one arrival in } (t,t+\Delta t]\}=o(\Delta t). \tag{5.5.7c}$$

For service completion (departure) $(n\ge 1)$:

$$P\{\text{one service completion in } (t,t+\Delta t]\}=\mu\Delta t+o(\Delta t), \tag{5.5.8a}$$

$$P\{\text{no service in } (t,t+\Delta t]\}=1-\mu\Delta t+o(\Delta t), \tag{5.5.8b}$$

$$P\{\text{more than one servie completion in } (t,t+\Delta t]\}=o(\Delta t). \tag{5.5.8c}$$

The notation $o(\Delta t)$ is read the "little o", meaning that:

$$\lim_{\Delta t\to 0}\frac{o(\Delta t)}{\Delta t}=0. \tag{5.5.9}$$

Relation (5.5.9) has been used to indicate insignificant contributions of other probabilities. Also, due to its 0 limiting value, the sum of probabilities in each case is 1.

For a transition from n to $n+1$, that is, $n+1$ arrivals in $t+\Delta t$ units of time, we have to have n arrivals during $(0,t]$ and 1 during Δt. Similarly, for a transition from n to $n-1$, that is, $n-1$ arrivals in $t+\Delta t$ unit of times, we have to have n arrivals during $(0,t]$ and 1 service completion during Δt. And, of course, the transition from n to more than $n+1$ or to less than $n-1$ occurs with almost zero probability. We summarize the activities involved in transitions leading to n tasks in the system at time $t+\Delta t$ as follows:

State of the System at time t	Activity in $(t,t+\Delta t)$	State of the System at time $t+\Delta t$
$n-1$ tasks	1 arrival	n tasks
$n+1$ tasks	1 service completion	n tasks
n tasks	0 arrival and 1 service completion	n tasks

Thus, from (5.5.7) and (5.5.8), we have:

$$
\begin{aligned}
p_{n,n+1}(\Delta t) &= [\lambda \Delta t + o(\Delta t)][1 - \mu \Delta t + o(\Delta t)] \\
&= \lambda \Delta t + o(\Delta t), \quad n = 0, 1, 2, \ldots,
\end{aligned}
\tag{5.5.10a}
$$

$$
\begin{aligned}
p_{n,n-1}(\Delta t) &= [1 - \lambda \Delta t + o(\Delta t)][\mu \Delta t + o(\Delta t)] \\
&= \mu \Delta t + o(\Delta t), \quad n = 1, 2, \ldots,
\end{aligned}
\tag{5.5.10b}
$$

$$
\begin{aligned}
p_{n,n}(\Delta t) &= [1 - \lambda \Delta t + o(\Delta t)][1 - \mu \Delta t + o(\Delta t)] \\
&= 1 - \lambda \Delta t - \mu \Delta t + o(\Delta t), \quad n = 1, 2, \ldots,
\end{aligned}
\tag{5.5.10c}
$$

$$
p_{n,m}(\Delta t) = o(\Delta t), \quad m \neq n-1, n, n+1.
\tag{5.5.10d}
$$

The transition probability matrix (5.3.2) for a birth–death process (sometimes called the generator matrix or Q-matrix) will now be of the form:

$$
\mathbf{P} =
\begin{pmatrix}
-\lambda & \lambda & 0 & 0 & 0 & 0 & \cdots \\
\mu & -(\lambda+\mu) & \lambda & 0 & 0 & 0 & \cdots \\
0 & \mu & -(\lambda+\mu) & \lambda & 0 & 0 & \cdots \\
0 & 0 & \mu & -(\lambda+\mu) & \lambda & 0 & \cdots \\
\vdots & \vdots & \vdots & \ddots & \ddots & \ddots & \cdots \\
\cdots & \cdots & \cdots & \cdots & \cdots & \cdots & \ddots
\end{pmatrix}
\tag{5.5.11}
$$

With the state of the system being the primary interest, we want to find distribution of $\xi(t)$, that is, we are looking for $P_n(t) \equiv P\{\xi(t) = n\}$. To do that, we will use the system of differential difference equations for $P_n(t)$. We note that finding such distribution would not involve the queue discipline. However, considering the waiting time, we will need the queue discipline.

Using the conditional and transitional probability properties, we present the absolute probabilities. Hence, to have n tasks in the system during the time $t + \Delta t$, there are four possibilities to consider: (1) $n + 1$ arrivals by t, no arrival and one service completion in $(t, t + \Delta t]$; (2) n arrivals by t, neither an arrival nor a service completion or one arrival and one service completion in $(t, t + \Delta t]$; (3) $n - 1$ arrivals by t, one arrival and no service completion in $(t, t + \Delta t]$; and (4) different cases than the three mentioned, that is, cases involving more than one arrival or one service completion. Thus, we will have:

$$
\begin{aligned}
P_n(t + \Delta t) = {}& P_{n+1}(t)\, P\{\text{no arrival and one service completion in } (t, t + \Delta t]\} \\
& + P_n(t)\, P\{\text{no arrival and no service completion or one} \\
& \quad \text{arrivaland one service completion in } (t, t + \Delta t]\} \\
& + P_{n-1}(t)\, P\{\text{one arrival and no service completion in } (t, t + \Delta t]\} \\
& + \text{terms containing probability of more than} \\
& \quad \text{one arrival or service completion in } (t, t + \Delta t].
\end{aligned}
$$

We could rewrite this expression in terms of the random variable $\xi(t)$ as follows. The notation $P_{-1}(\Delta t)$ would mean a service completion in Δt:

$$
\begin{aligned}
P_n(t+\Delta t) &= P\{\xi(t+\Delta t) = n\} \\
&= P\{\xi(t) = n+1, \xi(t+\Delta t) - \xi(t) = -1\} \\
&\quad + P\{\xi(t) = n, \xi(t+\Delta t) - \xi(t) = 0\} \\
&\quad + P\{\xi(t) = n-1, \xi(t+\Delta t) - \xi(t) = 1\} \\
&\quad + \sum_{k=2}^{n} P\{\xi(t) = n-k, \xi(t+\Delta t) - \xi(t) = k\} \\
&= P_{n+1}(t)\cdot P_{-1}(\Delta t) + P_n(t)\cdot P_0(\Delta t) + P_{n-1}(t)\cdot P_1(\Delta t) + o(\Delta t) \\
&= P_{n+1}(t)[[1-\lambda\Delta t + o(\Delta t)][\mu\Delta t + o(\Delta t)]] \\
&\quad + P_n(t)[[1-\lambda\Delta t + o(\Delta t)][1-\mu\Delta t + o(\Delta t)] + (\lambda\Delta t + o(\Delta t)(\mu\Delta t + o(\Delta t)] \\
&\quad + P_{n-1}(t)[[\lambda\Delta t + o(\Delta t)][1-\mu\Delta t + o(\Delta t)]] + o(\Delta t).
\end{aligned}
$$
(5.5.12)

To simplify (5.5.12), we note that $(\Delta t)^2 = o(\Delta t)$. Hence, we have:

$$
\begin{aligned}
P_n(t+\Delta t) &= P_{n+1}(t)\mu\Delta t + P_{n-1}(t)\lambda\Delta t \\
&\quad + P_n(t)(1-\lambda\Delta t - \mu\Delta t + o(\Delta t)), \quad n \geq 1.
\end{aligned}
$$
(5.5.13)

For $n = 0$, we argue similarly. Thus, similar to (5.5.13), we will have:

$$
P_0(t+\Delta t) = P_1(t)\mu\Delta t + P_0(t)(1-\lambda\Delta t) + o(\Delta t).
$$
(5.5.14)

Thus, we need to solve the system of difference equations (5.5.13) and (5.5.14). To do that, we start by multiplying out the right-hand sides of both (5.5.13) and (5.5.14) and move the terms to the left, which will obtain the following:

$$
\frac{P_n(t+\Delta t) - P_n(t)}{\Delta t} = \mu P_n(t) + \lambda P_{n-1}(t) - (\lambda+\mu) P_n(t) + \frac{o(\Delta t)}{\Delta t}, n \geq 1,
$$
(5.5.15)

and

$$
\frac{P_0(t+\Delta t) - P_0(t)}{\Delta t} = \mu P_1(t) - \lambda P_0(t) + \frac{o(\Delta t)}{\Delta t}.
$$
(5.5.16)

From (5.5.15) and (5.5.16), passing to the limit, therefore, we obtain the following system of differential difference equations:

$$
\begin{cases}
\dfrac{dP_0(t)}{dt} = -\lambda P_0(t) + \mu P_1(t) \\[2mm]
\dfrac{dP_n(t)}{dt} = -(\lambda+\mu) P_n(t) + \lambda P_{n-1}(t) + \mu P_{n+1}(t), \quad n \geq 1,
\end{cases}
$$
(5.5.17)

with $P_i(0) = \delta_{ij}$, where δ_{ij} is the *Kronecker delta*, which is defined as:

$$\delta_{ij} = \begin{cases} 1, & \text{if } i = j, \\ 0, & \text{if } i \neq j. \end{cases} \tag{5.5.18}$$

The system (5.5.17) is well known as the *forward Kolmogorov equations*. What (5.5.17) says is that $dP_n(t)/dt$ is the difference between [*flow rate of probability into state n at time t*] and [*flow rate of probability out of state n at time t*].

There are different methods available to solve the system (5.5.17). For instance, see Takács (1962), Gross et al. (2011), and Haghighi and Mishev (2013). The methods involve applying the Laplace transform, generating function, inverting of the Laplace transform, and choosing the coefficients of the Taylor expansion of the generating function. We now solve the system:

$$-2\left(\frac{\lambda}{\mu}\right)^2 \left[\sum_{n=0}^{\infty} a_n P_n\right]\left[\sum_{n=0}^{\infty} n a_n P_n\right]. \tag{5.5.19}$$

5.5.1. Transient Queue Length Distribution for *M/M/1*

In this section, our goal is to solve the differential–difference equation system adding to the first equation of (5.5.17) and as an application of what we have discussed in the previous chapters. The procedure of solving differential–difference equations is somewhat complicated, but it is quite instructive. The recursive method that is used in solving many systems does not work in solving (5.5.17) because of the presence of the differential part. After this example, we will present more models with stationary solutions of difference equations that are less complicated and where the iterative method sometimes works. Applying all the aforementioned leads to the following theorem for the time-dependent transition probabilities $P_{in}(t)$, where i is the initial state. However, keeping in mind that n could be $\geq i$ or $<i$, we just state these conditions and use the absolute probabilities of states as $P_n(t)$.

Theorem 5.5.1.1
For $n \geq i$ or $n < i$, $i = 0, 1, 2, \ldots$, the time-dependent (transient) distribution of the queue length for $M/M/1$ is:

$$P_n(t) = e^{-(\lambda+\mu)t}\left[\rho^{\frac{n-i}{2}} I_{n-i}\left(2 - \sqrt{\lambda\mu t}\right) + \rho^{\frac{n-i+1}{2}} I_{n+i+1}\left(2\sqrt{\lambda\mu t}\right),\right.$$
$$\left. + (1-\rho)\rho^n \sum_{k=n+i+2}^{\infty} \rho^{-k/2} I_k\left(2\sqrt{\lambda\mu t}\right), \tag{5.5.1.1}\right.$$

where $I_v(z)$, $v = 0, \pm1, \pm2, \ldots$, is the modified Bessel function of order v defined as:

$$I_v(z) = \sum_{k=0}^{\infty} \frac{\left(\frac{z}{2}\right)^{v+2k}}{k!\,\Gamma(v+k+1)}. \tag{5.5.1.2}$$

Proof

Let the generating function of $P_n(t)$, denoted by $G(z, t)$, be defined as:

$$G(z, t) = \sum_{n=0}^{\infty} P_n(t) z^n, \qquad (5.5.1.3)$$

and the Laplace transform of a function $f(t)$, denoted by $f^*(s)$, as:

$$f^*(s) = \int_0^\infty f(t) e^{-st} dt, \quad Re(s) > 0. \qquad (5.5.1.4)$$

In order to apply the generating function defined in (5.5.1.3) on the system of equation (5.5.17), we first multiply both sides of the first equation by $z^0 = 1$ that yields the same equation. Then, we multiply both sides of the nth ($n = 1$, $2, \ldots$) equation of the second set by z^n, add all these equations, and obtain the following:

$$\sum_{n=1}^{\infty} \frac{dP_n(t)}{dt} z^n = -(\lambda + \mu) \sum_{n=1}^{\infty} P_n(t) z^n + \lambda \sum_{n=1}^{\infty} P_{n-1}(t) z^n + \mu \sum_{n=1}^{\infty} P_{n+1}(t) z^n. \qquad (5.5.1.5)$$

Now performing some algebra on (5.5.1.5), addition to the first equation of (5.5.1.7) and from (5.5.1.3) using the fact that:

$$\sum_{n=1}^{\infty} \frac{dP_n(t)}{dt} z^n = \frac{\partial G(z, t)}{\partial t} - \frac{dP_0(t)}{dt} = \frac{\partial G(z, t)}{\partial t} - \mu P_1(t) + \lambda P_0(t),$$

we obtain:

$$\frac{\partial G(z, t)}{\partial t} = \frac{(1-z)}{z} [(\mu - \lambda z) G(z, t) - \mu P_0(t)]. \qquad (5.5.1.6)$$

Now we apply Laplace transform on both sides of (5.5.1.6). To simplify the expression obtained, we note that to evaluate the Laplace transform of the left-hand side, we use the integration by parts and use the initial condition $P_i(0) = \delta_{ij}$ to obtain:

$$\int_0^\infty \left(\frac{\partial G(z, t)}{\partial t} \right) e^{-st} dt = s G^*(z, s) - z^i. \qquad (5.5.1.7)$$

Using (5.5.1.7), the Laplace transform of (5.5.1.6), after some simplification, will be:

$$s G^*(z, s) - z^i = \frac{(1-z)}{z} [(\mu - \lambda z) G^*(z, s) = \mu P_0^*(s)]. \qquad (5.5.1.8)$$

Solving for $G^*(z, s)$ from (5.5.1.8), we obtain:

$$G^*(z, s) = \frac{z^{i+1} - \mu(1-z)P_0^*(s)}{(\lambda + \mu + s)z - \mu - \lambda z^2}. \qquad (5.5.1.9)$$

In order to find an expression for $P_0^*(s)$, we will use the well-known *Rouché theorem*. To do that, we note that the denominator of (5.5.1.9) has two zeros, say z_1 and z_2, given by:

$$z_1 = \frac{\lambda + \mu + s - \sqrt{(\lambda + \mu + s)^2 - 4\lambda\mu}}{2\lambda} \qquad (5.5.1.10)$$

and

$$z_2 = \frac{\lambda + \mu + s + \sqrt{(\lambda + \mu + s)^2 - 4\lambda\mu}}{2\lambda}, \qquad (5.5.1.11)$$

with $|z_1| < |z_2|$ and that:

$$z_1 + z_2 = \frac{\lambda + \mu + s}{\lambda} \quad \text{and} \quad z_1 z_2 = \frac{\mu}{\lambda}. \qquad (5.5.1.12)$$

Now letting $f(z) = (\lambda + \mu + s)z$ and $g(z) = -\mu - \lambda z^2$, the denominator of (5.5.1.10) has only one zero within the unit circle and $\mathrm{Re}(s) > 0$, and that must be z_1 since $|z_1| < |z_2|$. Hence, the numerator must have the same zero as the denominator. Thus, we will have:

$$P_0^*(s) = \frac{z_1^{i+1}}{\mu(1-z_1)}. \qquad (5.5.1.13)$$

Therefore, using (5.5.1.13), (5.5.1.9) reduces to an explicit expression for the Laplace transform of generating function of time-dependent probabilities $P_n(t), n \geq 0$ as:

$$G^*(z, s) = \frac{z^{i+1} - \dfrac{(1-z)z_1^{i+1}}{(1-z_1)}}{(\lambda + \mu + s)z - \mu - \lambda z^2}. \qquad (5.5.1.14)$$

To obtain $P_n(t), n \geq 0$, we need to invert (5.5.1.14). The denominator of (5.5.1.14) can be written as $\lambda(z - z_1)(z_2 - z)$, where z_1 and z_2 are given in (5.5.1.10) and (5.5.1.11). Thus, (5.5.1.14) can be rewritten as:

$$G^*(z, s) = \frac{z^{i+1} - \dfrac{(1-z)z_1^{i+1}}{(1-z_1)}}{\lambda(z - z_1)(z_2 - z)}. \qquad (5.5.1.15)$$

The expression in (5.5.1.15) can be expanded as follows:

$$G^*(z, s) = \frac{1}{\lambda z_2\left(1 - \dfrac{z}{z_2}\right)}\left[z^i\left(1 + \frac{z_1}{z} + \cdots + \left(\frac{z_1}{z}\right)^2\right) + \frac{z_1^{i+1}}{1 - z_1}\right]. \tag{5.5.1.16}$$

Using (5.5.1.12) in (5.5.1.16) leads to further simplification and we leave it as an exercise to show that for $n \geq i$, it reduces to:

$$P^*(s) = \frac{1}{\lambda}\left[z_2^{i-n-1} + \rho^{-1} z_2^{i-n-3} + \rho^{-2} z_2^{i-n-5} + \cdots + \rho^{-i} z_2^{-i-n-1} + \rho^{n+1}\sum_{k=n+i+2}^{\infty}(\rho z_2)^{-k}\right], \tag{5.5.1.17}$$

where $\rho = \lambda/\mu$.

To invert the Laplace transform, we note the following:

$$\mathcal{L}^{-1}\left(z_2^{-n}\right) = e^{-(\lambda+\mu)t} n\rho^{n/2} t^{-1} I_n\left(2\sqrt{\lambda\mu}t\right), \tag{5.5.1.18}$$

where $I_v(z)$, $v = 0, \pm 1, \pm 2, \ldots$, is the modified Bessel function of order v defined in (5.5.1.2). Moreover, note that the sum in (5.5.1.2) is convergent everywhere in the complex z-plan. Using the following two very useful properties of $I_v(z)$:

$$\frac{zv}{z} I_v(z) = I_{v-1}(z) - I_{v+1}(z) \quad \text{and} \quad I_v(z) = I_{-v}(z), \tag{5.5.1.19}$$

we will have (5.5.1.1) and, thus, the proof is complete.

For a different method of proof for Theorem 5.5.1.1, refer to Takács (1962, p. 24). We also note that using (5.5.1.2), it can be shown that:

$$P_n(t) = (1-\rho)\rho^n + e^{-(\lambda+\mu)t}\rho^n\sum_{m=0}^{\infty}\frac{(\lambda t)^m}{m!}\sum_{k=0}^{m+n+i+1}(m-k)\frac{(\mu t)^{k-1}}{k!}$$
$$+ e^{-(\lambda+\mu)t}\rho^n\sum_{m=0}^{\infty}(\lambda t)^{m+1}\frac{(\mu t)^{m+V}}{m!} \tag{5.5.1.20}$$
$$\times\left[\frac{(\lambda t)^{-U-1}}{(m+|i-n|)!} - \frac{(\mu t)^{U+1}}{(m+i+n+2)!}\right],$$

where $U = \min(i, n)$ and $V = \max(i, n)$. For proof see Jain et al. (2007).

5.5.2. Stationary Queue Length Distribution for *M/M/*1

In time-dependent process, as t approaches infinity, the steady-state distribution of the queue size can be obtained. Hence, we prove the following theorem.

Theorem 5.5.2.1

If $\rho = \lambda/\mu$, then the stationary distribution $P_n \equiv \lim_{t \to \infty} P_n(t)$, $n = 0, 1, 2, \ldots$, exists, is independent of the initial state, and:

$$P_n \equiv \lim_{t \to \infty} P_n(t) = (1-\rho)\rho^n, \quad \rho = \frac{\lambda}{\mu} < 1, \quad n = 0, 1, 2, \ldots. \qquad (5.5.2.1)$$

If $\rho = (\lambda/\mu) \geq 1$, the stationary distribution does not exist.

Proof

As we have discussed before, the stationary process (time-independent process) is the limiting value (as $t \to \infty$) of $P_n(t)$, denoted by P_n, that represents the portion of time that the system contains n tasks. Thus, the steady-state distribution (5.5.2.1) can be obtained directly from (5.5.1.1) by passing to the limit. However, we will prove the theorem directly from its own system of difference equations. Therefore, considering (5.5.17), we pass to the limit as time increases without bound and then drop the time parameter. When the probabilities are time independent, the derivative with respect to time will be zero. Thus, we will have:

$$\begin{cases} \lambda P_0 = \mu P_1 \\ (\lambda + \mu) P_n = \lambda P_{n-1} + \mu P_{n-1}, \quad n \geq 1, \end{cases} \qquad (5.5.2.2)$$

with $\sum_{n=0}^{\infty} P_n = 1$.

Once again we remind that by a state of the system we mean the total number of tasks in the system at any time, including in the service facility; moreover, that λ and μ represent infinitesimal transition rates in and out of the states of the system. The system (5.5.2.2) simply and succinctly says that *flow in equals flow out for state n*. The set of all such equilibrium equations for all states, that is, the system (5.5.2.2), is called a *balance equation*. Accordingly, the system (5.5.2.2) may be interpreted as:

(Long-time probability of being in state n)$\times$(Transition rates out of state n)

$$= \sum_{i=n-1, n+1} (\text{Long-time probability of being in state } i)$$

$\times$(Transition rate from state i to state n).

Now to solve (5.5.2.2), despite the complexity of (5.5.17), we can solve (5.5.2.2) by iterative method and use of mathematical induction to prove that the solution is indeed (5.5.2.1). This is because from the first equation of (5.5.2.2) we have:

$$P_1 = \frac{\lambda}{\mu} P_0 = \rho P_0.$$

From the second equation of (5.5.2.2) we have:

$$P_{n+1} = \frac{1}{\mu}[(\lambda+\mu)P_n - \lambda P_{n-1}], \quad n = 1, 2, \dots. \tag{5.5.2.3}$$

Thus, for $n = 1$, we have:

$$P_2 = \frac{1}{\mu}[(\lambda+\mu)P_1 - \lambda P_0] = \left(\frac{\lambda}{\mu}\right)^2 P_0.$$

Let us assume that for $n = k - 1$ we have:

$$P_k = \left(\frac{\lambda}{\mu}\right)^k P_0. \tag{5.5.2.4}$$

Then, from (5.5.2.3) and (5.5.2.4) for $n = k$ we will have:

$$P_{k+1} = \frac{1}{\mu}[(\lambda+\mu)P_k - \lambda P_{k-1}] = \frac{1}{\mu}\left[(\lambda+\mu)\left(\frac{\lambda}{\mu}\right)^k P_0 - \lambda\left(\frac{\lambda}{\mu}\right)^{k-1} P_0\right]$$

$$= \left(\frac{\lambda}{\mu}\right)^{k+1} P_0.$$

This proves the mathematical induction on parameter n that leads to:

$$P_n = \left(\frac{\lambda}{\mu}\right)^n P_0, \quad n = 0, 1, 2, \dots.$$

Then, from $\sum_{n=0}^n P_n = \sum_{n=0}^n (\lambda/\mu)^n P_0 = 1$, we solve for P_0. For this, we note that $\sum_{n=0}^n (\lambda/\mu)^n$ is a geometric series and it converges for $\rho = (\lambda/\mu) < 1$ to $1/1 - \rho$. Hence,

$$P_0 = \frac{1}{\sum_{n=0}^\infty \left(\frac{\lambda}{\mu}\right)^n} = 1 - \rho$$

and from this we obtain (5.5.2.1), and this completes the proof of the theorem.

As an alternative proof, suppose that the initial distribution of the queue size is denoted by $\{P_i(0)\}$ and is given by (5.5.2.1). Then we have to show that for all $t \geq 0$ we have:

$$P\{\xi(t) = i\} = P_i(0), \quad i = 0, 1, 2, \dots. \tag{5.5.2.5}$$

To show this, we note that:

$$P\{\xi(t) = i\} = \sum_{k=0}^{\infty} P_k(0) \cdot P_{ki}(t),$$

where $P_{ki}(t) = p_{ki}(t) + P_i(0)$. We leave it as an exercise to show that $\sum_{k=0}^{\infty} \rho^k p_{ki}(t) = 0$. Hence, (5.5.2.5) follows.

We note from (5.5.2.1) that the probability that the server is busy is a system performance measure. It is the *utility factor*, which is

$$1 - P_0 = \rho = \frac{\lambda}{\mu},$$

and that, in fact, is the *traffic intensity*. We also note that:

$$P\{\xi > k\} = \sum_{l=k+1}^{\infty} (1-\rho)\rho^l = \rho^{k+1} \tag{5.5.2.6}$$

and

$$P\{\xi \le k\} = 1 - P\{\xi > k\} = 1 - \rho^{k+1}. \tag{5.5.2.7}$$

For the average queue length in the stationary case, we denote the number of tasks in the system and the expected value of the number of tasks in the system by ξ and L, $L \equiv E\{\xi\}$, respectively. Thus, from (5.5.2.1) and the fact that $1/(1-b) = \sum_{n=0}^{\infty} b^n$, for $|b| < 1$, we have:

$$L = \sum_{n=0}^{\infty} nP_n = \sum_{n=1}^{\infty} n(1-\rho)\rho^n = (1-\rho)\rho \sum_{n=1}^{\infty} n\rho^{n-1} = (1-\rho)\rho \frac{d}{d\rho} \sum_{n=0}^{\infty} \rho^n$$

$$= (1-\rho)\rho \frac{d}{d\rho} \left(\frac{1}{1-\rho} \right) = (1-\rho)\rho \frac{1}{(1-\rho)^2} \tag{5.5.2.8}$$

$$= \frac{\rho}{1-\rho}.$$

For variance of ξ, we need the second moment of ξ. Hence,

$$E\{\xi^2\} = \sum_{n=0}^{\infty} n^2 P_n = \sum_{n=1}^{\infty} n^2 (1-\rho)\rho^n = (1-\rho) \sum_{n=1}^{\infty} [n(n-1) + n]\rho^n$$

$$= (1-\rho) \left[\sum_{n=1}^{\infty} n(n-1)\rho^n + \sum_{n=1}^{\infty} n\rho^n \right]$$

$$= (1-\rho) \left[\rho^2 \frac{d^2}{d\rho^2} \left(\frac{1}{1-\rho} \right) + \rho \frac{d}{d\rho} \left(\frac{1}{1-\rho} \right) \right] \tag{5.5.2.9}$$

$$= \frac{\rho(\rho+1)}{(1-\rho)^2}.$$

Thus, from (5.5.2.8) and (5.5.2.9) we have:

$$Var(\xi) = \frac{\rho(\rho+1)}{(1-\rho)^2} - \left(\frac{\rho}{1-\rho}\right)^2$$

$$= \frac{\rho}{(1-\rho)^2}. \tag{5.5.2.10}$$

Note that when ρ approaches 1, the mean and variance of the number of tasks in the system, (5.5.2.8) and (5.5.2.10), approach infinity. In other words, the stationary queue length is long when ρ is less than 1 but is near to it. Also, if $\rho \geq 1$, that is, the arrival rate is greater than or equal to the service rate, the queue length grows without bound, that is, the stationary distribution does not exist.

Now let us denote the number of tasks in the waiting line and its expected value by ξ_q and L_q, $L_q \equiv E\{\xi_q\}$, respectively. Then,

$$L_q = \sum_{n=0}^{\infty} (n-1) P_n = \sum_{n=1}^{\infty} n P_n - \sum_{n=1}^{\infty} P_n = L - (1 - P_0)$$

$$= L - \rho = \frac{\rho}{1-\rho} - \rho \tag{5.5.2.11}$$

$$= \frac{\rho^2}{1-\rho}.$$

We leave it as an exercise to find the variance of the number of tasks in the waiting line.

Note that from (5.5.2.9) and the penultimate line in (5.5.2.11), it can be seen that $L = L_q + \rho$, which means that in the stationary case, the expected number of tasks in the system is the sum of the expecting number in the waiting line and the expected number of tasks in service.

Summarizing what we found earlier, we have the following theorem:

Theorem 5.5.2.2
For the queueing system $M/M/1$, the stationary mean and variance of the number of tasks in the system and the mean number of tasks in waiting to be served with traffic intensity less than 1 are given, respectively, by (5.5.2.8), (5.5.2.10), and (5.5.2.11).

Example 5.5.2.1
Suppose a main frame server with a sufficiently large storage capacity receives programs for execution at a mean rate of three programs per minute. The computer has a mean execution time of 15 seconds per task (the mean service rate = 4 tasks per minute) on a first-come, first-served basis. We want to answer the following questions:

(i)   Write the system of difference equations governing the problem.

(ii)   Find the traffic intensity of the system.

(iii) Find the probability of the system being empty.

(iv)   Find the percent of time the service facility is idle.

(v)   Find the probability that there are at most two tasks in the system.

(vi)   Find the probability of more than nine tasks in the system.

(vii) Find the average number of tasks in the system.

(viii) Find the standard deviation of the number of tasks in the system.

(ix)   Find the average number of tasks waiting to be executed.

Answers

The system described can be modeled as an *M/M/*1 queueing process with parameters $\lambda = 3$ and $\mu = 4$. The states $0, 1, 2, \ldots$ represent the number of programs waiting to be executed in addition to the one being executed, if any.

(i)   From (5.5.2.2), the system of balance equations for this problem is:

$$\begin{cases} 3P_0 = 4P_1 \\ 7P_n = 3P_{n-1} + 4P_{n+1}, \quad n \geq 1. \end{cases}$$

(ii)   The traffic intensity of the system is $\rho = \lambda/\mu = 0.75$.

(iii) The probability that the system is empty is $P_0 = 1 - \rho = 1 - .75 = 0.25$.

(iv)   The stationary probability of the server facility being idle is the same as the probability that the system is empty. Hence, the percent of time the service facility is idle will be 25.

(v)   From (5.5.2.1), we have:

$$P_n = 0.25(0.75)^n, \quad n = 0, 1, 2, \ldots. \tag{5.5.2.12}$$

Hence, probability of having at most two tasks in the system means to have 0, 1, or 2 tasks in the system, that is, $P\{\xi \leq 2\}$. Thus, from (5.5.2.12) we have

$$P\{\xi \leq 2\} = P_0 + P_1 + P_2 = 0.25 + (0.25)(0.75) + (0.25)(0.75)^2 = 0.578125.$$

However, from (5.5.2.7), $P\{\xi \leq 2\} = 1 - (0.75)^3 = 0.578125$.

(vi)   From (5.5.2.6), the probability of having more than nine tasks in the system is $P\{\xi > 9\} = (0.75)^{10} = 0.056$.

(vii) The average number of programs in the system is $0.75/0.25 = 3$.

(viii) The standard deviation of the number of tasks in the system from (5.5.2.11) is $\sqrt{0.75/(0.25)^2} = 3.46$.

(ix)   The average number of programs waiting to be executed is $(0.75)^2/0.25 = 2.25$.

5.5.3. Stationary Waiting Time of a Task in *M/M/*1 Queue

Another important quantity of interest in a queueing theory is the waiting time of a task in the system and in the waiting line. The total time in the system is sometimes referred to as the *sojourn time*. We denote the means of these two quantities by W and W_q, respectively. Recall L and L_q that were denoted as the means of queue length and the length of the waiting line, respectively. The following is well known as *Little's formula*:

$$L = \lambda W. \tag{5.5.3.1}$$

Similarly,

$$L_q = \lambda W_q. \tag{5.5.3.2}$$

Equations (5.5.3.1) and (5.5.3.2) give relations between average queue lengths and average waiting times for the system and waiting line, respectively. Thus, for *M/M/*1, from (5.5.2.8) and (5.5.2.11), we have:

$$W = \frac{1}{\mu - \lambda} \tag{5.5.3.3}$$

and

$$W_q = \frac{\lambda}{\mu(\mu - \lambda)}. \tag{5.5.3.4}$$

We note that the average time spent in the system is the sum of the average time spent in the waiting line plus the average time spent in service. In fact, we can see this by adding the average service time $1/\mu$ to W_q given in (5.5.3.4) to obtain W given in (5.5.3.3).

Example 5.5.3.1
Consider the Example 5.5.2.1. Recall that we had the mean arrival rate, $\lambda = 3$ and the mean execution rate $\mu = 4$ programs per minute. We now want to find the average waiting time in the system and in the waiting line for a task.

Answer
In Example 5.5.2.1 we found $L = 3$ and $L_q = 2.25$. Hence, using Little's formula, we can find the average waiting times. However, directly from (5.5.3.3), we can have $W = 1/(4 - 3) = 1$ and $Wq = 3/4(4 - 3) = 0.75$.

5.5.4. Distribution of a Busy Period for *M/M/*1 Queue

As we earlier defined, by a *busy period* we mean the period of time during which the server is continuously busy. In other words, a busy period starts with an arrival and stops when the system is empty, that is, the system starts in state 1 and stops in state 0 the first time. From the point of view of the server, for a busy period, queue discipline is irrelevant.

The period for which the server is not busy is called the *idle period*. The idle period follows a busy period and together they make a *busy cycle*. As soon as a task arrives, the idle period ends. Hence, the idle period is the remaining interarrival time after the last task in the busy period leaves the system after its completion of service. The duration of successive idle periods and busy periods are independent random variables. Since the arrivals are Poisson, the interarrivals are exponential and, thus, memoryless. Hence, the idle periods will have an exponential distribution with the same parameter as the arrivals. If we denote the pdf of idle periods by $f_I(t)$, then:

$$fI(t) = \lambda e^{-\lambda t}, \quad t \geq 0. \tag{5.5.4.1}$$

There are different methods of finding the distribution of a busy period. We will choose the differential–difference equation method, the same approach as the transient solution of the queue length. For the busy period case, the state 0 is taken as an absorbing state and all others are transient. The transition probability matrix for the busy period is:

$$\mathbf{P} = \begin{pmatrix} 0 & 0 & 0 & 0 & 0 & 0 & \cdots \\ \mu & -(\lambda+\mu) & \lambda & 0 & 0 & 0 & \cdots \\ 0 & \mu & -(\lambda+\mu) & \lambda & 0 & 0 & \cdots \\ 0 & 0 & \mu & -(\lambda+\mu) & \lambda & 0 & \cdots \\ \vdots & \vdots & \vdots & \ddots & \ddots & \ddots & \cdots \\ \cdots & \cdots & \cdots & \cdots & \cdots & \cdots & \ddots \end{pmatrix}. \tag{5.5.4.2}$$

The system of differential–difference equations governing the probability of a busy period with the presence of absorbing state is:

$$\begin{cases} \dfrac{dP_0(t)}{dt} = \mu P_1(t) \\[2mm] \dfrac{dP_1(t)}{dt} = -(\lambda+\mu)P_1(t) + \mu P_2(t) \\[2mm] \dfrac{dP_n(t)}{dt} = -(\lambda+\mu)P_n(t) + \lambda P_{n-1}(t) + \mu P_{n+1}(t), \quad n \geq 2 \end{cases}, \tag{5.5.4.3}$$

with $P_1(0) = 1$ and $P_n(0) = 0$, for $n \neq 1$.

To solve (5.5.4.3), we do as we did with (5.5.17), that is, we will use probability generating functions and Laplace transforms. After performing similar manipulations, the Laplace transform of the generating function will be:

$$G^*(z, s) = \frac{z^2 - (1-z)(\mu - \lambda z)(z_1/s)}{-\lambda(z - z_1)(z - z_2)}, \tag{5.5.4.4}$$

where z_1 and z_2 are given by (5.5.1.10) and (5.5.1.11).

We define the Laplace transform of the busy period, that is, the length of a busy period $\leq t$, as $P_0^*(s)$, where $P_0(t)$ is the probability of the system being empty. Note that $P_0(t)$ can be obtained from the generating function given by (5.5.4.4) by substituting $z = 0$. The Laplace transform of $P_0(t)$ from (5.5.1.4) is $P_0^*(s)$, as desired. Thus, substituting $z = 0$ in (5.5.4.4) and use of z_2 is given by (5.5.1.11) yields:

$$P_0^*(s) = G^*(0, s) = \frac{\mu}{\lambda s z_2} \tag{5.5.4.5}$$

or

$$s P_0^*(s) = s G^*(0, s) = \frac{\mu}{\lambda z_2} = \frac{\lambda + \mu + s - \sqrt{(\lambda + \mu + s)^2 - 4\lambda\mu}}{2\lambda}. \tag{5.5.4.6}$$

It is well known that:

$$\mathcal{L}[P_0'(t)] = s P_0^*(s) - P_0(0) = s G^*(0, s). \tag{5.5.4.7}$$

Thus, inverting (5.5.4.6) using (5.5.4.7) and the fact that $P_0(0) = 0$, we obtain the probability distribution of the busy period $\leq t$ as:

$$P_0'(t) = \frac{\sqrt{\mu/\lambda}}{t} e^{-(\lambda+\mu)t} I_1\left(2\sqrt{\lambda\mu t}\right), \tag{5.5.4.8}$$

where $I_\nu(z)$, $\nu = 0, \pm 1, \pm 2, \ldots$, is the modified Bessel function of order ν defined by (5.5.1.2). Using combinatorial arguments as in Prabhu (2007), it can be shown that the probability distribution of the busy period can have an alternative form as:

$$P_0'(t) = e^{-(\lambda+\mu)t} \sum_{n=0}^{\infty} \frac{\lambda^{n-1} \mu^n t^{2n-2}}{n!(n-1)!}. \tag{5.5.4.9}$$

Let the length of the busy period be denoted by the random variable B. To find the mean and variance of B, note from (5.5.1.4) that if $f(t)$ is the pdf of a random variable T and then $f^*(s)$ is its Laplace transform, then:

$$E(T) = -f^{*\prime}(0) \quad \text{and} \quad E(T^2) = -f^{*\prime\prime}(0). \tag{5.5.4.10}$$

Thus, we leave it as an exercise to show that from (5.5.4.6), using (5.5.4.10), we have:

$$E(B) = \frac{1}{\mu(1-\rho)} \quad \text{and} \quad Var(B) = \frac{1+\rho}{\mu^2 (1-\rho)^3}, \quad (5.5.4.11)$$

for $\rho = (\lambda/\rho) < 1$.

The earlier discussion was based on the case that the system starts with an idle period. If $\xi(0) = 0$, then the system starts with an idle period and every busy period has the same distribution. However, the system may start with a busy period, that is, the busy period could be defined as the period initiating with i, $i \geq 1$, tasks in the system, that is, $\xi(0) = i$, $i \geq 1$, and stops when the number of tasks in the system becomes 0 or $i - 1$ for the first time. The latter will be discussed later.

Suppose $\xi(0) = i$, $i \geq 1$. Let the distribution of the busy periods following the initial one be denoted by $T(t)$. Then, the distribution function of the initial busy period is $T_i(t)$, where $T_i(t)$ is the i-fold convolution of $T(t)$ with itself. Let B_i be a random variable representing the length of the busy period started by i tasks in the system. We leave it as an exercise to show that the probability distribution, mean, and variance of the length of the busy period are, respectively, as follows:

$$P_0'(t) = \frac{i\sqrt{\mu/\lambda}}{t} e^{-(\lambda+\mu)t} I_i\left(2\sqrt{\lambda\mu t}\right), \quad (5.5.4.12)$$

$$E(B_i) = \frac{i}{\mu(1-\rho)} \quad \text{and} \quad Var(B_i) = \frac{i(1+\rho)}{\mu^2 (1-\rho)^3}. \quad (5.5.4.13)$$

5.6. FINITE BUFFER SINGLE-SERVER MARKOVIAN QUEUE: *M/M/1/N*

In this section, we assume that the capacity of the system is finite, say N. That is, the buffer size is $N - 1$. Hence, the incoming tasks after the system is full will be lost. All other conditions and properties of $M/M/1$ remain the same. Hence, an *M/M/1/N queueing system* is a single-server queueing system with Markovian arrival and Markovian service time processes with infinite source of arrivals and finite capacity N.

As before, let us define by $\xi(t)$ the state of the system, that is, the number of tasks in the system at time t with state space $\{0, 1, 2, \ldots, N\}$. Hence, $\{\xi(t)\}$ is a Markov process. Again, assuming that $P\{\xi(0) = i\} = 1$, the transition probabilities will be defined as follows:

$$p_{in}(t) = P\{\xi(t) = n | \xi(0) = i\}, \quad i, n = 0, 1, 2, \ldots, N. \quad (5.6.1)$$

The transition probability matrix, in this case, will be:

$$
\mathbf{P} = \begin{array}{c} \\ 0 \\ 1 \\ 2 \\ 3 \\ \vdots \\ N-1 \\ N \end{array}
\begin{array}{c}
\begin{array}{ccccccccc} 0 & 1 & 2 & 3 & 4 & \ldots & N-1 & N \end{array} \\
\begin{pmatrix}
-\lambda & \lambda & 0 & 0 & 0 & 0 & \cdots & 0 \\
\mu & -(\lambda+\mu) & \lambda & 0 & 0 & 0 & \cdots & 0 \\
0 & \mu & -(\lambda+\mu) & \lambda & 0 & 0 & \cdots & 0 \\
0 & 0 & \mu & -(\lambda+\mu) & \lambda & 0 & \cdots & 0 \\
\vdots & \vdots & \vdots & \vdots & \ddots & \ddots & \ddots & \cdots & 0 \\
\vdots & \vdots & \vdots & \vdots & \ddots & \ddots & \ddots & -(\lambda+\mu) & \lambda \\
\cdots & \cdots & \cdots & \cdots & \cdots & \cdots & \mu & -\mu
\end{pmatrix}
\end{array}. \quad (5.6.2)
$$

The differential–difference equations governing the distribution of the queue length of $M/M/1/N$ system is as follows:

$$
\begin{cases}
\dfrac{dP_0(t)}{dt} = -\lambda P_0(t) + \mu P_1(t), \\[2mm]
\dfrac{dP_n(t)}{dt} = -(\lambda+\mu)P_n(t) + \lambda P_{n-1}(t) + \mu P_{n+1}(t), \, 1 \le n \le N-1, \\[2mm]
\dfrac{dP_N(t)}{dt} = -\mu P_N(t) + \lambda P_{N-1}(t),
\end{cases}
\quad (5.6.3)
$$

with $P_i(0) = \delta_{ij}$, where δ_{ij} is the *Kronecker delta*, which is defined as in (5.5.18). We note that from state n, only departure is allowed.

As in the case of $M/M/1$, there are different methods for finding the distribution of the queue length for $M/M/1/N$. For instance, Takács (1962, p. 13) uses the Chapman–Kolmogorov equation:

$$
P_{i,k}(t+s) = \sum_{j=0}^{N} P_{ij}(t) P_{jk}(s), \quad t, s > 0 \quad (5.6.4)
$$

and eigenvalue method to prove the following theorem. We use our notation $P_n(t)$ that stands for the queue length distribution, assuming the initial queue size i.

Theorem 5.6.1
If

$$
\rho = \frac{\lambda}{\mu} \ne 1,
$$

then:

$$
P_n(t) = \frac{1-\rho}{1-\rho^{N+1}} \rho^n + \frac{2}{N+1} \sum_{j=1}^{N} \frac{e^{-(\lambda+\mu)t + 2t\sqrt{\lambda\mu}\cos[j\pi/(N+1)]} \rho^{(n-i)/2}}{\left[1 - 2\sqrt{\rho}\cos\dfrac{\pi j}{N+1} + \rho \right]}
$$
$$
\cdot \left[\sin\frac{ij\pi}{N+1} - \rho^{1/2}\sin\frac{(i+1)j\pi}{N+1} \right]\left[\sin\frac{nj\pi}{N+1} - \rho^{1/2}\sin\frac{(n+1)j\pi}{N+1} \right],
$$

$$
(5.6.5)
$$

and if $\rho = 1$, then:

$$P_n(t) = \frac{1}{N+1} + \frac{1}{N+1} \sum_{j=1}^{N} \frac{e^{-2\lambda t + 2\lambda t \cos[j\pi/(N+1)]}}{\left(1 - 2\cos\frac{\pi j}{N+1}\right)} \cdot$$

$$\left[\sin\frac{ij\pi}{N+1} - \sin\frac{(i+1)j\pi}{N+1}\right] \cdot \left[\sin\frac{nj\pi}{N+1} - \sin\frac{(n+1)j\pi}{N+1}\right]. \tag{5.6.6}$$

We note that if $N \to \infty$, we will obtain the transient solution of the *M/M/1* case. However, instead of the Bessel function we had previously used, this time we will use the trigonometric function. It is as follows:

$$P_n(t) = Q_n(t) + \begin{cases} (1-\rho)\rho^n & \text{if } \rho < 1, \\ 0, & \text{if } \rho \geq 1, \end{cases} \tag{5.6.7}$$

where

$$Q_n(t) = \frac{2}{\pi} e^{-(\lambda+\mu)t} \rho^{\frac{n-1}{2}} \int_0^\pi \frac{e^{2\sqrt{\lambda\mu}t \cos y}}{1 + \rho - 2\rho^{\frac{1}{2}}\cos y}$$

$$\times \left[\sin iy - \rho^{\frac{1}{2}}\sin(i+1)y\right] \times \left[\sin ny - \rho^{\frac{1}{2}}\sin(n+1)y\right] dy. \tag{5.6.8}$$

Passing to limit as $t \to \infty$, from (5.6.7) and (5.6.8), we obtain the stationary distribution of *M/M/1* given by (5.5.2.1).

The stationary distribution of the queue length of *M/M/1/N*, that is, when t approaches infinity, can be easily obtained from Theorem 5.6.2 as stated in the next paragraph. See also Takács (1962, p. 21).

Theorem 5.6.2

For the system *M/M/1/N*, the limiting probability distribution of the queue length, denoted by P_n, $P_n = \lim_{t \to \infty} P_{in}(t)$, $n = 0, 1, 2, \ldots N$, exists and is independent of the initial state of the system. The Pn, $n = 0, 1, 2, \ldots, N$ is:

$$P_n = \begin{cases} \dfrac{1-\rho}{1-\rho^{N+1}}\rho^n, & \text{if } \rho \neq 1, \\[3mm] \dfrac{1}{N+1}, & \text{if } \rho = 1. \end{cases} \tag{5.6.9}$$

If the initial queue size, $\xi(0)$, is a random variable with distribution function

$$P\{\xi(0) = i\} = P_i, \quad i = 0, 1, 2, \ldots,$$

defined by (5.6.9), then:

$$P\{\xi(t) = i\} = P_i \quad i = 0, 1, 2, \ldots, \quad t \geq 0.$$

Hence, $\{\xi(t)\}$ is the stationary process for the queue length.

For the expected queue length, denoted by L_N, we have:

$$
\begin{aligned}
L_N &= \sum_{k=0}^{N} k \frac{(1-\rho)\rho^k}{1-\rho^{N+1}} = \frac{(1-\rho)\rho}{1-\rho^{N+1}} \sum_{k=0}^{N} k\rho^{k-1} \\
&= \frac{(1-\rho)\rho}{1-\rho^{N+1}} \sum_{k=0}^{N} \frac{d\rho^k}{d\rho} = \frac{(1-\rho)\rho}{1-\rho^{N+1}} \frac{1-(N+1)\rho^N + N\rho^{N+1}}{(1-\rho)^2} \\
&= \frac{\rho}{1-\rho} - \frac{(N+1)\rho^{N+1}}{1-\rho^{N+1}}, \quad \rho \neq 1,
\end{aligned}
\tag{5.6.10}
$$

and

$$
L_N = \sum_{k=0}^{N} k P_k = \sum_{k=0}^{N} \frac{k}{N+1} = \frac{N}{2}, \quad \rho = 1. \tag{5.6.11}
$$

Note

The following results are very clear. Hence, we leave them as exercises to explain, if necessary.

(1) The expected number of tasks attending the system $M/M/1/N$ in a unit time, called *effective arrival rate*, denoted by $\lambda^{(1)}$, is:

$$
\lambda^{(1)} = \lambda(1 - P_N) = \begin{cases} \dfrac{\lambda(1-\rho^N)}{1-\rho^{N+1}}, & \text{if } \rho \neq 1, \\[3mm] \dfrac{\lambda N}{N+1}, & \text{if } \rho = 1. \end{cases}
\tag{5.6.12}
$$

(2) The mean interarrival times attending the system is $1/\lambda^{(1)}$. From (5.6.12), it can be seen that $1/\lambda^{(1)} < 1/\lambda$.

(3) The expected number of possible tasks lost due to the finiteness of the system is:

$$
\lambda - \lambda(1 - P_N) = \lambda P_N = \begin{cases} \dfrac{\lambda(1-\rho)\rho^N}{1-\rho^{N+1}}, & \text{if } \rho \neq 1, \\[3mm] \dfrac{\lambda}{N+1} & \text{if } \rho = 1. \end{cases}
\tag{5.6.13}
$$

(4) For $M/M/1/N$, the *utilization factor* is ρ_1 given by:

$$
\rho_1 = \frac{\lambda(1)}{\mu} = \rho(1 - P_N) = \begin{cases} \dfrac{\rho(1-\rho^N)}{1-\rho^{N+1}}, & \text{if } \rho \neq 1, \\[3mm] \dfrac{\rho N}{N+1}, & \text{if } \rho = 1. \end{cases}
\tag{5.6.14}
$$

(5) The expected number of tasks being served, for *M/M/*1/*N*, in a unit of time is:

$$\mu\rho_1 = \begin{cases} \dfrac{\lambda(1-\rho^N)}{1-\rho^{N+1}}, & \text{if } \rho \neq 1, \\[2ex] \dfrac{\lambda N}{N+1}, & \text{if } \rho = 1. \end{cases} \tag{5.6.15}$$

(6) By Little's formula, the expected waiting time in the system is $L_N/\lambda^{(1)}$.

(7) Let us denote the probability that an effective arrival, that is, an arrival who can attend the queue, finds n tasks in the system within an infinitesimal time interval of $(t, t + \Delta t)$, at its epoch of arrival, be denoted by q_n. Then, using Bayes' theorem (see Chapter 1), we have:

$$q_n = P\{n \text{ in the system} \,|\, \text{an arrival is about to occur}\}$$

$$= \frac{P\{\text{an arrival is about to accur} \,|\, n \text{ is the system}\} \times P_n}{\sum_{k=0}^{N-1} P\{\text{an arrival is about to occur} \,|\, k \text{ in the system}\}}$$

$$= \lim_{\Delta t \to 0} \frac{[\lambda\Delta t + o(\Delta t)P_n]}{\sum_{k=0}^{N-1}[\lambda\Delta t + o(\Delta t)P_k]} = \frac{\lambda P_n}{\sum_{k=0}^{N-1} \lambda P_k}$$

$$= \frac{P_n}{1 - P_N}, \quad n = 0, 1, 2, \cdots, N - 1.$$

5.7. *M/M/*1 QUEUE WITH FEEDBACK

Consider the *M/M/*1 queueing system described in Section 5.5. Suppose that after completion of service, there are two possibilities: the task needs to return to the system for further service with probability $p, 0 < p < 1$, or exit the system with probability of $q = 1 - p$. If returned, the task should join the tail of the queue and wait for another service as a new arrival. Eventually, after completion of all needed services, the task will leave the system. Hence, the rate of departure from the system will be $q\mu$. This rate, that is, the rate at which tasks actually leave the system and a change in state occurs, is called the *effective service rate*. In contrast, the rate of return, that is, the *feedback*, will be $p\mu$. Of course, a feedback does not make a change in the state of the system. Thus, we have what is known as the *single-sever queue with feedback*. Figure 5.7.1 illustrates such a system.

With description of the system as given earlier, the system of differential–difference equations for an *M/M/*1 feedback queue will be the same as for *M/M/*1, except that $q\mu$ replaced the μ, that is,

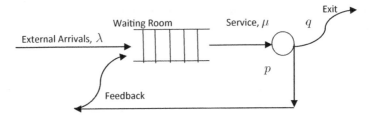

Figure 5.7.1. Traffic flow diagram for a single-server feedback queue.

$$\begin{cases} \dfrac{dP_0(t)}{dt} = -\lambda P_0(t) + q\mu P_1(t) \\ \dfrac{dP_n(t)}{dt} = -(\lambda + q\mu)\,P_n(t) + \lambda P_{n-1}(t) + q\mu P_{n+1}(t), \quad n \geq 1, \end{cases} \tag{5.7.1}$$

with $P_i(0) = \delta_{ij}$, where δ_{ij} is the *Kronecker delta*, which is defined as in (5.5.18). A similar system can be obtained for the stationary $M/M/1$ with feedback, $M/M/1/N$ with feedback, and stationary $M/M/1/N$ with feedback. We can also find similar performance measure as for $M/M/1$ for $M/M/1$ with feedback.

For instance, suppose computer programs arrive for execution to a computer server with an average of λ per minute. On any one trip through the process, the probability that a program leaves the server after execution is 0.1, that is, a program may return for further service with a probability of 0.9. Thus, if the average execution rate is 10 programs per minute, then on the average, the departure rate of programs would be $0.1 \times 10 = 1$ per minute.

5.8. SINGLE-SERVER MARKOVIAN QUEUE WITH STATE-DEPENDENT BALKING

Definition 5.8.1

Customer impatience is a very important feature of a queue that splits into three forms: balking, reneging, and jockeying. An *impatient* task may arrive at the system, but it may not attend (or may be reluctant to attend) the queue due to a variety of reasons such as length of the queue or waiting time. In this case, we say that the task *balks*. On the other hand, a task may attend the system, wait for a while, and then leave the system before receiving service, that is, reluctant to remain in the queue after joining. In this case, we say the task has *reneged*. In case of a multiple-server system, the impatience appears as *jockeying* between parallel service lines.

Note that both concepts of balking and reneging could be considered as the same concept since both are about leaving the system before receiving service. However, one leaves the system immediately at the arrival by observing the size of the queue, for instance, and the other leaves it after staying in the queue for a while. Because of this commonality, some authors use reneging for both concepts; we distinguish them.

The notion of balking and reneging queue is applicable in almost all areas of parallel computing systems wherein buffers are finite and/or queue length is too long. It is also applicable in the military, when shipments of troops, goods, and ammunition situate under enemy attack and minimal loss is desirable. Balking and reneging in the military could be thought of as the case that an aircraft needs to land on a ship for emergency needs while all spaces of the ship are occupied or cargo must leave a location that is under bombardment of the enemy. "The supercarrier, too valuable to risk losing, and able to hurl its ordnance beyond the horizon, is made to stand away from battle, sometimes hundreds of miles away." "A *Nimitz*-class carrier operates with a crew of more than 3200 trained and specialized young men and women enlisted and officers" Robbins (2003). In operating the orders in a production system, balking and reneging could be interpreted as not receiving an order because the manufacturer is backlogged or cancellation of an order before or after an order was placed. Consideration of zero balking rate may be desirable since a vessel that does not enter the warfare area fails its mission. Thus, consideration of the model without balking will help decision making during wartime.

In this section, we only consider a single-server Markovian queue with state-dependent balking. Other kinds of balking and reneging queues will be considered in the multiserver section, Section 5.11. Each of these cases will serve as an example of the application of the differential–difference system of equations solved in stationary cases.

We now consider an $M/M/1$ queueing system. Suppose that an arriving task attends the queue with probability a_n, $n \geq 0$, when there are n tasks in the system at its arrival epoch. That is, the attendance of a task is state dependent. This means, an arriving task becomes discouraged when it sees a long number of tasks in the system when it arrives or it calculates the estimate of the time it has to wait in line before its service starts. Thus, the arriving rate is $\lambda a_n, n \geq 0$. Note that a_n is a monotonically decreasing function of the system size, that is, $0 \leq a_{n+1} \leq a_n \leq 1, n > 0$, and $a_0 = 1$.

Let $\xi_n(t)$ denote the state of the system at time t when the nth task arrives, that is, there are $n - 1$ tasks ahead of it. We also let the stationary state of the system at the epoch of arrival of the nth task be denoted by the random variable ξ_n. We further denote by $P_n(t)$ the probability that at time t the system is in state n, and by P_n the stationary probabilities of $P_n(t)$, that is,

$$P_n = \lim_{t \to \infty} P_n(t). \tag{5.8.1}$$

Based on a description of the model, the transient system of balance differential–difference equations for the distribution of the number of tasks in the system $M/M/1$ with random balking will be the same as $M/M/1$ except that λ is replaced with λb_n, that is,

$$\begin{cases} P_0'(t) = -\lambda a_0 P_0(t) + \mu P_1(t), \\ P_n'(t) = -(\lambda a_n + \mu) P_n(t) + \lambda a_{n-1} P_{n-1} + \mu P_{n+1}(t), \quad n \geq 1. \end{cases} \qquad (5.8.2)$$

From (5.8.2), passing to the limit as t approaches infinity, and using (5.8.1), the system of balance difference equations for the stationary distribution of the number of tasks in the system will be as follows:

$$\begin{cases} \lambda a_0 P_0 = \mu P_1, \\ (\lambda a_n + \mu) P_n = \lambda a_{n-1} P_{n-1} + \mu P_{n+1}, \quad n \geq 1. \end{cases} \qquad (5.8.3)$$

Solving (5.8.3), using the recursive method (leaving out the intermediate steps as exercise), we obtain:

$$P_0 = \frac{1}{1 + \sum_{n=1}^{\infty} \left[\prod_{i=0}^{n-1} a_i \left(\frac{\lambda}{\mu} \right)^{i+1} \right]}, \qquad (5.8.4)$$

$$P_n = \left[\prod_{i=0}^{n-1} a_i \left(\frac{\lambda}{\mu} \right)^{i+1} \right] P_0, \quad n = 1, 2, \dots. \qquad (5.8.5)$$

It could be instructive to use the generating function to obtain the result rather than using the recursive method that we used earlier. Then, we will be able to find the mean and variance. However, we would still need the distribution to find the values of these measures of effectiveness. To do that, let

$$Q_n = a_n P_n, \quad n = 0, 1, 2, \dots, \quad Q_0 = P_0. \qquad (5.8.6)$$

Then, using (5.8.6), the system (5.8.3) can be rewritten as:

$$\begin{cases} \mu P_1 = \lambda Q_0, \\ \mu P_{n+1} = \mu P_n + \lambda Q_n - \lambda Q_{n-1}, \quad n \geq 1. \end{cases} \qquad (5.8.7)$$

We now define two probability generating functions P_n and Q_n, respectively, as follows:

$$G(z) = \sum_{n=0}^{\infty} P_n z^n, Q(z) = \sum_{n=0}^{\infty} Q_n z^n, |z| < 1, G(1) = 1. \qquad (5.8.8)$$

Now, as we have previously done, we multiply both sides of the equations in system (5.8.7) by the appropriate powers of z (first equation by zero power) and add. Hence, after some algebra simplification, we will have the following:

$$\mu G(z) - \lambda z Q(z) = \mu P_0. \qquad (5.8.9)$$

From (5.8.8) when $z = 1$, (5.8.9) yields:

$$P_0 = 1 - \lambda \sum_{n=0}^{\infty} a_n P_n.$$
(5.8.10)

Hence,

$$G(z) = \frac{\lambda}{\mu} z Q(z) + 1 - \lambda \sum_{n=0}^{\infty} a_n P_n.$$
(5.8.11)

We note that to find the distribution, instead of using the recursive method that we used earlier and obtained (5.8.5), we could use the generating function obtained in (5.8.11). To do that, we can let $q_n = a_n P_n = r^n P_n$. Then, we will have $Q(z) = \sum_{n=0}^{\infty} r^n P_n z^n = G(rz)$. Thus, from (5.8.11) we will have:

$$G(z) = \frac{\lambda}{\mu} z G(rz) + 1 - \lambda \sum_{n=0}^{\infty} a_n P_n,$$
(5.8.12)

from which, taking the corresponding coefficients, we will obtain (5.8.5).

Now from (5.8.11), finding the first and the second derivatives evaluated at $z = 1$ will give us the first and the second moments from which the mean and variance can be found in terms of a_n. Once the values of a_n, $n = 0, 1, 2, \ldots$, is given, since P_n's are given by (5.5.4) and (5.8.5), the mean and variance will be obtained. Here are the mean and variance, respectively:

$$E[\xi_n] = G'(1) = \frac{\lambda}{\mu} \sum_n (1 + n) a_n P_n$$
(5.8.13)

and

$$\begin{aligned}
Var[\xi_n] &= G''(1) + G'(1) - [G'(1)]^2 \\
&= \frac{\lambda}{\mu} \sum_{n=0}^{\infty} a_n P_n - \left(\frac{\lambda}{\mu}\right)^2 \left[\sum_{n=0}^{\infty} a_n P_n\right]^2 + 3\left(\frac{\lambda}{\mu}\right) \sum_{n=0}^{\infty} n a_n P_n \\
&\quad - \left(\frac{\lambda}{\mu}\right)^2 \left(\sum_{n=0}^{\infty} n a_n P_n\right)^2 + \frac{\lambda}{\mu} \sum_{n=0}^{\infty} n(n-1) a_n P_n \\
&\quad - 2\left(\frac{\lambda}{\mu}\right)^2 \left[\sum_{n=0}^{\infty} a_n P_n\right]\left[\sum_{n=0}^{\infty} n a_n P_n\right].
\end{aligned}$$
(5.8.14)

5.9. MULTISERVER PARALLEL QUEUE

In the last three sections we have discussed single-server queues of different types. When there are more than one, say m, servers in the service station, set in parallel and that work independently of each other, we say that the system is a *parallel queueing system*. Such a system is sometimes referred to as an *m-server queueing system, many-server queueing system, multichannel queueing system*, or a *multiprocessor system*.

The structure of a multiserver queue is as follows. Tasks arrive at a system according to a Poisson distribution with parameter λ at epochs $\tau_0, \tau_1, \ldots, \tau_n$ with $\tau_0 = 0$ and form one waiting line if all servers are busy. If an arriving task finds an idle server at the time of its arrival, its service will start immediately. Thus, the interarrival times $\tau_n - \tau_{n-1}, n = 1, 2, \ldots$, are independent identically distributed positive random variables with a common distribution function $F(x) = 1 - e^{-\lambda x}, x \geq 0$. There are m service stations, set in parallel, each with a single-server. This is the same as having one station with m independent individual servers. Queue discipline is FIFO. The service times are independent identically distributed positive random variables with a common distribution function $H(x) = 1 - e^{-\mu x}, x \geq 0$, independent of the arrival processes. The number of tasks in the system, denoted by $\xi(t)$, determines the state of the system, that is, as before, if the initial state of the system is E_i and

$$P_n(t) = P\{\xi(t) = n \mid \xi(0) = i\}, \quad i, n \geq 0,$$

where the system is in state E_n at time t if there are n tasks in the system, that is, in the waiting line and in service.

5.9.1. Transient Queue Length Distribution for *M/M/m*

We develop the differential–difference equations based on the given structure of the system similar to what we did for the single-server case as follows. The case $m = 1$ is the *M/M/1* that we discussed in Section 5.5. We leave the case $m = 2$ as an exercise. Hence, we will develop the case for $m \geq 3$, although the results will hold for $m \geq 2$ with some modification.

Probability of exactly one arrival during the time interval $(t, t + \Delta t)$

$$= 1 - e^{-\lambda \Delta t} = \lambda \Delta t - \frac{(\lambda \Delta t)^2}{2!} + \cdots = \lambda \Delta t + o(\Delta t). \tag{5.9.1.1}$$

Probability that no service is completed during the interval $(t, t + \Delta t)$

$$= e^{-\mu \Delta t} = 1 - \mu \Delta t + \frac{(\mu \Delta t)^2}{2!} + \cdots = 1 - \mu \Delta t + o(\Delta t). \tag{5.9.1.2}$$

Probability of a transition from state E_n to state E_n

$$= \begin{cases} 1 - (\lambda + n\mu)\Delta t + o(\Delta t), & \text{if } 0 \leq n \leq m-1, \\ 1 - (\lambda + m\mu)\Delta t + o(\Delta t), & \text{if } n \geq m. \end{cases} \tag{5.9.1.3}$$

Probability of a transition from state E_k to state E_{k-1} during the interval $(t, t+\Delta t)$

$$= \begin{cases} n\mu\Delta t + o(\Delta t), & \text{if} \quad n \leq m, \text{ since only n servers are busy,} \\ m\mu\Delta t + o(\Delta t), & \text{if} \quad n \geq m, \text{ since there is no server idle.} \end{cases}$$

$$(5.9.1.4)$$

Probability of a transition from state E_n to state E_{n+1} during the interval $(t, t+\Delta t)$

$$= [\lambda\Delta t + o(\Delta t)][1 - \mu\Delta t) + o(\Delta t)] = \lambda\Delta t + o(\Delta t).$$

$$(5.9.1.5)$$

The following should be noted to see the first part of (5.9.1.4) better: suppose that there are $n, n = 1, 2, \ldots$, servers that are busy at a time. Then, the probability that any one of the n busy servers completes service within the interval $(t, t + \Delta t]$, using binomial distribution, is:

$$\binom{n}{1}[\mu\Delta t + o(\Delta t)][1 - \mu\Delta t + o(\Delta t)]^{n-1} = n\mu\Delta t + o(\Delta t).$$

Similarly, the probability that any k of the n busy servers completes service within the interval $(t, t + \Delta t]$, is

$$\binom{n}{k}[\mu\Delta t + o(\Delta t)]^k[1 - \mu\Delta t + o(\Delta t)]^{n-k} = o(\Delta(t).$$

That is why when there are n servers that are busy, the only way to have a transition from state E_k to state E_{k-1} during the interval $(t, t+\Delta t)$ is $n\mu\Delta t + o(\Delta t)$.

Thus, as before, we have:

$$P_0(t+\Delta t) = [1 - \lambda\Delta t + o(\Delta t)]P_0(t) + [\mu\Delta t + o(\Delta t)]P_1(t), \tag{5.9.1.6}$$

$$P_n(t+\Delta t) = [1 - (\lambda + n\mu)\Delta t + o(\Delta t)]P_n(t) + [(n+1)\mu\Delta t + o(\Delta t)]P_{n+1}(t)$$
$$+ [\lambda\Delta t + o(\Delta t)]P_{n-1}(t), \quad 1 \leq n < m, \tag{5.9.1.7}$$

$$P_n(t+\Delta t) = [1 - (\lambda + m\mu)\Delta t + o(\Delta t)]P_n(t) + [m\mu\Delta t + o(\Delta t)]P_{n+1}(t)$$
$$+ [\lambda\Delta t + o(\Delta t)]P_{n-1}(t), \quad n \geq m. \tag{5.9.1.8}$$

Again, as we did before, after some algebra and passing to limit, we will have the differential–difference equations as follows:

$$\begin{cases} P_0'(t) = -\lambda P_0(t) + \mu P_1(t), \\ P_n'(t) = -(\lambda + n\mu)P_n(t) + \lambda P_{n-1}(t) + (n+1)\mu P_{n+1}(t), & 1 \leq n < m, \\ P_n'(t) = -(\lambda + m\mu)P_n(t) + \lambda P_{n-1}(t) + m\mu P_{n+1}(t), & n \geq m, \end{cases} \tag{5.9.1.9}$$

with $P_i(0) = \delta_{ij}$, where δ_{ij} is the Kronecker delta, which is defined in (5.5.18).

Now, as before, we solve the system (5.9.1.9) by performing some algebra and then applying the generating function and the Laplace transform. Thus, let the generating function of $P_n(t)$ for the multiserver system, denoted by $G(z, t)$, be defined as in (5.5.1.3) and the Laplace transform of a function $f(t)$, denoted by $f*(s)$, be defined as in (5.5.1.4). Then, we leave it as an exercise to complete the work and show that the Laplace transform of the generating function of the queue length, denoted by $G*(z, s)$ is:

$$G^*(z, s) = \frac{z^{i+1} - \mu(1-z)\sum_{n=0}^{m-1}(m-n)z^n P_n^*(s)}{sz - (1-z)(m\mu - \lambda z)}. \qquad (5.9.1.10)$$

We will now again use the well-known Rouché theorem. Since $G*(z, s)$ converges inside and on $|z| = 1$ for $\mathrm{Re}(s) > 0$, the zeros of the denominator of (5.9.1.10) inside and on $|z| = 1$ must also be zeros of the numerator. The denominator of (5.9.1.10) is a quadratic equation in z, having two roots,

$$z_j(s) = \frac{\lambda + m\mu + s \pm \sqrt{(\lambda + m\mu + s)^2 - 4\lambda m\mu}}{2\lambda}, j = 1, 2, \qquad (5.9.1.11)$$

one of which, by Rouche's theorem, is inside the disk $|z| \le 1$. This is because if we let $f(z) = (\lambda + m\mu + s)z$ and $g(z) = \lambda z^2 + m\mu$, then, on the unit circle $|z| = 1$ will have $|f(z)| = |\lambda + m\mu + s| > |\lambda + m\mu| = |g(z)|$, since $\lambda, \mu, m > 0$. Hence, $|g(z)| < |f(z)|$ on $|z| = 1$. Thus, $f(z)$ and $f(z) + g(z)$, that is, the denominator of (5.9.1.10), have the same number of zeros within $|z| = 1$. However, $z = 0$ is the only root of $f(z)$. Therefore, the denominator of (5.9.1.10) has a single zero within $|z| = 1$. Let $z_1(s)$ and $z_2(s)$ be roots of the denominator of (5.9.1.10), with negative and positive square root, respectively. Then, since $|z_1(s)| < |z_2(s)|$, the zero inside $|z| = 1$ must be $z_1(s)$ and there is no zero on $|z| = 1$. We drop the (s) from z's since there will not be any confusion.

Now, since

$$z_1 + z_2 = \frac{\lambda + m\mu + s}{\lambda}, z_1 + z_2 = \frac{m\mu}{\lambda} \quad \text{and} \quad s = -\lambda(1 - z_1)(1 - z_2), \qquad (5.9.1.12)$$

$z_1(s)$ must be a root of the numerator, which means it has $(z - z_1)$ as a factor. We therefore have the following:

$$\sum_{n=0}^{n-1}(m-n)z_1^n P_n^*(s) = \frac{z_1^{i+1}}{\mu(1 - z_1)}. \qquad (5.9.1.13)$$

Substituting (5.9.1.13) in (5.9.1.10), after some algebra, we obtain:

$$G^*(z, s) = \frac{(1-z)z_1^{i+1} - z^{i+1}(1 - z_1)}{(1 - z_1)[\lambda z^2 - (\lambda + m\mu + s)z - m\mu]}. \qquad (5.9.1.14)$$

From (5.9.1.14) we have to find $P_n^*(s)$. Then, to find $P_n(t)$ we need to invert the Laplace transform using

$$P_n(t) = \frac{1}{2\pi i} \int_{a-i\infty}^{a+i\infty} e^{st} P_n^*(s)\, ds \tag{5.9.1.15}$$

for any positive number t, where $i^2 = -1$. To make this possible, we have to find $P_n^*(s)$ for $0 \le n < m - 1$, and then we can find $P_n^*(s)$ for $n \ge m$ explicitly. Hence, we need m equations. The first $m - 1$ of (5.9.1.9) and (5.9.1.13) make the m equations needed. The first $m - 1$s of (5.9.1.9) are:

$$\begin{cases} P_0'(t) = -\lambda P_0(t) + \mu P_1(t) \\ P_n'(t) = -(\lambda + n\mu) P_n(t) + \lambda P_{n-1}(t) + (n+1)\mu P_{n+1}(t), \quad 1 \le n \le m-2. \end{cases} \tag{5.9.1.16}$$

Note from (5.9.1.16) that the condition $n \ge 3$ is necessary for the second equation to hold. From this system, we should be able to obtain $P_n^*(s)$ for $0 \le n < m - 1$ in terms of $P_{m-1}^*(s)$. To do that, we define a new generating function as follows:

$$Q(z, t) = \sum_{n=0}^{m-2} P_n(t) z^n. \tag{5.9.1.17}$$

We assume $\xi(0) = i$, $i > m - 2$. Now we go through the standard generating function method and obtain:

$$-\frac{\partial Q(z, t)}{\partial t} + \mu(1-z)\frac{\partial Q(z, t)}{\partial t} = \lambda(1-z)Q(z, t) + \lambda z^{m-1} P_{m-2}(t) \tag{5.9.1.18}$$
$$- \mu(m-1) z^{m-2} P_{m-1}(t).$$

Equation (5.9.1.18) is a first-order partial differential equation. Hence, the associate equations are:

$$\frac{dt}{-1} = \frac{dz}{\mu(1-z)}$$
$$= \frac{dQ(z, t)}{\mu(1-z)Q(z, t) + \lambda z^{m-1} P_{m-2}(t) - \mu(m-1) z^{m-2} P_{m-1}(t)}. \tag{5.9.1.19}$$

Solving the first part of (5.9.1.19), which is a linear differential equation, we obtain $t - \ln C + \ln(1 - z)$, which gives $-\mu t = -\mu \ln C + \ln(1 - z)$ and, thus,

$$C_1 e^{\mu t} = 1 - z, \, C_1 = (1-z)e^{-\mu t} \quad \text{and} \quad z = 1 - C_1 e^{\mu t}, \tag{5.9.1.20}$$

where C_1 is the constant of integration. Substituting the value of z from (5.9.1.20) in the second part of (5.9.1.19) and taking the term thus obtained equal t, after some algebra, we obtain:

$$\frac{\partial Q(z,t)}{\partial t} + \mu C_1 e^{\mu t} Q(z,t) = \mu(m-1)(1-C_1 e^{\mu t})^{m-2} P_{n-1}(t)$$

$$- \lambda(1-C_1 e^{\mu t})^{m-1} P_{n-2}(t).$$

(5.9.1.21)

The solution of the linear nonhomogeneous differential equation (5.9.1.21) is:

$$Q(z,t) = e^{-\frac{\lambda}{\mu}C_1 e^{\mu t}} \int_0^t \left\{ \mu(m-1)(1-C_1 e^{\mu x})^{m-2} P_{n-1}(x) \right.$$

$$\left. - \lambda(1-C_1 e^{\mu x})^{m-1} P_{n-2}(x) e^{\frac{\lambda}{\mu}C_1 e^{\mu x}} \right\} dx.$$

(5.9.1.22)

Notes

1. Since $Q(z, 0) = 0$, the second integration constant $C_2 = f(C_1) = f[(1 - z)e^{-\mu t}] = 0$ for some $f(C_1)$.
2. In the earlier argument, we based our discussion on the assumption that $i > m - 2$. The same argument can be applied for the case $i \leq m - 2$. However, in the latter case, C_2 will not be zero since then $Q(z, 0) = z^i$ and $f(1 - z) = z^i e^{\lambda/\mu(1-z)}$. Hence, letting $y = 1 - z$ implies that:

$$z = 1 - y, \ f(y) = (1-y)^i e^{\frac{\lambda}{\mu}y} \ \text{and} \ f[(1-z)e^{-\mu t}] = [1-(1-z)e^{-\mu t}]^i e^{\frac{\lambda}{\mu}(1-z)e^{-\mu t}}.$$

Thus, substituting C_1 from (5.9.1.20), equation (5.9.1.22) can be rewritten as:

$$Q(z,t) = e^{-\frac{\lambda}{\mu}(1-z)} \int_0^t \left\{ \mu(m-1)[1-(1-z)e^{\mu(t-x)}]^{m-2} P_{n-1}(x) \right.$$

$$\left. - \lambda[1-(1-z)e^{-\mu(t-x)}]^{m-1} P_{n-2}(x) e^{\frac{\lambda}{\mu}(1-z)e^{\mu}(t-x)} \right\} dx$$

(5.9.1.23)

$$+ e^{-\frac{\lambda}{\mu}(1-z)(1-e^{-\mu t})} [(1-(1-z)e^{-\mu t}]^i.$$

As it can be seen, the last term of equation (5.9.1.23) is the second integration constant for the case $i \leq m - 2$.

We now return to the case $i > m - 2$. Let

$$f(t) = (1-e^{-t})^{\tau-1} (1-\lambda e^{-t})^{a e - t}, \quad Re(\tau) > 0, \quad |arg(1-\lambda)| < \pi.$$

(5.9.1.24)

It is known (see for instance Erdelyi, 1954, p. 147, formula 40) that the Laplace transform of $f(t)$ defined in (5.9.1.24) is:

$$Lf(t) = L\left[\left(1-e^{-t}\right)^{\tau-1}\left(1-\lambda e^{-t}\right)^{ae^{-t}}\right] = \frac{\Gamma(\tau)\Gamma(v)}{\Gamma(\tau+v)}\Phi_1(v,\mu,\tau,\lambda,a),$$

$$Re(v) > 0,$$

(5.9.1.25)

where $\Phi_1(v, \mu, \tau, \lambda, a)$ is the generalized hypergeometric function defined by:

$$\Phi_1(v,\mu,\tau,\lambda,a) = \sum_{j=0}^{\infty}\sum_{k=0}^{\infty}\frac{(v)_{k+j}(\mu)_j}{(\tau)_{k+j}\,k!\,j!}\lambda^k a^j,$$

where $(\alpha)_0 \equiv 1$, and $(\alpha)_n \equiv \alpha(\alpha + 1) \ldots (\alpha + n + 1)$. We now choose $\tau = 1$, $\mu = -(m - 2)$, and $y = \mu w$. Hence, for $Re(s) > 0$ we have:

$$\int_0^t e^{-sw}\left[1-(1-z)e^{\mu\omega}\right]^{m-2}e^{\frac{\lambda}{\mu}(1-z)e-\mu w}\,dw$$

$$= \int_0^{\infty} e^{-(s/\mu)y}\left[(1-z)e^{-y}\right]^{m-2}e^{\frac{\lambda}{\mu}(1-z)e^{-y}}\frac{dy}{\mu}$$

(5.9.1.26)

$$= \frac{1}{\mu}\frac{\Gamma\left(\frac{s}{\mu}\right)}{\Gamma\left(\frac{s}{\mu}+1\right)}\Phi_1\left[\frac{s}{\mu},1-(m-2),1,(1-z),\frac{\lambda}{\mu}(1-z)\right].$$

Now, applying the Laplace transform on (5.9.1.23) and using (5.9.1.26), we obtain:

$$Q^*(z,s) = e^{-\frac{\lambda}{\mu}(1-z)}\left\{(m-1)P_{m-1}^*(s)\right.$$

$$\times\frac{1}{\mu}\frac{\Gamma\left(\frac{s}{\mu}\right)}{\Gamma\left(\frac{s}{\mu}+1\right)}\Phi_1\left[\frac{s}{\mu},1-(m-2),1,(1-z),\frac{\lambda}{\mu}(1-z)\right]$$

$$\left.-\frac{\lambda}{\mu}P_{m-2}^*(s)\frac{\Gamma\left(\frac{s}{\mu}\right)}{\Gamma\left(\frac{s}{\mu}+1\right)}\times\Phi_1\left[\frac{s}{\mu},-(m-1),1,(1-z),\frac{\lambda}{\mu}(1-z)\right]\right\}.$$

(5.9.1.27)

From the definition of the generating function, $P_n(t)$ is the coefficient of z^n in $Q(z, t)$. Thus, $P_{m-2}^*(s)$ can be obtained, in terms of $P_{m-1}^*(s)$ by $m - 2$ times differentiating $Q^*(z, s)$ and evaluating at $z = 0$. That is,

$$P_{m-2}^*(s) = \frac{1}{(m-2)!} \frac{\partial \delta^{m-2} Q^*(z, s)}{\partial z^{m-2}} \bigg|_{z=0}. \tag{5.9.1.28}$$

Then, substituting $P_{m-2}^*(s)$ from (5.9.1.28) in (5.9.1.27), we obtain $P_n^*(s)$ for $0 \le n \le m - 2$ in terms of $P_{m-1}^*(s)$. Substituting the expressions obtained in (5.9.1.13) yields $P_{m-1}^*(s)$.

Now, let

$$g(z) = \Phi_1 \left[\frac{s}{\mu}, -(m-2), 1, (1-z), \frac{\lambda}{\mu}(1-z) \right] \quad \text{and} \quad f(z) = e^{-\frac{\lambda}{\mu}(1-z)}. \tag{5.9.1.29}$$

From the Leibniz differentiation theorem, we obtain:

$$\frac{1}{(m-2)!} \frac{\partial^{m-2}}{\partial z^{m-2}} f(z)g(z) \bigg|_{z=0} = \frac{1}{(m-2)!} \sum_{k=0}^{m-2} \binom{m-2}{k} \left(\frac{\lambda}{\mu} \right)^{m-k-2} e^{-\frac{\lambda}{\mu}}$$
$$\times \frac{d^n}{dz^n} \Phi_1 \left[\frac{s}{\mu}, -(m-2), 1, (1-z), \frac{\lambda}{\mu}(1-z) \right]_{z=0}. \tag{5.9.1.30}$$

Hence, from (5.9.1.30), (5.9.1.13), and after some algebra and simplification, we will have:

$$P_{m-2}^*(s) = \frac{\Gamma\left(\frac{s}{\mu}\right)(m-1)}{(m-2)!\Gamma\left(\frac{s}{\mu}+1\right)} P_{m-1}^*(s) \sum_{k=0}^{m-2} \binom{m-2}{k} \left(\frac{\lambda}{\mu} \right)^{m-k-2} e^{-\frac{\lambda}{\mu}}$$

$$\times \left\{ \frac{\frac{dk}{dz^k}\left(\Phi_1\left[\frac{s}{\mu}, -(m-2), 1, (1-z), \frac{\lambda}{\mu}(1-z)\right]\right)_{z=0}}{1 + \frac{\lambda}{\mu} \frac{\Gamma\left(\frac{s}{\mu}\right)}{(m-2)!\Gamma\left(\frac{s}{\mu}+1\right)} \sum_{k=0}^{m-2} \binom{m-2}{k} \left(\frac{\lambda}{\mu} \right)^{m-k-2} e^{-\frac{\lambda}{\mu} \frac{d^k \Phi_1}{dz^k}}\bigg|_{z=0}} \right\}. \tag{5.9.1.31}$$

For $n \le m - 2$:

$$
P_n^*(s) = \frac{\Gamma\left(\frac{s}{\mu}\right)}{n!\,\Gamma\left(\frac{s}{\mu}+1\right)} \Bigg\{ (m-1) P_{m-1}^*(s) \sum_{k=0}^{n} \binom{n}{k}\left(\frac{\lambda}{\mu}\right)^{n-k} e^{-\frac{\lambda}{\mu}}
$$

$$
\times \frac{d^n}{dz^n} \Phi_1\left[\frac{s}{\mu}, -(m-n), 1, (1-z), \frac{\lambda}{\mu}(1-z)\right]_{z=0}
\tag{5.9.1.32}
$$

$$
-\frac{\lambda}{\mu} P_{m-2}^*(s) \sum_{k=0}^{n} \binom{n}{k}\left(\frac{\lambda}{\mu}\right)^{n-k} e^{-\frac{\lambda}{\mu}}
$$

$$
\times \frac{d^n}{dz^n} \Phi_1\left[\frac{s}{\mu}, (m-1), 1, (1-z), \frac{\lambda}{\mu}(1-z)\right]_{z=0} \Bigg\},
$$

$$
P_{m-1}^*(s) = \frac{\text{Numerator}}{\text{Denominator}},
\tag{5.9.1.33}
$$

where

$$
\text{Numerator} = \frac{z_2^{i+1}}{1-z_2},
$$

and

$$
\text{Denominator} = z_2^{m-1} + \frac{\Gamma\left(\frac{s}{\mu}\right)}{\Gamma\left(\frac{s}{\mu}+1\right)} \sum_{k=0}^{m-3} (m-k) z_2^k \frac{1}{k!} \times \Bigg\{ (m-1) \sum_{j=0}^{k} \binom{k}{j}\left(\frac{\lambda}{\mu}\right)^{k-j} e^{-\frac{\lambda}{\mu}}
$$

$$
\times \frac{d^n}{dz^n} \Phi_1\left[\frac{s}{\mu}, -(m-2), 1, (1-z), \frac{\lambda}{\mu}(1-z)\right]_{z=0}
$$

$$
-\frac{\lambda}{\mu} \frac{P_{m-2}^*(s)}{P_{m-1}^*(s)} \sum_{j=0}^{k} \binom{k}{j}\left(\frac{\lambda}{\mu}\right)^{k-j} e^{-\frac{\lambda}{\mu}} \frac{d^j \Phi_1}{dz^j}\Bigg]_{z=0} \Bigg\} + 2 z_2^{m-2} \frac{P_{m-2}^*(s)}{P_{m-1}^*(s)}.
\tag{5.9.1.34}
$$

$P_n^*(s)$ for $n \ge m$ can be obtained as the coefficients of z^n from the generating function and again by applying the Leibniz theorem. Hence, we have the following:

For $n \ge m$:

$$
P_n^*(s) = \frac{\mu}{\lambda} \left[\sum_{k=0}^{m-1} (m-k) P_k^*(s) \frac{\lambda}{\mu\mu z_2^{n-k-2}} \frac{1 - \left(\dfrac{z_2}{z_1}\right)^{n-k+1}}{1 - \dfrac{z_2}{z_1}} \right.
$$

$$
\left. - (m-k) P_k^*(s) \frac{\lambda}{\mu\mu z_z^{n-i-3}} \frac{1 - \left(\dfrac{z_2}{z_1}\right)^{n-k}}{1 - \dfrac{z_2}{z_1}} \right],
$$

(5.9.1.35)

where the expression for $P_k^*(s)$ when $0 \le k \le m-1$ should be used from what was previously found.

5.9.2. Stationary Queue Length Distribution for *M/M/m*

The limiting distribution of the number of tasks in the system at time t, $\xi(t)$, can be obtained from the system of differential–difference equation (5.9.1.9) by passing to limit and dropping t from the terms involved. Hence, derivatives with respect to t become zero and by moving terms around, we obtained the system of stationary balance equations. The solution of such a system is the distribution we are looking for and we state this in the following theorem.

Theorem 5.9.2.1
If $\lambda < m\mu$, then the process $\{\xi(t), 0 \le t < \infty\}$ has a unique stationary distribution $\lim_{t\to\infty} P\{\xi(t) = n\} = P_n$, $n = 0, 1, 2, \dots$, given by:

$$
P_n = \begin{cases} \dfrac{\left(\dfrac{\lambda}{\mu}\right)^n}{n!} P_0, & \text{if, } 1 \le n \le m \\[4mm] \dfrac{\left(\dfrac{\lambda}{m\mu}\right)^n m^m}{m!} P_0, & \text{if, } n \ge m \end{cases},
$$

(5.9.2.1)

where P_0 is given by:

$$
P_0 = \frac{1}{\displaystyle\sum_{n=0}^{m-1} \left[\dfrac{\left(\dfrac{\lambda}{\mu}\right)^n}{n!} \right] + \dfrac{\left(\dfrac{\lambda}{\mu}\right)^m}{m!\left(1 - \dfrac{\lambda}{m\mu}\right)}}.
$$

(5.9.2.2)

If $\lambda > m\mu$, then the stationary distribution does not exist and $P_n = 0$, $\forall n$.

Note that (5.9.2.1) holds for $m \geq 1$. For instance, if $m = 1$, then (5.9.2.1) reduces the stationary queue length for $M/M/1$ as $(1 - \rho)\rho^n, n = 0, 1, 2, \dots$, where $\rho = \lambda/\mu$ (see 5.5.2.1).

Proof

Using the system of equation (5.9.1.9), passing to the limit as $t \to \infty$, and moving terms around, we will have:

$$\begin{cases} \lambda P_0 = \mu P_1 \\ (\lambda + n\mu) P_n = \lambda P_{n-1} + (n+1)\lambda P_{n+1}, & 1 \leq n < m, \\ (\lambda + m\mu) P_n = \lambda P_{n-1} + m\mu P_{n+1}, & n \geq m, \end{cases} \quad (5.9.2.3)$$

with $P_i(0) = \delta_{ij}$, where δ_{ij} is the Kronecker delta, which is defined in (5.5.18).

The system (5.9.2.3) can easily be solved by recursive method as we did in the $M/M/1$ case. We leave the detail of the iterative method and completion of the proof as an exercise.

Notes:

1. Let $\rho(\lambda/m\mu)$. Then, ρ is the traffic intensity for the system $M/M/m$.
2. When $n \geq m$, the system $M/M/m$ may be regarded as an $M/M/1$ with service rate $m\mu$.
3. A task has to wait for service only if all servers are busy. The probability of such an event is $\sum_{n=m}^{\infty} P_n$. Thus, we have:

$$P\{\text{a task waits for service}\} = \frac{\left(\dfrac{\lambda}{\mu}\right)^m}{m!} \times \frac{1}{1 - \dfrac{\lambda}{m\mu}} \times \frac{1}{\left[\displaystyle\sum_{n=0}^{m-1} \frac{\left(\dfrac{\lambda}{\mu}\right)^n}{n!}\right] + \dfrac{\left(\dfrac{\lambda}{\mu}\right)^m}{m!\left(1 - \dfrac{\lambda}{m\mu}\right)}}.$$

$$(5.9.2.4)$$

Relation (5.9.2.4) is called *Erlang's delay formula* or *Erlang's second formula*.

Now let L and L_q, as before, denote the mean stationary queue length, that is, the number in the system (including in services) and in the waiting line (only), respectively. Then, from (5.9.2.1), (5.9.2.2), geometric series and its derivative, and some algebra, we have:

$$L = \sum_{n=0}^{\infty} n P_n = \left[\sum_{n=1}^{m} n \frac{\left(\frac{\lambda}{\mu}\right)^n}{n!} + \sum_{n=m+1}^{\infty} n \left(\frac{\lambda}{m\mu}\right)^{n-m} \frac{\left(\frac{\lambda}{\mu}\right)}{m!} \right] P_0$$

$$= \left[\frac{\lambda}{\mu} \sum_{n=1}^{m} \frac{\left(\frac{\lambda}{\mu}\right)^{n-1}}{(n-1)!} + \frac{\left(\frac{\lambda}{\mu}\right)^m}{m!} \sum_{n-m=1}^{\infty} n \left(\frac{\lambda}{m\mu}\right)^{n-m} \right] P_0$$

$$= \left[\frac{\lambda}{\mu} \sum_{k=0}^{m-1} \frac{\left(\frac{\lambda}{\mu}\right)^k}{k!} + \frac{\left(\frac{\lambda}{\mu}\right)^m}{m!} \sum_{k=1}^{\infty} (k+m) \left(\frac{\lambda}{m\mu}\right)^k \right] P_0$$

$$= \left[\frac{\lambda}{\mu} \sum_{k=0}^{m-1} \frac{\left(\frac{\lambda}{\mu}\right)^k}{k!} + \frac{\left(\frac{\lambda}{\mu}\right)^m}{m!} \left(\frac{\frac{\lambda}{m\mu}}{\left(1-\frac{\lambda}{n\mu}\right)^2} + \frac{\frac{\lambda}{\mu}}{\left(1-\frac{\lambda}{m\mu}\right)} \right) \right] P_0$$

$$= \frac{\left(\frac{\lambda}{m\mu}\right)\left(\frac{\lambda}{\mu}\right)^m P_0}{m! \left(1-\frac{\lambda}{m\mu}\right)^2} + \frac{\lambda}{\mu} \left[\sum_{k=0}^{m-1} \frac{\left(\frac{\lambda}{\mu}\right)^k}{k!} + \frac{\left(\frac{\lambda}{\mu}\right)^m}{m! \left(1-\frac{\lambda}{m\mu}\right)} \right] P_0$$

$$= \frac{\lambda}{\mu} + \frac{\left(\frac{\lambda}{m\mu}\right)\left(\frac{\lambda}{\mu}\right)^m}{m! \left(1-\frac{\lambda}{m\mu}\right)^2} P_0,$$

(5.9.2.5)

and

$$L_q = \sum_{n=m+1}^{\infty} (n-m) \frac{\left(\frac{\lambda}{\mu}\right)}{m!} \left(\frac{\lambda}{m\mu}\right)^{n-m} P_0$$

$$= \frac{\left(\frac{\lambda}{\mu}\right)^m}{m!} P_0 \sum_{n-m=1}^{\infty} (n-m) \left(\frac{\lambda}{m\mu}\right)^{n-m} = \frac{\left(\frac{\lambda}{\mu}\right)^m}{m!} P_0 \sum_{k=1}^{\infty} k \left(\frac{\lambda}{m\mu}\right)^k$$

(5.9.2.6)

$$= \frac{\left(\frac{\lambda}{\mu}\right)^m \frac{\lambda}{m\mu}}{m! \left(1-\frac{\lambda}{m\mu}\right)^2} P_0,$$

where P_0 is given by (5.9.2.2).

5.9.3. Stationary Waiting Time of a Task in *M/M/m* Queue

Chan and Lin (2003) recently presented a new method of analysis of waiting time distribution for the transient *M/M/m* queueing system case. They have shown that the conditional waiting time obeys an Erlang distribution with rate $m\mu$, where μ is the service rate of a server. The authors obtained an explicit closed-form solution by means of the probability density function of the Erlang distribution. The derivation of the result proved to be very simple. They have pointed out the significance of Khintchine's method and its close relation to their proposed method. They have also shown that the waiting time distribution can be obtained from Takács's waiting time distribution for the *G/M/m* queue as a special case. This reveals some insight into the significance of Takács's more general but rather complex result.

This method and other properties of the transient *M/M/m*'s waiting time distribution has been discussed in Haghighi and Mishev (2013). Hence, we discuss only the stationary case here, that is, the waiting time of a task as $t \to \infty$. In doing that, we note that there will only be a waiting time if the number of tasks in the system at the arrival of a task is the same as the number of servers or more. In that case, the interdeparture times are independent and identically distributed random variables with exponential are distributed with parameter $m\mu$.

As before, we recognize two random variables T and T_q to represent the waiting time in the system and in the waiting line, respectively. Of course, the waiting time of a task in the system includes the service time of the task, and is referred to as the sojourn time. Thus, let the functions $F(t) = P\{T \le t\}$ and $F_q(t) = P\{T_q \le t\}$ be the probability distributions functions of T and T_q, respectively. If no task is waiting in line, it would mean that the number of tasks in the system is less than the number of servers. In other words,

$$F_q(0) = P\{T_q = 0\} = P\{\text{number of tasks in the system} < m\}$$

$$= \sum_{n=0}^{m-1} P_n = \sum_{n=0}^{m-1} \left[\frac{\left(\frac{\lambda}{\mu}\right)^n}{n!} \right] P_0. \tag{5.9.3.1}$$

Substituting P_0 from (5.9.2.2) into (5.9.3.1) we will have:

$$F_q(0) = 1 - \frac{\left(\frac{\lambda}{\mu}\right)^m}{m!\left(1 - \frac{\lambda}{m\mu}\right)} P_0. \tag{5.9.3.2}$$

Instead of the distribution of the waiting time $F(t) = P\{T \le t\}$, it is more convenient to compute $P\{T > t\} = 1 - F(t)$. To compute $P\{T > t\}$, we consider the conditional probability of $P\{T > t\}$ given that an arriving task at epoch τ_n finds the system in state n, that is, $P\{T > t|\xi(\tau_n) = n\}$, where $\xi(t)$ is the random variable denoting the number of tasks in the system at time t. Then, using the law of total probability and properties of Poisson process and some algebra, it can be shown (see Haghighi and Mishev, 2013) that:

$$P\{T > t\} = \sum_{n=0}^{\infty} P_n P\{T > t|\xi(\tau_n) = 0\}$$

$$= \sum_{n=m}^{\infty} \left[\frac{\left(\dfrac{\lambda}{m\mu}\right)^n}{m!\,m^{n-m}} \right] P_0 \int_{y=t}^{\infty} \left[\frac{(m\mu)^{n-m+1}\, y^{n-m}}{(n-m)!} \right] e^{-m\mu y}\, dy$$

$$\tag{5.9.3.3}$$

$$= m\mu P_m \int_{y=t}^{\infty} e^{-(m\mu-\lambda)y}\, dy$$

$$= \left(\frac{mP_m}{m - \dfrac{\lambda}{m\mu}} \right) e^{-(m\mu-\lambda)t}, \quad t \geq 0.$$

To find the mean waiting times in line and in the system, we may apply Little's formula again (see 5.5.3.1 and 5.5.3.2). Hence,

$$W = \frac{L}{\lambda} = \frac{1}{\mu} + \frac{\left(\dfrac{\lambda}{\mu}\right)^m}{m\mu m!\left(1 - \dfrac{\lambda}{m\mu}\right)^2} P_0 \tag{5.9.3.4}$$

and

$$W_q = \frac{L_q}{\lambda} + \frac{\left(\dfrac{\lambda}{\mu}\right)^m}{m\mu m!\left(1 - \dfrac{\lambda}{m\mu}\right)^2} P_0, \tag{5.9.3.5}$$

where P_0 is given by (5.9.2.2).

It would be instructive to see that (5.9.3.5) may be obtained differently (see Bhat, 2008) as follows. If a task finds n, $n = m, m + 1, \ldots$, tasks in the waiting line at its arrival, then those n tasks must complete their services before the new arrival's service starts. That is, we have to have n departures. But interdeparture times are exponential, that is, the departures follow the Poisson process with parameter $m\mu t$. Hence, we have:

$$dF_q = \sum_{n=m}^{\infty} P_n \frac{(m\mu t)^{n-m}}{(n-m)!} m\mu e^{-m\mu t} dt$$

$$= P_m e^{-m\mu t} \sum_{n=m}^{\infty} \left(\frac{\lambda}{m\mu}\right)^{n-m} \frac{(m\mu t)^{n-m}}{(n-m)!} m\mu t dt \qquad (5.9.3.6)$$

$$= m\mu P_m e^{-m\mu\left(1-\frac{\lambda}{m\mu}\right)t} dt.$$

From the iterative method driving (5.9.2.1), we can have:

$$P_n = \left(\frac{\lambda}{m\mu}\right)^{n-m} P_m, \quad n \geq m. \qquad (5.9.3.7)$$

Thus, from (5.9.3.6), (5.9.3.7), and (5.9.3.3), we have:

$$W_q = \int_0^{\infty} t dF_q(t) = \int_0^{\infty} m\mu P_m t e^{-m\mu\left(1-\frac{\lambda}{m\mu}\right)t} dt$$

$$= \frac{P_m}{m\mu\left(1-\frac{\lambda}{m\mu}\right)^2} = \frac{\left(\frac{\lambda}{\mu}\right)^m}{m\mu m!\left(1-\frac{\lambda}{m\mu}\right)^2} P_0, \qquad (5.9.3.8)$$

where P_0 is given by (5.9.2.2).

Example 5.9.3.1

For a system with three servers in parallel, if $\lambda = 3$ and $\mu = 4$, then the probability that the system is empty at any time is $P_0 = 0.47$ and $F_q(0) = 0.96$. This means the probability that an arriving task not to wait in line and its service starts immediately as it arrives is 96%. Of course, there is also the probability that there is at most one server that is busy at any time. The average waiting time for a task in the waiting line is 0.005 unit of time. The average number of tasks in the system and in the waiting line, at any time, is 0.77 and 0.02, respectively. Note that with the given rate values, the traffic intensity is only 0.25 and hence, one should expect the system not to be busy. For instance, the probability of an arriving task finding two tasks in the system (i.e., being served) is 0.13, while finding five tasks (i.e., two tasks in the waiting line) is 0.002.

5.10. MANY-SERVER PARALLEL QUEUES WITH FEEDBACK

5.10.1. Introduction

We consider the same model described in Section 5.9, except for the following. After being served, a task may either leave the system permanently with probability q or immediately rejoin the waiting line (if there is one, otherwise it will join anyone of the idle servers) with probability p, where $p + q = 1$. It is assumed that the return of a task is an event independent of any other event involved in the system and, in particular, independent of the number of its previous returns. We denote the sojourn time of the nth task by θ_n.

Our main interest is finding the distribution and the first two moments of θ_n when $m = 2$ in the case of a stationary process, that is, in the case when θ_n has the same distribution for every $n = 1, 2, \ldots$.

As before, the system is in state E_n at time t if there are n tasks in the system (i.e., in the waiting line and in service). Again, under conditions described, the stochastic process $\{\xi(t), t \geq 0\}$ is a homogeneous Markov process with denumerable state space $S = \{0, 1, 2, \ldots\}$.

5.10.2. Stationary Distribution of the Queue Length

Using the Chapman–Kolmogorov equation, the system of difference equations for the system is as follows:

$$\begin{cases} \lambda P_0 = q\mu P_1 \\ (\lambda + nq\mu) P_n = \lambda P_{n-1} + (n+1)q\mu P_{n+1}, & 1 \leq n < m, \\ (\lambda + mq\mu) P_n = \lambda P_{n-1} + mq\mu P_{n+1}, & n \geq m, \end{cases} \quad (5.10.2.1)$$

with $P_i(0) = \delta_{ij}$, where δ_{ij} is the Kronecker delta, which is defined in (5.5.18). As it can be observed, the system (5.10.2.1) is the same as (5.9.2.1), except μ has been replaced by $q\mu$. Hence, the solution can be obtained as:

Theorem 5.10.2.1
If $\lambda < mq\mu$, then the process $\{\xi(t), t \geq 0\}$ has a unique stationary distribution $\lim_{t \to \infty}\{\xi(t) = n, n = 0, 1, \ldots\} = P_n$, given by:

$$P_n = \begin{cases} \dfrac{\left(\dfrac{\lambda}{q\mu}\right)}{n!} P_0, & \text{if, } 1 \leq n \leq m, \\[4mm] \dfrac{\left(\dfrac{\lambda}{mq\mu}\right)^n m^m}{m!} P_0, & \text{if, } n \geq m, \end{cases} \quad (5.10.2.2)$$

where P_0 is given by:

$$P_0 = \cfrac{1}{\left[\sum_{n=0}^{m-1} \cfrac{\left(\dfrac{\lambda}{q\mu}\right)^n}{n!}\right] + \cfrac{\left(\dfrac{\lambda}{q\mu}\right)^m}{m!\left(1 - \dfrac{\lambda}{mq\mu}\right)}}.$$ (5.10.2.3)

If $\lambda \geq mq\mu$, then stationary distribution does not exist and $P_0 = 0$, $\forall n$.

5.10.3. Stationary Waiting Time of a Task in Many-Server Queue with Feedback

The argument for the waiting time distribution for the case of feedback multiserver is similar to the one for the multiserver without feedback. Hence, we will eliminate part of the argument and get to the results (see Haghighi and Mishev, 2013). Suppose that the nth arriving task finds ζ_n tasks in the system. Thus, if $\zeta_n = j < m$, then its service begins immediately. However, if $\zeta_n = j \geq m$, then the task has to wait not only for all tasks in line in front of it to complete their services, but for one server to be idle to take it to serve, that is, it has to wait for $j + 1 - m$ tasks to complete their services and exit the service station before its service starts. Again, as before, since interdeparture times are iid exponential, the outputs from the service station follow a Poisson process with an average rate of $m\mu$. Thus, the distribution of $j + 1 - m$ successive outputs will be the $(j + 1 - m)$th-fold convolution of $1 - e^{-m\mu t}$ with itself, which is a gamma distribution with density function $[(m\mu t)^{j-m}/(j - m)!]m\mu e^{-m\mu t}$.

Let the two random variables T and T_n represent the waiting time in the system and the nth arriving task, respectively. Of course, the waiting time of a task in the system means waiting for $j + 1 - m$ successive outputs. We also let the functions $F(t) = P\{T \leq t\}$ and $F_n(t) = P\{T_n \leq t\}$ be the probability distributions functions of T and T_n, respectively. Note that $\lim_{n \to \infty} P\{T_n < t\} = P\{T \leq t\} = F(t)$. Thus, we will have the following equation and the theorem that follows it:

$$T(t) = \sum_{j=0}^{m-1} P_j + \sum_{j=m}^{\infty} P_j \int_0^t \frac{(m\mu x)^{j-m}}{(j-m)!} m\mu e^{-m\mu x} dx, \quad \frac{\lambda}{m\mu q} < 1.$$ (5.10.3.1)

Theorem 5.10.3.1

If $\lambda < mq\mu$, then:

1.

$$T(t) = 1 - \cfrac{\left(\dfrac{\lambda}{mq\mu}\right)^m P_0}{m!\left(1 - \dfrac{\lambda}{mq\mu}\right)} e^{-m\mu\left(1 - \frac{\lambda}{mq\mu}\right)t},$$ (5.10.3.2)

which is independent of the order services, and P_0 is given by:

$$P_0 = \cfrac{1}{\left[\displaystyle\sum_{n=0}^{m-1} \cfrac{\left(\cfrac{\lambda}{q\mu}\right)^n}{n!}\right] + \cfrac{\left(\cfrac{\lambda}{q\mu}\right)^m}{m!\left(1-\cfrac{\lambda}{mq\mu}\right)}}. \tag{5.10.3.3}$$

If $\lambda \geq mq\mu$, then stationary distribution does not exist and $P_0 = 0$, $\forall n$. The mean waiting time of the arriving task is:

2.

$$E(T_n) = \frac{1}{m\mu}\sum_{j=m}^{\infty}(j+1-m)P(T_n = j) = \cfrac{\left(\cfrac{\lambda}{q\mu}\right)^m P_0}{m\mu m!\left(1-\cfrac{\lambda}{mq\mu}\right)^2}. \tag{5.10.3.4}$$

3. The mean sojourn time of the first cycle for the nth task, denoted by $E\{\theta_n^{(1)}\}$, is the sum of its waiting time and its first service time. Thus, the mean sojourn time for the first cycle equals the expected value of the waiting time plus the expected value of the service time. Then,

$$E\{\theta_n^{(1)}\} \cfrac{\left(\cfrac{\lambda}{q\mu}\right)^m P_0}{m\mu m!\left(1-\cfrac{\lambda}{mq\mu}\right)^2} + \frac{1}{\mu}. \tag{5.10.3.5}$$

5.11. MANY-SERVER QUEUES WITH BALKING AND RENEGING

In this section, we will study a many-server queueing model with balking and reneging. We will obtain the stationary distribution of the number of tasks in the system.

Multiple-server queues with balking and reneging have been studied in the past four decades. See for instance, Montazer-Haghighi (1976, 1981) and Montazer-Haghighi et al. (1986). Balking and reneging in operating the orders in a production system could be interpreted as cancellation of some orders before or after those orders were placed.

We start with two special cases in Section 5.11.1 and then the general case in Section 5.11.2.

5.11.1. Priority *M/M/*2 with Constant Balking and Exponential Reneging

We want to consider the following model. There are two identical service stations, each consisting of one server, set in parallel. Each station has a finite

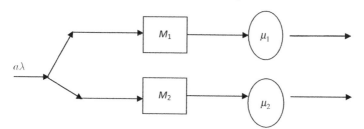

Figure 5.11.1. A parallel two-station single-server system.

buffer with capacities $M_1 - 1$ and $M_2 - 1$, respectively, so that the total capacities of the stations are M_1, and M_2, respectively, and the total capacity of the system, hence, is $M = M_1 - M_2$.

Tasks arrive to the system from an infinite source probabilistically according to the Poisson distribution with mean λ. The service times at each station are *iid* random variables having negative exponential distributions with means $1/\mu_i, i = 1, 2$, where $\mu_i, i = 1, 2$, are positive real numbers. At each station, units will be served on a FIFO basis.

The system may lose some of its potential arrivals because tasks may balk or renege. The balking and reneging tasks, coupled with those who attend and finally obtain service, constitute potential arrivals. If an arriving task finds both servers busy at the time of its arrival, it will attend the system with a constant probability $a, 0 \le a \le 1$, that is, it will balk with probability $b = 1 - a$. See Figure 5.11.1.

While a task is waiting for service, it may decide to leave the system; that is, it may renege. The length of time a task may wait in a station before it reneges is assumed to be a random variable. Distributions of the waiting times of reneging tasks are negative exponential distribution with means $1/\alpha_i, i = 1, 2$, where α_i, $i = 1, 2$, are positive real numbers. Reneging cannot occur for the tasks in service. Thus, the average number of reneged tasks from the system while servers are busy will be $\alpha_1 + \alpha_2$. If a task leaves the system due to balking, reneging, or completing its service and rejoining the system, it will be considered as a new arrival, independent of the previous balking, reneging, or being served.

Let us assume that station 1 has the *priority of receiving the arrivals*. That is, if a task arrives, it must join station 1 unless there are already M_1 in that station. In that case, the arriving task will try attending station 2 unless there are already M_2 in that station. In this case, the arriving task will be lost. Thus, loss to the system will be due to balking, reneging, or both stations being full.

Due to the priority restriction, it is possible that at some point, any of the two servers becomes idle, and naturally the buffer of the station with the idle server must be empty while there is a task in the other buffer waiting to be served. Thus, there are two options:

(1) let the system be as described and not bother with this idleness or
(2) avoid this situation and let the next task move to the other station with the idle server. Hence, if at any time there are two or more tasks in the

system, neither of the servers will be idle. Note that this case is different from jockeying. Also note that in this case, states $(2, 0)$ and $(0, 2)$ are the same as $(1, 1)$.

We choose option (2) and set up the model mathematically. Let $P_{m1,m2}$ denote the stationary probability of having m_1 tasks in the first station and m_2 in the second.

Before we discuss the model in general, we note a very special case in which the service rates are equal, neither station has a buffer, and there is neither balking nor reneging. That is, $\mu_1 = \mu_2 = \mu$, $M_1 = M_2 = 1$, $a = 1$, and $r_1 = r_2 = 0$. Hence, the model is an *M/M/2 loss system*. We leave it as an exercise to show that by letting the traffic intensity be denoted by $\rho = \lambda/\mu$, the distribution of the queue size in this very particular case is as follows:

$$P_{0,0} = \frac{2}{\rho^2 + 2\rho + 2}, \qquad P_{0,1} = \frac{\rho^2}{(\rho^2 + 2\rho + 2)(1+\rho)},$$

$$P_{1,0} = \frac{\rho(2+\rho)}{(\rho^2 + 2\rho + 2)(1+\rho)}, \qquad P_{1,1} = \frac{\rho^2}{\rho^2 + 2\rho + 2}.$$

We now go back to our model with unequal service rates, capacities as described, and priority as mentioned. For the number of tasks in the system decrease, we either have to have a service completed or a reneging occurred. The system of balance equations is as follows:

1. $\lambda P_{0,0} = \mu_1 P_{1,0} + \mu_2 P_{0,1}$,
2. $(\lambda + \mu_1)P_{1,0} = \lambda P_{0,0} + \mu_2 P_{1,1}$,
3. $(\lambda + \mu_2)P_{0,1} = \mu_1 P_{1,1}$,
4. $(a\lambda + \mu_1 + \mu_2)P_{1,1} = \lambda P_{0,1} + (\alpha_1 + \mu_1 + \mu_2)P_{2,1} + (\mu_1 + \alpha_2 + \mu_2)P_{1,2}$,
5. $(a\lambda + \mu_1 + (m_2 - 1)\alpha_2 + \mu_2)P_{1,m_2} = (\alpha_1 + \mu_1)P_{2,m_2}$
   $\quad + (\mu_1 + m_2\alpha_2 + \mu_2)P_{1,m_2+1}, \quad 2 \le m_2 \le M_2 - 1$,
6. $[a\lambda + \mu_1 + (M_2 - 1)\alpha_2 + \mu_2]P_{1,M_2} = (\alpha_1 + \mu_1)P_{2,M_2}$,
7. $[a\lambda + (m_1 - 1)\alpha_1 + \mu_1 + \mu_2]P_{m_1,1} = (a\lambda P_{m_1-1,1} + (\mu_1\alpha_1 + \mu_1 + \mu_2)P_{m_1+1,1}$
   $\quad + (\alpha_2 + \mu_2)P_{m_1,2}, \quad 2 \le m_1 \le M_1 - 1$,
8. $[a\lambda + (m_1 - 1)\alpha_1 + \mu_1 + (m_2 - 1)\alpha_2 + \mu_2]P_{m_1,m_2}$
   $\quad = a\lambda P_{m_1-1,m_2} + (m_1\alpha_1 + \mu_1)P_{m_1+1,m_2} + (m_2\alpha_2 + \mu_2)P_{m_1,m_2+1}$,
   $\quad 2 \le m_1 \le M_1 - 1, \quad 2 \le m_2 \le M_2 - 1$,
9. $[a\lambda + m_1\alpha_1 + \mu_1 + (M_2 - 1)\alpha_2 + \mu_2)P_{m_1,M_2}] = a\lambda P_{m_1-1,M_2}$
   $\quad + (m_1\alpha_1 + \mu_1)P_{m_1+1,M_2}, \quad 2 \le m_1 \le M_1 - 1$,
10. $[a\lambda + (M_1 - 1)\alpha_1 + \mu_1 + \mu_2)P_{M_1,1} = a\lambda P_{M_1-1,1} + (\alpha_2 + \mu_2)P_{M_1,2}$,

11. $[a\lambda + (m_1 - 1)\alpha_1 + \mu_1 + (M_2 - 1)\alpha_2 + \mu_2]P_{M_1,m_2} = a\lambda P_{M_1-1,m_2}$

$\qquad + a\lambda P_{M_1,m_2-1} + (m_2\alpha_2 + \mu_2)P_{M_1,m_2+1}, \quad 2 \le m_2 \le M_2 - 1,$

12. $[(M_1 - 1)\alpha_1 + \mu_1 + (M_2 - 1)r_2 + \mu_2]P_{M_1,M_2} = a\lambda P_{M_1-1,M_2} + a\lambda P_{M_1,M_2-1},$

$$(5.11.1.1)$$

with

$$\sum_{m_1=0}^{M_1} \sum_{m_2=0}^{M_2} P_{m_1,m_2} = 1. \qquad (5.11.1.2)$$

Since the system (5.11.1.1) is a finite system of equations, we can solve it using matrices. We leave the general solution as an exercise. However, in the next section we will solve a special case.

5.11.1.1. Special Case: $M_1 = 1$, $M_2 \ge 3$
We now consider the case where there is no buffer in station 1, that is, at any time, there may be no task or one task in that station and hence there will not be any reneging from station 1. Thus, the system of balance equation (5.11.1.1) reduces to:

1. $\lambda P_{0,0} = \mu_1 P_{1,0} + \mu_2 P_{0,1},$
2. $(\lambda + \mu_1)P_{1,0} = \lambda P_{0,0} + \mu_2 P_{1,1},$
3. $(\lambda + \mu_2)P_{0,1} = \mu_1 P_{1,1},$
4. $(a\lambda + \mu_1 + \mu_2)P_{1,1} = \lambda P_{0,1} + \lambda P_{1,0} + (\mu_1 + \alpha_2 + \mu_2)P_{1,2},$
5. $[a\lambda + \mu_1 + (m_2 - 1)\alpha_2 + \mu_2]P_{1,m_2} = a\lambda P_{1,m_2-1}$

$\qquad + (\mu_1 + m_2\alpha_2 + \mu_2)P_{1,m_2+1}, \quad 2 \le m_2 \le M_2 - 1,$

6. $[\mu_1 + (M_2 - 1)\alpha_2 + \mu_2]P_{1,M_2} = a\lambda P_{1,M_2-1},$

$$(5.11.1.3)$$

with

$$P_{0,0} + P_{0,1} + \sum_{m_2=0}^{M_2} P_{1,m_2} = 1. \qquad (5.11.1.4)$$

Note that the system (5.11.1.3) reduces to:

$$\begin{cases} -a\lambda P_{1,1} + (\mu_1 + \alpha_2 + \mu_2)P_{1,2} = 0, \\ -[a\lambda + \mu_1 + (m_2 - 1)\alpha_2 + \mu_2]P_{1,m_2} + a\lambda P_{1,m_2-1} \\ \quad + (\mu_1 + m_2\alpha_2 + \mu_2)P_{1,m_2+1} = 0, \quad 2 \le m_2 \le M_2 - 1, \\ -[\mu_1 + (M_2 - 1)\alpha_2 + \mu_2]P_{1,M_2} + a\lambda P_{1,M_2-1} = 0. \end{cases} \qquad (5.11.1.5)$$

We refer the reader to Haghighi and Mishev (2013) for the algorithmic solution, unless one wants to try matrix solution.

5.11.1.2. Special Case: $M_1 = 1$, $M_2 \geq 3$ with No Reneging   Now suppose there is no reneging in the system described in the special case considered in 5.11.1.1, that is, $r_1 = 0$ and $r_2 = 0$. The probability distribution of the length of the queue for the system indicating an appropriate station for this case is:

$$P_{0,0} = \frac{1}{S} X_{0,0}, \quad P_{0,1} = \frac{1}{S} X_{0,1}, \quad P_{1,0} = \frac{1}{S} X_{1,0},$$

$$P_{1,n} = \frac{1}{S} X_{1,n} = \frac{1}{S} \left(\frac{a\lambda}{\mu_1 + \mu_2} \right)^{n-1}, \quad n = 1, 2, \cdots, M_2,$$

$$X_{0,0} = \frac{\mu_1 \mu_2 (2\lambda + \mu_1 + \mu_2)}{\lambda^2 (\lambda + \mu_2)}, \quad X_{0,1} = \frac{\mu_1}{\lambda + \mu_2} \quad X_{1,0} = \frac{\mu_2 (2\lambda + \mu_1 + \mu_2)}{\lambda(\lambda + \mu_2)},$$

$$X_{1,1} = 1, \text{ and } S = X_{0,0} + X_{0,1} + X_{1,0} + \sum_{n=1}^{M_2} X_{1,n}.$$

5.11.2. *M/M/m* with Constant Balking and Exponential Reneging

In this section we will study a multiserver Markovian queueing system with constant balking rate and exponential reneging with constant parameter. We will obtain the steady-state distribution of the number of tasks in the system. Further, for a particular case under the condition that a task cannot leave the system once it is in the system, we will obtain an expression for the average loss of tasks during a fixed period of time.

We consider the system with m-servers sitting in parallel. Tasks arrive from a single infinite source in accordance with the Poisson process of parameter λ. If, at its arrival, a task finds all the m-servers busy, it will balk with a constant probability of b and attend with a constant probability a, $a + b = 1$. Thus, $b\lambda$ is the instantaneous balking rate and $a\lambda$ is the arrival rate when all servers are busy. Furthermore, a task may also renege if its service does not begin by a certain time. The length of time that it will wait is a random variable having an exponential distribution with parameter α.

We note that it appears reasonable to assume in certain situations that the balking probability is constant and is not dependent on queue length. Such situations may arise when the queue length is not observable by the task and it has no knowledge of the queue length. For example, one may consider a service facility, such as a telephone exchange, where requests for connections are received over the phone.

Tasks are served on a FIFO basis by one of the m-servers; the service times by each server are *iid* exponential random variables with parameter μ, that is, service rates are equal for all servers. If a task balks or reneges and decides to return later, it will be considered as a new arrival, independent of its previous balking or reneging.

5.11.3. Distribution of the Queue Length for *M*/*M*/*m* System with Constant Balking and Exponential Reneging

Without loss of generality, we suppose that the system starts operating at time $t = 0$ when there is no task waiting in the waiting lines; however, some servers may be busy. As before, we denote the transient state probability that there are n in the system by $P_n(t)$. Then, based on the standard argument we have been offering for different models, the differential–difference system of equations governing the model is as follows:

$$\begin{cases} P_0'(t) = -\lambda P_0(t) + \mu P_1(t), \\ P_n'(t) = -(\lambda + n\mu) P_n(t) + \lambda P_{n-1}(t) + (n+1)\mu P_{n+1}(t), \quad 1 \le n < m, \\ P_m'(t) = -(a\lambda + m\mu) P_m(t) + \lambda P_{m-1}(t) + (m\mu + \alpha) P_{n+1}(t), \\ P_n'(t) = -[a\lambda + m\mu + (n-m)\alpha] P_n(t) + a\lambda P_{n-1}(t) \\ \qquad + [m\mu + (n+1-m)\alpha] P_{n+1}(t), \quad n > m. \end{cases} \tag{5.11.2.1}$$

Assume that the stationary probability distribution, $P_n = \lim_{t\to\infty} P_n(t)$, exists, it will be the solutions of the following system of difference equations:

$$\begin{cases} \lambda P_0 = \mu P_1, \\ (\lambda + n\mu) P_n = \lambda P_{n-1} + (n+1)\mu P_{n+1}, \quad 1 \le n \le m-1, \\ (a\lambda + m\mu) P_m = \lambda P_{m-1} + (m\mu + \alpha) P_{m+1}, \\ a\lambda + m\mu + (n-m)\alpha] P_n = a\lambda P_{n-1} \\ \qquad + [m\mu + (n+1-m)\alpha] P_{n+1}, \quad n \ge m. \end{cases} \tag{5.11.2.2}$$

Using our standard argument and denoting $\varepsilon(t)$ as the queue length at time t, we will have the following theorem:

Theorem 5.11.2.1
If $\lambda < m\mu$, then:

$$P_n = \frac{1}{n!}\left(\frac{\lambda}{\mu}\right)^n P_0, \quad n \le m, \tag{5.11.2.3a}$$

$$P_{m+1} = \frac{1}{m!}\left(\frac{\lambda}{\mu}\right)^m \left(\frac{a\lambda}{m\mu + \alpha}\right) P_0, \tag{5.11.2.3b}$$

$$P_{m+2} = \frac{1}{m!}\left(\frac{\lambda}{\mu}\right)^m \frac{(a\lambda)^2}{(m\mu + \alpha)(m\mu + 2\alpha)} P_0, \tag{5.11.2.3c}$$

$$P_n = \frac{1}{m!}\left(\frac{\lambda}{\mu}\right)^m \frac{(a\lambda)^{n-m}}{\prod_{k=1}^{n-m}(m\mu + k\alpha)} P_0, \quad n > m, \tag{5.11.2.3d}$$

$$P_0 = \left[\sum_{j=0}^{m} \frac{1}{j!} \left(\frac{\lambda}{\mu} \right)^j + \frac{1}{m!} \left(\frac{\lambda}{\mu} \right)^m \sum_{j=1}^{\infty} \frac{(a\lambda)^j}{\prod_{k=1}^{j} (m\mu + k\alpha)} \right]^{-1}. \qquad (5.11.2.3e)$$

$E\{\text{Number of tasks attended}\}$

$$= \left[\left(\frac{\lambda}{\mu} \right)^m \frac{a\lambda}{(m-1)!} \frac{(m+1)\mu - a\lambda}{(m\mu + a\lambda)^2} + \sum_{n=1}^{m} \frac{1}{(n-1)!} \left(\frac{\lambda}{\mu} \right)^n \right] P_0. \qquad (5.11.2.3f)$$

Proof

From the first two equations of (5.11.2.2), we get (5.11.2.3a). From (5.11.2.3a), we find P_{m-1}, P_m and then use the third equation of (5.11.2.2) to find (5.11.2.3b). Putting $n = m + 1$ in the last equation of (5.11.2.2) and using (5.11.2.3a) and (5.11.2.3b), we obtain (5.11.2.3c). Hence, in general for $n > m$, we will have (5.11.2.3d). P_0, (5.11.2.3e), can be obtained from the normalizing condition $\sum_{n=0}^{\infty} P_n = 1$. We leave it as an exercise to show (5.11.2.3f).

The case of queueing with just balking alone can be treated as a particular case of the general case considered with $\alpha \to 0$ or $(1/\alpha) \to \infty$. Letting $\rho = \lambda/m\mu$, we will then have:

$$P_n = \begin{cases} \dfrac{(m\rho)^n}{n!} P_0, & 0 \leq n \leq m \\[3mm] \dfrac{1}{m!} \left(\dfrac{m}{a} \right)^m (a\rho)^n P_0, & n > m \end{cases} \qquad (5.11.2.4)$$

and

$$P_0 = \left[\sum_{j=0}^{m} \frac{(c\rho)^j}{j!} + \frac{am^m \rho^{m+1}}{m!(1 - a\rho)} \right]^{-1}. \qquad (5.11.2.5)$$

5.12. SINGLE-SERVER MARKOVIAN QUEUEING SYSTEM WITH SPLITTING AND DELAYED FEEDBACK

Haghighi et al. (2011b) recently considered a Poisson single-processor model with splitter and feedback with two buffers as waiting rooms, one before the service station and one before the feedback station. In this section, we will discuss this model due to two important features after completion of service of a task, namely splitting and returning with delay to the tail of the queue.

5.12.1. Description of the Model

Let us consider the single-processor model illustrated in Figure 5.12.1. There are two types of inputs: external arrivals and inputs from feedbacks. External tasks arrive according to a Poisson distribution with parameter λ to a waiting

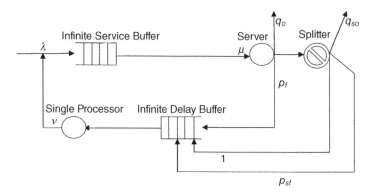

Figure 5.12.1. Single processor with splitting and delayed feedback.

buffer with infinite capacity in front of a single server. The service distribution is exponential with parameter μ. We refer to this server and its buffer as the *main service station*. After a task leaves the main service station, it does one of the following three things:

1. leaves the system with probability q_o, $0 \leq q_o \leq 1$;
2. returns to the end of waiting buffer for further service with probability p_f through a delay station; or
3. proceeds to a splitter with probability p_s, $q_o + p_f + p_s = 1$.

The splitter splits the task into two subtasks: one returns to the end of waiting buffer through the delay buffer and the other either leaves the system with probability q_{so}, or returns to the end of waiting buffer for further service with probability p_{sf}, through the delay buffer, $q_{so} + p_{sf} = 1$. The splitting is immediate, that is, it does not require any delay time.

The *delay station* consists of an infinite capacity buffer and a single delay processor. The movement process from the delay station to the tail of the queue, that is, the main service station, is by batches with exponential distribution with parameter v. The maximum number of tasks in a batch is N, where N is a positive integer. The delay processing starts only if there are at least k tasks in the delay buffer, where k is a known fixed number, $1 \leq k \leq N$, called the *threshold*. Tasks forming a batch feedback into the service buffer in their order of arrival to the delay station.

After the processing at the delay station is completed, the processed batch joins the queue tail of the main service station. We assume that the capacities of both buffers (one in the main service station and one in the delay station) are infinite. Hence, there is no possibility of blocking in the system. It is also assumed that the return of a task is an event independent of any other event involved and, in particular, independent of the number of its returns. Discipline in the main service station is that tasks will be served one at a time in the order that they arrive.

Here, focusing on the case $k = N = 1$, we have a tandem queue with two infinite buffers, Poisson external arrival, single servers at each station, exponential service distribution with parameter μ at the main service station and v at the delay station, and splitter device after the main service station. We will report our findings for the case $1 \le k \le N$ elsewhere.

5.12.2. Analysis

Let the random variables $\xi_1(t)$ and $\xi_2(t)$ denote the number of tasks, at time t, in the main service station, including the one being served, and at the delay station, including those being processed, respectively. Further, let the joint probability of m, $m \ge 0$, tasks at the main service station, including the one being served, and n, $n \ge 0$, tasks at the delay station, including those being processed, at time t be denoted by $\Psi_{m,n}(t)$, that is, $\Psi_{m,n}(t) = P\{\xi_1(t) = m, \xi_2(t) = n\}$. Thus, $P\{\xi_1(t) = m, \xi_2(t) = n\}$ is a finite irreducible Markov chain. Further, let $\Psi_{m,n}(t) = \lim_{t \to \infty} \Psi_{m,n}(t)$ be the steady-state probability of having m tasks in the main service station and n tasks in the delay station. Our first goal is to find the steady-state probability $\Psi_{m,n}$. It is known that our system is not Markovian. See, for example, Brémaud (1978).

Based on the description of the model and from Figure 5.12.1, we see that we have external and internal arrivals to the service station and internal arrivals to the delay station. Let us denote by λ_1 and λ_2 the average arrival rates to the service station and delay station, respectively. Then, since when the system is busy the arrival rates in and out of the delay station are the same, from Figure 5.12.1, it can be seen that:

$$\begin{cases} \lambda_1 = \lambda + \lambda_2, \text{ and} \\ \lambda_2 = \lambda_1 [p_f + p_s(1 + p_f)], \end{cases}$$

from which we will have:

$$\lambda_1 = \frac{\lambda}{1 - [p_f + p_s(1 - p_{sf})]} \qquad (5.12.1)$$

and

$$\lambda_2 = \frac{\lambda [p_f + p_s(1 + p_{sf})]}{1 - [p_f + p_s(1 + p_{sf})]}. \qquad (5.12.2)$$

We can find the distribution of the number of tasks in the system from the following theorem. Theorem 5.12.1 shows the existence of the solution, theoretically. However, the solution of the functional equation mentioned in the theorem is beyond the scope of this book and we refer the readers to Haghighi et al. (2011a). Nevertheless, we will use the theorem to find the expected values of the number of tasks in each station.

Theorem 5.12.1

Let $a = q_o\mu$, $b = (p_f + p_s q_{so})\mu$, $c = p_s p_{sf}\mu (a + b + c = \mu)$ and $d = q_o - p_s p_{sf}$. If $d > 0$, $(\lambda_1/\mu) < 1$, and $(\lambda_2/v) < 1$, then for $k = N = 1$, the stationary joint distribution function of the number of tasks in each station exists and can be obtained by the coefficient of the Maclaurin series expansion of the solution of the functional equation for the generating function of the distribution function:

$$A(w, z)G_1(w, z) = B_1(w, z)G_1(0, z) + B_2(w, z)G_1(w, 0), \quad (5.12.3)$$

where A, B_1, B_2 and $G_1(w, z)$ are defined as:

$$A(w, z) = cz^3 + bz^2 + \left[a + \lambda w^2 - (\lambda + \mu + v)w\right]z + vw^2, \quad (5.12.4)$$

$$B_1(w, z) = cz^3 + bz^2 + (a - \mu w)z, \quad (5.12.5)$$

$$B_2(w, z) = vw^2 - vwz, \quad (5.12.6)$$

and

$$G_1(w, z) = \sum_{m=0}^{\infty}\sum_{n=0}^{\infty} \Psi_{m,n} w^m z^n, \quad |z| < 1, |w| < 1 \quad (5.12.7)$$

is the generating function of the joint distribution function of the number of tasks in each station. The index of G indicates the value of k, which in this case is 1. Furthermore, the expected number of tasks at each station is found using the algorithm given in the proof.

Proof

From the description of the model, the system of stationary balance difference equations in this case is as follows:

(i) $\quad \lambda\Psi_{0,0} = a\Psi_{1,0}$ (5.12.8)

(ii) $\quad (\lambda + v)\Psi_{0,1} = a\Psi_{1,1} + b\Psi_{1,0}$

(iii) $\quad (\lambda + v)\Psi_{0,n} = a\Psi_{1,n} + b\Psi_{1,n-1} + c\Psi_{1,n-2}, n \geq 2$

(iv) $\quad (\lambda + \mu)\Psi_{m,0} = \lambda\Psi_{m-1,0} + a\Psi_{m+1,0} + v\Psi_{m-1,1}, m \geq 1$

(v) $\quad (\lambda + \mu + v)\Psi_{m,1} = \lambda\Psi_{m-1,1} + a\Psi_{m+1,1} + b\Psi_{m+1,0} + v\Psi_{m-1,2}, m \geq 1$

(vi) $\quad (\lambda + \mu + v)\Psi_{m,n} = \lambda\Psi_{m-1,n} + a\Psi_{m+1,n} + b\Psi_{m+1,n-1} + c\Psi_{m+1,n-2} + v\Psi_{m-1,n+1}$,
$\quad m \geq 1, n \geq 1$

(vii) $\quad \sum_{m=0}^{\infty}\sum_{n=0}^{\infty} \Psi_{m,n} = 1.$

We note that by letting $q_o = 1$, the system of equation (5.12.8) reduces to that of $M/M/1$. Moreover, by letting $p_s = 0$, it reduces to the one by Kleinrock and Gail (1996), with the exception of the delayed distribution. Furthermore, if

$p_s = 0$ the system (5.12.8) reduces to the one by Jackson (1957), with the exception of allowing exit in deciding to feedback.

Now, we go back to the proof. We define

$$G_1(w, z) = \sum_{m=0}^{\infty} \sum_{n=0}^{\infty} \Psi_{m,n} w^m z^n \qquad (5.12.9)$$

as the generating function of the joint steady-state probability distribution $\Psi_{m,n}$. Let:

$$F_m(z) = \sum_{n=0}^{\infty} \Psi_{m,n} z^n, \quad m = 1, 2, \ldots. \qquad (5.12.10)$$

Then, the generating function defined in (5.12.9) can be rewritten as:

$$G_1(w, z) = \sum_{m=0}^{\infty} F_m(z) w^m. \qquad (5.12.11)$$

We now apply (5.12.10) on (5.12.8). Thus, from (iii) of system (5.12.8) we will have:

$$(\lambda + v)[F_0(z) - \Psi_{0,0} - z\Psi_{0,1}] = a[F_1(z) - \Psi_{1,0} - z\Psi_{1,1}]$$
$$+ bz[F_1(z) - \Psi_{1,0}] + cz^2 F_1(z)$$

or

$$(\lambda + v)F_0(z) - (a + bz + cz^2)F_1(z) = v\Psi_{0,0}. \qquad (5.12.12)$$

Now from (vi) of (5.12.8), after some algebra, we will have:

$$(\lambda + \mu + v)[F_m(z) - \Psi_{m,0} - z\Psi_{m,1}] = \lambda[F_{m-1}(z) - \Psi_{m-1,0} - z\Psi_{m-1,1}]$$
$$+ a[F_{m+1}(z) - \Psi_{m+1,0} - z\Psi_{m+1,1}]$$
$$+ bz[F_{m+1}(z) - \Psi_{m+1,0}]$$
$$+ cz^2 + vF_{m+1}(z)$$
$$+ \frac{v}{z}[F_{m-1}(z) - \Psi_{m-1,0} - z\Psi_{m-1,1} - z^2\Psi_{m-1,2}],$$
$$m \geq 1. \qquad (5.12.13)$$

Relation (5.12.13) can be reduced to:

$$\left(\lambda + \frac{v}{z}\right)F_{m-1}(z) - (\lambda + \mu + v)F_m(z) + (a + bz + cz^2)F_{m+1}(z)$$
$$= \frac{v}{z}\Psi_{m-1,0} - v\Psi_{m,0}, \quad m \geq 1. \qquad (5.12.14)$$

Summing (5.12.14) over m gives:

$$\left[\left(\lambda+\frac{v}{z}\right)w-(\lambda+\mu+v)+\frac{1}{w}(a+bz+cz^2)\right]G_1(w,z)$$

$$=\left[\frac{1}{w}(a+bz+cz^2)-\mu\right]F_0(z) \tag{5.12.15}$$

$$+\left(\frac{vw}{z}-v\right)(\Psi_{0,0}+\Psi_{1,0}w+\Psi_{2,0}w^2+\cdots).$$

Since $G_1(0,z)=\sum_{n=0}^{\infty}\Psi_{0,n}z^n=F_0(z)$ and $G_1(w,0)=\sum_{m=0}^{\infty}\Psi_{m,0}w^m$, equation (5.12.15) can be rewritten as:

$$\left[\left(\lambda+\frac{v}{z}\right)w-(\lambda+\mu+v)+\frac{1}{w}(a+bz+c^2)\right]G_1(w,z)$$

$$=\left[\frac{1}{w}(a+bz+cz^2)-\mu\right]G_1(0,z)+\left(\frac{vw}{z}-v\right)G_1(w,0) \tag{5.12.16}$$

or

$$\{cz^3+bz^2+[q+\lambda w^2-(\lambda+\mu+v)w]z+vw^2\}G_1(w,z)$$

$$=[cz^3+bz^2+(a-\mu w)z]G_1(0,z)+(vw^2-vwz)G_1(w,0), \tag{5.12.17}$$

from which we have (5.12.3).

We note that $A(w,z)$ in (5.12.5) is a third-degree polynomial function with respect to z and a second-degree polynomial with respect to w. Since $A(w,z)$ is a cubic nonsingular polynomial with respect to z, it is an elliptic curve and, thus, of *genus one* (roughly speaking, genus is the number of holes in a surface; e.g., a sphere has genus 0, a torus has genus 1).

To prove the remaining part of the theorem, we substitute $z=1$ in (5.12.5)–(5.12.7) and use the fact that $a+b+c=\mu$ to obtain:

$$A(w,1)=c+b+a+\lambda w^2-(\lambda+\mu+v)w+vw^2$$

$$=(1-w)(\mu-\lambda w-vw), \tag{5.12.18}$$

$$B_1(w,1)=\mu(1-w), \tag{5.12.19}$$

and

$$B_2(w,1)=-vw(1-w). \tag{5.12.20}$$

From (5.12.3) and (5.12.18)–(5.12.20), we have:

$$(\mu-\lambda w-vw)G_1(w,1)=\mu G_1(0,1)-vwG_1(w,0). \tag{5.12.21}$$

Now substituting $w = 1$ in (5.12.21) we have:

$$\mu G_1(0, 1) - v G_1(1, 0) = \mu - \lambda - v. \tag{5.12.22}$$

In a similar manner, substituting $w = 1$ in (5.12.4)–(5.12.6), we have:

$$A(1, z) = (z - 1)[cz(z + 1) + bz - v], \tag{5.12.23}$$
$$B_1(1, z) = (z - 1)[cz(z + 1) + bz], \tag{5.12.24}$$

and

$$B_2(1, z) = -v(z - 1), \tag{5.12.25}$$

From (5.12.3) and (5.12.23)–(5.12.25), we have:

$$[cz(z + 1) + bz - v]G_1(1, z) = (z + 1) + bz]G_1(1, 0) - v G_1(1, 0). \tag{5.12.26}$$

Now substituting $z = 1$ in (5.12.26) and using the fact that $G_1(1, 1) = 1$, we have:

$$(2c + b)G_1(0, 1) - v G_1(1, 0) = 2c + b + v. \tag{5.12.27}$$

Thus, from (5.12.22) and (5.12.27), we have the following system of equations:

$$\begin{cases} \mu G_1(0, 1) - v G_1(1, 0) = \mu - \lambda - v, \\ (2c + b)G_1(0, 1) - v G_1(1, 0) = 2c + b - v. \end{cases} \tag{5.12.28}$$

System (5.12.28) is a nonhomogeneous system of two equations with two unknowns. We solve it by the elimination method and obtain its solutions as:

$$G_1(0, 1) = 1 - \frac{\lambda_1}{\mu} \quad \text{and} \quad G_1(1, 0) = 1 - \frac{\lambda_2}{v}, \tag{5.12.29}$$

Since $0 < G_1(0, 1) < 1$ and $0 < G_1(1, 0) < 1$, from (5.12.29), we will have:

$$0 < \frac{\lambda_1}{\mu} < 1 \quad \text{and} \quad 0 < \frac{\lambda_2}{v} < 1.$$

Now, from the assumption of the theorem,

$$\frac{\lambda}{d\mu} = \frac{\lambda_1}{\mu},$$

and, hence, $d > 0$. Thus, all conditions of the Theorem 5.12.1 are satisfied.

The joint probability distribution could be obtained from the Maclaurin expansion of (5.12.7). In contrast, substituting 1 for each w and z, separately, in (5.12.7) and evaluating the derivative with respect to the other variable at 1 gives the expected number of tasks in each station. This completes the proof of Theorem 5.12.1.

5.12.3. Computation of Expected Values of the Queue Length and Waiting Time at each Station, Algorithmically

In this section, we offer an algorithm to find expected values of the queue length and waiting time at each station, as well as a numerical example illustrating the workability of the algorithm. From the statement of Theorem 5.12.1, recall that $a = q_o\mu$, $b = (p_f + p_s q_{so})\mu$, $c = p_s p_{sf}\mu\ (a + b + c = \mu)$, $d = q_o - p_s p_{sf}$, $d > 0$, $\lambda_1/\mu < 1$, and $\lambda_2/v < 1$. Moreover, from the description of the model, recall that p_s, $q_o + p_f + p_s = 1$ and $q_{so} + p_{sf} = 1$.

Before giving the steps, we need a brief discussion of $G(0, z)$. Let $z(w)$ be an algebraic curve, say L, defined over the complex plane by $A(w, \zeta(w)) = 0$ such that:

$$B_1(w,\zeta(w))G(0, \zeta(w)) + B_2(w, \zeta(w))G(w, 0) = 0. \qquad (5.12.30)$$

Similarly, let $w(z)$ be an algebraic curve defined over the complex plane by $A(w(z), z) = 0$ such that:

$$B_1(w(z), z)G(0, z) + B_2(w(z), z)G(\omega(z), 0) = 0. \qquad (5.12.31)$$

Now from (5.12.30), we have:

$$G(w, 0) = -\frac{B_1(\omega(z), z)}{B_2(\omega(z), z)}G(0, z), \quad w \in \left[w_1^*, w_2^8\right], \quad z \in L, \qquad (5.12.32)$$

where $\left[w_1^*, w_2^*\right]$ is a cut on the real axis. Thus, if we find $G(0, z)$, we will have $G(w, 0)$ and in turn $G(w, z)$.

Note that since $G(w, 0)$ is real on the cut $\left[w_1^*, w_2^*\right]$, it implies that the right-hand side of (5.12.32) is real on the cut. We wish to move away from the cut $\left[w_1^*, w_2^*\right]$ and get on the contour L, and particularly on the unit circle L_o. This can be done by *conformal mapping*, as we will see later.

For $z \in L$, let $\Psi(z)$ be a complex function of complex variable z, defined by:

$$\Psi(z) = u(z) + iv(z) \equiv G(0, z). \qquad (5.12.33)$$

Also, let $B(z)$ be a complex function of complex variable z, defined by:

$$B(z) = b(z) + ia(z) \equiv \frac{B_1(\omega(z), z)}{B_2(\omega(z), z)}, \qquad (5.12.34)$$

where $b(z) = \Re(B(z))$ and $a(z) = \Im(B(z))$. Then, relation (5.12.32) can be rewritten as:

$$G(w, 0) = B(z)\Psi(z)$$
$$= [b(z)u(z) - a(z)v(z)] + i[a(z)u(z) + b(z)v(z)], \quad z \in L. \tag{5.12.35}$$

Thus, for t such that $w_1^* \le t \le w_2^*$, we must have:

$$\Im[B(z)\Psi(z)] = a(z)u(z) + b(z)v(z) = 0, \quad \text{for } z \in L. \tag{5.12.36}$$

Relation (5.12.36) is a *homogenous Hilbert boundary value problem* (see Gakhov, 1966) or Problem H_0, with index χ that needs to be solved. The solution will be the unknown function $G(0, z)$ analytic in and continuous on L. It is well known that Problem H_0 has a unique solution if and only if $\chi = 0$, as in our case (see Smirnov, 1964, Gakhov, 1966, or Muskhelishvili, 1992).

Now, dividing the boundary conditions (5.12.36) by $\sqrt{[a(z)]^2 + [b(z)^2]}$, it will be reduced to the case when $[a(z)]^2 + [b(z)]^2 = 1$, where

$$a_1(z) = \frac{a(z)}{\sqrt{[a(z)]^2 + [b(z)]^2}} \quad \text{and} \quad b_1(z) = \frac{b(z)}{\sqrt{a[z]^2 + [b(z)]^2}}. \tag{5.12.37}$$

Thus, (5.12.36) can be rewritten as:

$$\Im[B_0(z)\Psi(z)] = a_1(z)u(z) + b_1(z)v(z) = 0, \quad \text{for } z \in L. \tag{5.12.38}$$

where $B_0(z) = B(z)/\sqrt{[a(z)]^2 + [b(z)]^2}$, and $B(z) \equiv -(B_1(\omega(z), z)/B_2(\omega(z), z))$, for $z \in L$.

The standard way to solve this type of boundary value problem is to transform the boundary condition (5.12.37) on the unit circle (see Cohen and Boxma, 1983 and Muskhelishvili, 1992) and then transform back on the region L^+. Thus, we introduce the conformal mapping $t = f(z): L^+ \to L_o^+$ onto its inverse $z = f^{-1}(t): L_o^+ \to L^+$, where the $+$ sign indicates the region interior of the curve. Using these mappings, we can reduce the Riemann–Hilbert problem on L to the same on the unit circle. Thus, the following:

New Problem H_0

Determine a function $\Psi_0(t)$ such that:

$$\Im\left[\tilde{B}_0(t)\Psi_0(t)\right] = \tilde{a}_1(t)u(t) + \tilde{b}_1(t)v(t) = 0, \quad \text{for } t \in L_o, \tag{5.12.39}$$

where

$$\tilde{B}_0(t) = B_0\left(f^{-1}(t)\right), \tag{5.12.40}$$

$$\tilde{a}_1(t) = a_1\left(f^{-1}(t)\right) = \frac{a\left(f^{-1}(t)\right)}{\sqrt{\left[a\left(f^{-1}(t)\right)\right]^2 + \left[b\left(f^{-1}(t)\right)\right]^2}}, \tag{5.12.41}$$

and

$$\tilde{b}_1(t) = b_1\left(f^{-1}(t)\right) = \frac{b\left(f^{-1}(t)\right)}{\sqrt{\left[a\left(f^{-1}(t)\right)\right]^2 + \left[b\left(f^{-1}(t)\right)\right]^2}}. \tag{5.12.42}$$

Hence, the solution of (5.12.38) is:

$$\Psi(z) = \Psi_0(f(z)), \quad z \in L^+. \tag{5.12.43}$$

Therefore, the solution of (5.12.39) is:

$$\Psi_0(t) = c e^{\sigma(t)}, \quad t \text{ is within the unit circle}, \tag{5.12.44}$$

where

$$\sigma(t) = \frac{1}{2\pi i} \int_L \frac{1}{s}\left\{\arctan \frac{\tilde{b}_1(s)}{\tilde{a}_1(s)}\right\} \frac{s+t}{s-t} ds, \tag{5.12.45}$$

where $\tilde{a}_1(t)$ and $\tilde{b}_1(t)$ are given by (5.12.41) and (5.12.42), respectively (see Cohen and Boxma, 1983, p.58).

Algorithm

Step 1. Choosing Data. Choose probabilities such that all conditions mentioned earlier are satisfied. Also, choose values for parameters so that $\lambda > 0$, $\mu \geq \lambda/d > 0$, and $v \geq \lambda(1-d)/d > 0$. If we let $r = 2c + b/\mu$, then $0 < v < \lambda r/(1-r)$.

Step 2. From equation (5.12.5), solve $A(w, z) = 0$ for w. This gives two roots as functions of z, say $w_1(z)$ and $w_2(z)$.

Step 3. For z on the unit circle, choose the unique root $w_i(z)$, $i = 1$ or 2, of $A(w, z) = 0$ such that $w_i(z) \leq 1$. To do this, choose $w_0 = 1/2 + \left(\sqrt{3}/2\right)i$ so that $|z| = 1$. Then, try the values of roots mentioned in Step 2 at this z-value; choose the one with absolute value less than or equal to 1 and call it $\omega(z)$.

Step 4. From equation (5.12.34), find $b(z) = \Re(B(z))$ and $a(z) = \Im(B(z))$.

Step 5. From Step 4 and (5.12.37), find $a_1(z)$ and $b_1(z)$.

Step 6. Find $\theta(v_k)$, which is the last iteration of the following process (see Step 6.4). To find $\theta(v_k)$, we do perform the following (for more detail, see Cohen and Boxma, 1983, IV.1.3):

Step 6.1. Solve:

$$A(w, z) = (\lambda z + v)w^2 - \left[(\lambda + \mu + v)z\right]w + \left[az + bz^2 + cz^3\right] = 0$$

for w, say, that gives:

$$w_{1,2} = \frac{(\lambda+\mu+v)z \pm \sqrt{D(z)}}{2(\lambda z+v)}, \qquad (5.12.46)$$

where $D(z) = (\lambda+\mu+v)^2 z^2 - 4(\lambda z+v)(az+bz^2+cz^3)$. From (5.12.46), we have:

$$w_1+w_2 = \frac{(\lambda+\mu+v)}{\lambda z+v} = 2\Re(\omega(z)) \qquad (5.12.47)$$

and

$$w_1 w_2 = \frac{az+bz^2+cz^3}{\lambda z+v} = |\omega(z)|^2. \qquad (5.12.48)$$

We note that $D(z)$ must be negative since (5.12.48) shows that the roots are complex conjugates.

Now, solving (5.12.48) for z, we obtain:

$$z = \frac{2v\Re(\omega(z))}{(\lambda+\mu+v)-2\lambda\Re(\omega(z))} = \frac{2v\delta}{\lambda+\mu+v-2\lambda\delta}, \qquad (5.12.49)$$

where $\delta \equiv \Re(\omega(z))$.

Substituting z from (5.12.49) into (5.12.48), we obtain:

$$\begin{aligned} H(\delta) \equiv |\omega(z)|^2 = &\frac{c}{\lambda}\left(\frac{2\delta v}{\lambda+\mu+v-2\delta\lambda}\right)^2 \\ &+ \left(\frac{b}{\lambda}-\frac{cv}{\lambda^2}\right)\left(\frac{2\delta v}{\lambda+\mu+v-2\delta\lambda}\right) + \left(\frac{a}{\lambda}-\frac{bv}{\lambda^2}+\frac{cv^2}{\lambda^3}\right) \\ &+ \left(-\frac{av}{\lambda}+\frac{bv^2}{\lambda^2}-\frac{cv^3}{\lambda^3}\right)\left(\frac{2\delta\lambda v}{\lambda+\mu+v-2\delta\lambda}+v\right)^{-1}. \end{aligned} \qquad (5.12.50)$$

It is clear from (5.12.50) that $H(\delta) > 0$.

Now we draw a circle with an embedded triangle $\triangle OAB$ as follows:

$$\overline{OA} = \Re(\omega(z)) = \delta, \overline{OB} = |\omega(z)|$$

and from

$$\triangle OAB : \cos(\theta_0(v_k)) = \overline{OA}/\overline{OB}.$$

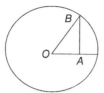

From $\triangle OAB$ and (5.12.50) we have:

$$\cos(\theta_0(v_k)) = \frac{\delta}{|\omega(z)|} = \frac{\delta}{\sqrt{H(\delta)}}. \tag{5.12.51}$$

Therefore,

$$\delta - \cos(\theta_0(v_k))\sqrt{H(\delta)} = 0, \tag{5.12.52}$$

where δ depends on $\theta_0(v_k)$, that is, $\delta = \delta(\theta_0(v_k))$, $k = 0, 1, \ldots, K$.

Step 6.2. Set $\theta_0(v_k) = v_k = 2\pi k/K$, $k = 0, 1, \ldots K$. Using Newton's method, solve equation (5.12.25), that is,

$$\delta(\theta_0(v_k)) = \cos(\theta_0(v_k))\sqrt{H(\delta(\theta_0(v_k)))} = 0, \tag{5.12.53}$$

for $\delta(\theta_0(v_k))$, $k = 0, 1, \ldots, K$.

Note that Step 6.2 could alternatively be done (as we do in the numerical example) using the system of equation (5.12.54) and solving for three unknowns: $\delta \equiv \Re(\omega(z))$, $|\omega(z)|$, and z:

$$\begin{cases} 2\Re(\omega(z)) = \dfrac{(\lambda + \mu + v)z}{\lambda z + v}, \\[2ex] |\omega(z)|^2 = \dfrac{az + bz^2 + cz^3}{\lambda z + v}, \\[2ex] \dfrac{\Re(\omega(z))}{|\omega(z)|} = \cos(\theta_0(v_k)), \quad k = 0, 1, \ldots, K. \end{cases} \tag{5.12.54}$$

To solve (5.12.54), we follow Step 6.2. Alternatively, see the following (for more detail, see Cohen and Boxma, 1983, IV.1.3).

Step 6.2. Alternative: Set $\theta_0(v_k) = v_k = 2\pi k/K$, $k = 0, 1, \ldots, K$, K should be divisible by 4. Using Newton's method, solve equation (5.12.55) for $\delta(\theta_0(v_k))$, $k = 0, 1, \ldots, K$:

$$\sqrt{\delta(\theta_0(v_k))} - \cos(\theta_0(v_K))\sqrt{H(\delta(\theta_0(v_k)))} = 0, \quad k = 1, 2, \ldots, K, \tag{5.12.55}$$

where

$$H_1(\delta) = \frac{2}{\lambda + \mu + v}\left[a + b\left(\frac{2\delta v}{\lambda + \mu + v - 2\lambda\delta}\right) + c\left(\frac{2\delta v}{\lambda + \mu + v - 2\lambda d}\right)^2 \right], \tag{5.12.56}$$

and $\delta \equiv \Re(\omega(z))$.

We should point out that the proper choice of a starting point is very important in Newton's method since it is possible that the method fails to converge.

For values of z on the cut, we have to find $\Re(\omega(z)) = \delta((\lambda + \mu + v)z/2(\lambda z + v))$ at the end points of the cut. Then we choose the initial value for the Newton method, a value between the two values of δ just found. This will assure the Newton's approximation to result in positive values.

Step 6.3. Using the Trapezoidal rule, find $\theta_1(v_k)$ for each $k = 0, 1, \ldots, K$ from:

$$\theta_1(v_k) = v_k - \int_0^{2\pi} \ln[\rho(\theta_0(\tau))]\cos\left[\frac{1}{2}(\tau - v_k)\right]d\tau, \qquad (5.12.57)$$

where $\rho(\theta_0(\tau)) = \delta(\theta_0(\tau))/\cos(\theta_0(\tau))$.

The integrand in (5.12.57) has a singularity at $\tau = v_k$. To deal with this singularity, we can rewrite the integrand as:

$$\{\ln[\rho(\theta_0(\tau))] - \ln[\rho(\theta_0(v_k))]\}\cot\left[\frac{1}{2}(\tau - v_k)\right] + \ln[\rho(\theta_0(v_k))]\cot\left[\frac{1}{2}(\tau - v_k)\right].$$

$$(5.12.58)$$

The integral of the last term equals zero; the value of the first integrand at $\tau = v_k$ can easily be determined by L'Hôspital's rule. Therefore, the resulting integral can be evaluated using a standard numerical integration procedure, that is, the trapezoidal rule. Thus, we can rewrite (5.12.57) as:

$$\theta_1(v_k) = v_k - \int_0^{2\pi} \{[\rho(\theta_0(\tau))] - \ln[\rho(\theta_0(v_k))]\}\cot\left(\frac{\tau - v_k}{2}\right)d\tau. \qquad (5.12.59)$$

In applying the trapezoidal rule, denote the integrand of the integral in (5.12.57) by $h(\tau)$, that is,

$$h(\tau) = \{\ln[\rho(\theta_0(\tau))] - \ln[\rho(\theta_0(v_k))]\}\cot\left(\frac{\tau - v_k}{2}\right). \qquad (5.12.60)$$

Then,

$$\int_0^{2\pi} h(\tau)d\tau \approx \frac{\pi}{K}[h(v_0) + 2h(v_1) + \cdots + 2h(v_k) + \cdots + 2h(v_{K-1}) + h(v_K)],$$

$$(5.12.61)$$

where $v_k = 2\pi k/K$, $k = 0, 1, \ldots, K$. To find $h(v_k)$, we can rewrite $h(\tau)$ as:

$$h(\tau) = \frac{\ln\left[\dfrac{\rho(\theta_0(\tau))}{\rho(\theta_0(v_k))}\right]}{\tan\left(\dfrac{\tau - v_k}{2}\right)}.$$ (5.12.62)

However,

$$\lim_{\tau \to v} \frac{\ln\left[\dfrac{\rho(\theta_0(\tau))}{\rho(\theta_0(v_k))}\right]}{\tan\left(\dfrac{\tau - v_k}{2}\right)} \cdot = \frac{2}{\rho(\theta_0(v_k))} \frac{d}{d\tau}\rho(\theta_0(\tau))\Big|_{\tau=v}.$$ (5.12.63)

Therefore, for this case, we have:

$$h(v_k) = \frac{2}{\rho(\theta_0(v_k))} \frac{d}{d\tau}\rho(\theta_0(\tau))\Big|_{\tau=v}.$$ (5.12.64)

Step 6.4. For $n = 1, 2, \ldots$, repeat Steps 6.1–6.3 to find $\theta_n(v_k)$ and $\theta_{n+1}(v_k)$, $k = 0, 1, \ldots, K$, until the condition $\max_{0 \le n \le N-1}|\theta_{n+1}(v_k) - \theta_n(v_k)| < 10^{-5}$ is satisfied. This condition may be revised in practice. The last iteration in this process will be $\theta(v_k)$, $k = 0, 1, \ldots, K$.

Step 7. Find $f(1)$ from:

$$f^{-1}(t) = te^{\frac{1}{2\pi}\int_0^{2\pi} \ln[\rho(\theta(u))]\frac{e^{iv}+2}{e^{iv}-2}dv}, \ |t| < 1.$$

To do this, use Newton's method and solve $f^{-1}(\eta) = 1$ in $[0, 1]$. Thus, $f(1) = \eta$. Make sure that $\eta < 1$.

Step 8. Find $f'(1)$. To do this, use the trapezoidal rule for the integral on the right-hand side of:

$$f'(1) = \frac{1}{\dfrac{d}{dt}[f^{-1}(t)]\Big|_{t=\eta}}.$$

Step 9. From:

$$f^{-1}(e^{iv_k}) = \frac{\delta(\theta(v_k))}{\cos(\theta(v_k))}e^{i\theta(v_k)}, \ k = 0, 1, \ldots, K-1,$$

find $f^{-1}(e^{iv_k})$.

Step 10. Using Trapezoidal rule, find:

$$\frac{d}{dz}G(0,z)\Big|_{z=1}$$

from

$$\frac{d}{dz}G(0,z)\Big|_{z=1} = G(0,1)\frac{i}{\pi}\int_0^{2\pi}\left\{\tan^{-1}\left[\frac{b_1\left(f^{-1}\left(e^{iv}\right)\right)}{a_1\left(f^{-1}\left(e^{iv}\right)\right)}\right]\right\}\frac{f'(1)}{\left[e^{iv}-f(1)\right]^2}\,dv.$$

Step 11. Differentiating both sides of (5.12.26) with respect to z, setting $z = 1$ and solving for

$$\frac{d}{dz}G(1,z)\Big|_{z=1},$$

we obtain:

$$\frac{\partial}{\partial z}G(1,z)\Big|_{z=1} = \frac{1}{v-(2c+b)}\left[\frac{\lambda(3c+b)}{a-c}-(2c+b)\frac{\partial}{\partial z}G(0,z)\Big|_{z=1}\right]. \quad (5.12.65)$$

It should be noted that since expected values for the model must be nonnegative, in the numerical evaluation we should have one of the following two cases:

1. $v-(2c+b)>0$ and $\dfrac{\lambda(3c+b)}{a-c}-(2c+b)\dfrac{\partial}{\partial z}G(0,z)\Big|_{z=1}>0$

    or

2. $v-(2c+b)>0$ and $\lambda(3c+b)/(a-c)-(2c+b)(\partial/\partial z)G(0,z)\big|_{z=1}<0$

Step 12. Find:

$$\frac{\partial}{\partial w}G(w,1)\Big|_{w=1}$$

from

$$\frac{\partial}{\partial w}G(w,1)\Big|_{w=1} = \frac{1}{a-c-\lambda}\left[\frac{\lambda a}{a-c}+(a-c)\frac{\partial}{\partial z}G(0,z)\Big|_{z=1}\right]-\frac{\partial}{\partial z}G(1,z)\Big|_{z=1}.$$

$$(5.12.66)$$

Here is how we obtained (5.12.65). We put $w = z = \zeta$ in (5.12.3). For this choice of w, $B_2(w,z) = 0$. Hence, (5.12.3) reduces to:

$$A(\zeta,\zeta)G(\zeta,\zeta) = B_1(\zeta,\zeta)G(0,\zeta). \quad (5.12.67)$$

Differentiating both sides of (5.12.66) with respect to ζ, solving for

$$\frac{dG(\zeta, \zeta)}{d\zeta},$$

$$\frac{dG(\zeta, \zeta)}{d\zeta} = \frac{\partial G}{\partial w} + \frac{\partial G}{\partial z},$$

simplifying, and applying L'Hôspital's rule as $\zeta \to 1$, we obtain:

$$\frac{\partial}{\partial w} G(w, 1)\Big|_{w=1} + \frac{\partial}{\partial z} G(1, z)\Big|_{z=1} = \frac{1}{a-c-\lambda}\left[\frac{\lambda a}{a-c} + (a-c)\frac{\partial}{\partial z} G(0, z)\Big|_{z=1}\right].$$

$$(5.12.68)$$

This is a relationship between the two expected values, from which (5.12.65) follows.

Step 13. Calculate the average waiting time at each station using $W_i = L_i/\lambda_i$, $i = 1, 2$.

5.12.4. Numerical Example

This example is to illustrate how the algorithmic steps work. We choose the number of iterations for approximations of integrals by trapezoidal rule as $K = 30$, for approximations of derivatives by Newton's method as $N = 10$, and the number of implementation of the approximation as $R = 10$. We also use the following notations and conditions:

1. $a = q_o\mu$, $b = (p_f + p_s q_{so})\mu$, $c = p_s p_{sf}\mu$, $a + b + c = \mu$, $d = q_o - p_s q_{so} > 0$
2. $\lambda > 0$, $\mu \geq \lambda/d$, and $v \geq \lambda(1 - d)/d$
3. λ_1 and λ_2 are as found in (5.12.1) and (5.12.2), respectively: $\lambda_1/\mu < 1$ and $\lambda_2/v < 1$
4. $0 < G_1(0, 1) < 1$ and $0 < G_1(1, 0) < 1$

To check the validity of the result, not only do we compare the special cases with those that exist in the literature, but we will also simulate the system and compare the analytical and simulated results.

For different values of the model parameters, we refer to Table 5.12.1 and Table 5.12.2, which include the expected values of the queue lengths at each station (including those in service and weighting for departure from the feedback buffer).

Remark on Special Cases

1. If $q_o = 1$ and $v = 0$, then the system (5.12.8) reduces to the system for stationary $M/M/1$ queue. In this case, if X represents the number of tasks in the system, then the expected value of the queue length E_{mml} is:

TABLE 5.12.1. $q_s = 0.9$, $q_o = 0.6$, $p_s = 0.05$, $p_{sf} = 0.1$, $p_f = 0.35$, $q_o + q_f + p_s = 1$, $q_{so} + p_{sf} = 1$, $a = q_o\mu$, $b = (p_f + p_s q_{so})\mu$, $c = p_s p_{sf}\mu$, $d = 0.595$, and $r = (2c + b)/\mu = 0.405$

λ	μ	v	λ_1	ρ_1	λ_2	ρ_2	ρ	$E(X_1)$	$E(X_2)$	$E(W_1)$	$E(W_2)$
1	1.8	28	1.6807	0.93371	0.68067	0.024310	0.95802	0.19927	0.669670	0.1186	0.9838
2	3.6	32	3.3613	0.93371	1.36130	0.042542	0.97625	13.25300	0.124870	3.9428	0.0917
2	4.0	28	3.3613	0.84034	1.36130	0.048619	0.88896	2.07740	0.705270	0.6180	0.5181
20	36.0	300	33.6130	0.93371	13.61300	0.045378	0.97908	14.24300	0.004702	0.4237	0.0003
30	55.0	400	50.4200	0.91673	20.42000	0.051050	0.96778	6.78040	2.826800	0.1345	0.1384

$$E_{mml}(X) = \frac{\rho_{mml}}{1 - \rho_{mml}}, \qquad (5.12.69)$$

where $\rho_{mml} = \lambda/\mu$.

2. If $p_s = 0$ in the system (5.12.8), then the system described reduces to a stationary $M/M/1$ queue with random delayed feedback with a single return facility discussed in Kleinrock and Gail (1996). In this case, the expected values of the queue lengths for the stations are:

$$E_T(S_1) = \frac{\rho_T}{1 - \rho_T} \quad \text{and} \quad E_T(S_2) = \frac{p_f\mu}{v}\rho_T, \qquad (5.12.70)$$

where $\rho_T = \lambda/q_o\mu$. Note that the expected value of the number of tasks in the first station is the same as the expected queue length for $M/M/1$ without feedback, that is, (5.12.69). See Table 5.12.3 for numerical values and comparison.

5.12.5. Discussion and Conclusion

It is important to observe all conditions of stabilities mentioned throughout the chapter. It seems the system does not respond well when traffic intensities at each station is very low. Moreover, due to the stability of the system, we have chosen the service and feedback rates depending on the external arrival rate. This can be seen quite consistently throughout the numerical example.

Note that $q_o - p_s q_{so} > 0$, and denoting

$$\rho_0 = \frac{\lambda}{(q_o - p_s q_{so})\mu} \quad \text{and} \quad \rho_2 = \frac{(1 - q_o + p_s q_{so})\lambda}{(q_o - p_s q_{so})v},$$

we should have:

$$\frac{\rho_1}{\rho_2} > 1. \qquad (5.12.71)$$

TABLE 5.12.2. $q_s = 0.6$, $q_o = 0.6$, $p_s = 0.05$, $p_{sf} = 0.4$, $p_f = 0.35$, and $q_{so} + p_{sf} = 1$, $a = q_o\mu$, $b = (pf + p_s q_{so})\mu$, $c = p_s p_s \mu$, $d = 0.595$, and $r = (2c + b)/\mu = 0.42$

λ	μ	v	λ_1	ρ_1	λ_2	ρ_2	ρ	$E(X_1)$	$E(X_2)$	$E(W_1)$	$E(W_2)$
0.605	1.09	31	1.0431	0.95698	0.43810	0.014132	0.97111	21.1020000	0.051588	20.2301	0.1178
0.800	1.43	31	1.3703	0.96455	0.57931	0.018687	0.98324	10.2370000	0.392940	7.4706	0.6783
1.000	1.77	31	1.7241	0.97409	0.72414	0.023359	0.99745	23.8720000	0.308410	13.8461	0.4259
1.000	1.80	28	1.7241	0.95785	0.72414	0.025872	0.98371	17.7810000	0.203910	10.3132	0.2816
2.000	3.60	50	3.4483	0.95785	1.44830	0.028966	0.98682	16.0860000	0.474230	4.6649	0.3274
10.000	18.00	200	17.2410	0.95785	7.24140	0.036207	0.99406	21.3710000	0.546980	1.2395	0.0755
14.000	26.00	530	24.1380	0.92838	10.13800	0.019128	0.94751	11.5680000	0.819750	0.4792	0.0809
17.000	32.00	530	29.3100	0.91595	12.31000	0.023227	0.93918	2.3423000	4.749000	0.0799	0.3858
20.000	40.00	400	34.4830	0.86207	14.48300	0.036207	0.89828	0.0085305	4.522400	0.0002	0.3123
26.000	47.00	525	44.8280	0.95378	18.82800	0.035862	0.98964	11.7740000	4.587300	0.2626	0.2436
30.000	55.00	500	51.7240	0.94044	21.72400	0.043448	0.98389	16.2580000	0.063853	0.3143	0.0029

TABLE 5.12.3. $q_s = 0$, $q_o = 0.6$, $p_s = 0$, $p_{sf} = 0$, $p_f = 0.4$, $a = q_o\mu$, $b = (p_f + p_s q_{so})\mu$, $c = p_s p_{sf}\mu$, $d = 0.6$, and $r = (2c + b)/\mu = 0$

λ	μ	v	λ_1	ρ_1	λ_2	ρ_2	ρ ρ_T	$E(X_1)$ $E_T(X_1)$	$E(X_2)$ $E_T(X_2)$
0.605	1.05	29	1.0083	0.96032	0.40333	0.013908	0.97423	22.2370	0.0460570
							0.96032	24.2000	0.0080137
1.000	2.00	30	1.6667	0.83333	0.66667	0.022222	0.85556	4.1987	0.1144200
							0.83333	5.0000	0.0111110
2.000	3.50	40	3.3333	0.95238	1.33330	0.033333	0.98571	14.2620	0.3910200
							0.95238	20.0000	0.0190480
10.000	17.50	200	16.6670	0.95238	6.66670	0.033333	0.98571	12.8370	1.8166000
							0.95238	20.0000	0.0190480
20.000	37.00	400	33.3330	0.90090	13.33300	0.033333	0.93423	7.7132	0.8332100
							0.90090	9.0909	0.0180180
30.000	53.00	500	50.0000	0.94340	20.00000	0.040000	0.98340	12.7940	2.1312000
							0.94340	16.6670	0.0226420

We then rewrite (5.12.64) as:

$$\left.\frac{\partial}{\partial z}G(1,z)\right|_{z=1} = \frac{1}{\left(\dfrac{\rho_1}{\rho_2}-1\right)}\left\{\rho_1\left[\frac{p_s p_{sf}}{1-q_o+p_s p_{sf}}+\frac{1-q_o+p_s p_{sf}}{q_o-p_s p_{sf}}\right]-\left.\frac{\partial}{\partial z}G(0,z)\right|_{z=1}\right\}$$

(5.12.72)

and (5.12.67) as:

$$\left.\frac{\partial}{\partial w}G(w,1)\right|_{w=1} + \left.\frac{\partial}{\partial z}G(1,z)\right|_{z=1} = \frac{1}{1-\rho_1}\left[\frac{\rho_1 q_o}{q_o-p_s p_{sf}}+\left.\frac{\partial}{\partial z}G(0,z)\right|_{z=1}\right].$$

(5.12.73)

All relations (5.12.70)–(5.12.72) must be satisfied and both (5.12.71) and (5.12.72) must be positive in order for the expected values to make sense, and thus will help choosing the parameters λ, μ, and v.

What we did in this section was that we considered a Poisson single-processor model with splitter and feedback with two buffers as waiting rooms, one before the service station and one before a delayed feedback station. We focused on feedback batch size equal to 1. Applying generating function on the steady-state system of differential–difference equations, we obtained a functional equation. Through an algorithm, we offered a step-by-step method of obtaining the mean queue lengths at each station.

EXERCISES

5.1. Prove that if $\mathbf{P}(t)$ is a stochastic matrix, then prove that $\mathbf{P}^n(t)$ is also a stochastic matrix for any positive integer n.

5.2. Find $\mathbf{P}^n$ if the transition probability matrix for a discrete-time Markov chain is:

$$\mathbf{P} = \begin{pmatrix} \dfrac{1}{3} & \dfrac{2}{3} \\[2mm] \dfrac{2}{5} & \dfrac{3}{5} \end{pmatrix}.$$

5.2. Let $\{X_n, n = 0, 1, 2, \ldots\}$ be a Markov chain with state space $\{S = 0, 1, 2 \ldots\}$ and transition matrix $\mathbf{P}$:

$$\mathbf{P} = \begin{pmatrix} \dfrac{1}{2} & 0 & \dfrac{1}{2} \\[2mm] \dfrac{2}{5} & \dfrac{1}{5} & \dfrac{2}{5} \\[2mm] 0 & \dfrac{2}{5} & \dfrac{3}{5} \end{pmatrix}.$$

Find:

a. $P(X_2 = 3 | X_1 = 0, X_0 = 1)$.
b. $P(X_2 = 3, X_1 = 0 | X_0 = 1)$.
c. $P(X_2 = 3, X_1 = 0 | X_0 = 0)$.
d. $P(X_{n+1} = 3, X_n = 0 | X_{n-1} = 0)$, $n = 1, 2, \ldots$

5.3. Write the one-step transition matrix for a Bernoulli counting process.

5.4. Let $\{X_n, n = 0, 1, 2, \ldots\}$ be a Markov chain with state space $S = 0, 1, 2, \ldots$ and transition matrix $\mathbf{P}$:

$$\mathbf{P} = \begin{pmatrix} \dfrac{1}{5} & \dfrac{3}{10} & \dfrac{1}{2} \\[2mm] \dfrac{4}{5} & \dfrac{1}{5} & 0 \\[2mm] \dfrac{3}{5} & 0 & \dfrac{2}{5} \end{pmatrix}.$$

Find:

a. the matrix of three-transition probabilities
b. $P(X_3 = 0)$ and $P(X_0 = 0, X_1 = 1, X_3 = 2)$, given the initial distribution

$$P(X_0 = i) = \frac{1}{3}, \quad i = 0, 1, 2, \dots .$$

5.5. Find the stationary distribution for a Markov chain with a state space $S = \{0, 1, 2, 3, 4\}$ and transition matrix:

$$\mathbf{P} = \begin{pmatrix} 0 & \dfrac{1}{5} & \dfrac{4}{5} & 0 & 0 \\[2mm] 0 & 0 & 0 & \dfrac{9}{10} & \dfrac{1}{10} \\[2mm] 0 & 0 & 0 & \dfrac{1}{10} & \dfrac{9}{10} \\[2mm] 1 & 0 & 0 & 0 & 0 \\[2mm] 1 & 0 & 0 & 0 & 0 \end{pmatrix}.$$

5.6. A copier is working. It has a probability q of failure (total breakdown) each time it is used. Let $p = 1 - q$ be the probability that it functions without breakdown on a given use. Each time the copier fails, a technician is called for repair and the copier cannot be used until it is completely repaired. What proportion of the time can we expect the copier in functioning condition?

5.7. Consider a birth-and-death process with transition probabilities $\lambda_n = \lambda$ and $\mu_n = n\mu$, $n \geq 0$.

a. Denote the probability of population size to be n at time t by $p_n(t)$, $n = 0, 1, \dots$ Find the differential–difference equation for the process.

b. Let $G(z, t)$ define a generating function for $p_n(t)$, $n = 0, 1, \dots$ Find the partial difference equation that $G(z, t)$ should satisfy.

c. Let $P_0(0) = 1$ be the initial condition for the process, Show that the generating function of solution to the partial difference equation mentioned in part (a) is $G(z, t) = e^{(\lambda/\mu)\left(1 - e^{-\mu}\right)(z-1)}$

5.8. Consider a birth-and-death process with transition probabilities

$$\lambda_n = \begin{cases} (N-n)\lambda, & n \leq N \\ 0, & n \geq N \end{cases}, \quad \mu_n = \begin{cases} n\mu, & n \leq N \\ 0, & n \geq N \end{cases}.$$

Denote the probability of population size to be n at time t by $p_n(t)$, $n = 0, 1, \dots, N$. Find the differential–difference equation for the process.

5.9. Consider a birth-and-death process with transition probabilities

$$\lambda_0 = 1, \quad \lambda_n = \frac{1}{1 + \left(\dfrac{n}{1+n}\right)^2}, \quad \mu_n = 1 - \lambda_n, \quad n = 1, 2, \dots .$$

Show that this is a transient process.

5.10. Suppose that the number of automobiles passing a traffic light daily within an afternoon during rush hour, 4:00–6:00 P.M., follows a homogeneous Poisson process with mean $\lambda = 40$ per half an hour. Among these automobiles, 80% disregard the red light. What is the probability that at least one automobile disregards the red light between 8:00 and 9:00 A.M.?

5.11. A Geiger counter is struck by radioactive particles according to a homogeneous Poisson process with parameter $\lambda = 1$ per 12 seconds. On the average, the counter only records 8 out of 10 particles.

a. What is the probability that the Geiger counter records at least three particles per minute?

b. What are the mean and variance (in minutes) of the random variable Y, representing time, between the occurrences of two successively recorded particles?

5.12. Consider a stationary $M/M/1$ queue with arrival and service rates λ and μ, respectively. Suppose a customer balks (depart without service) within an interval of length Δt with probability $\alpha \Delta t + o(\Delta t)$, where o is the "little o."

a. Express P_{n+1}, the probability that there are $n + 1$ in the system, in terms of P_n, $n = 0, 1, 2, \ldots$

b. Solve the differential balance equations found in part (a) for P_n, $n = 0$, $1, 2, \ldots$, when $\alpha = \mu$.

5.13. Consider a stationary $M/M/1/N$ queue with arrival and service rates λ and μ, respectively. Express P_{n+1}, the probability that there are $n + 1$ in the system, in terms of P_n, $n = 0, 1, 2, \ldots, N$.

5.14. Consider a machine that may be up and running perfectly, up and running inefficiently, or down (not operating). Assume that state changes can only occur after a fixed time unit of length 1. The machine may stay in the same state for while, that is, it may work perfectly for a while, stay down for a while until it is repaired, or work for a while, but not efficiently. Let X_n be a random variable denoting the state of the machine at time n. Also assume that the sequence $\{X_n, n = 0, 1, \ldots\}$ forms a Markov chain with transition matrix

$$P = \begin{pmatrix} \dfrac{4}{5} & \dfrac{1}{10} & \dfrac{1}{10} \\[2mm] 0 & \dfrac{3}{5} & \dfrac{2}{5} \\[2mm] \dfrac{4}{5} & 0 & \dfrac{1}{5} \end{pmatrix}.$$

a. Find the stationary state probabilities.

b. Suppose that if the machine is in perfect working condition, it will generate $800 profit per unit time. If it is in inefficient working condition, it will generate $500 profit per unit time. However, if it is in the "down" condition it will generate a $100 loss. What is the expected profit per unit time if the machine is to work for a sufficiently long time?

5.15. Find the waiting time distribution for stationary $C/M/1/N$.

5.16. Show that the process $\{X(t), t \geq 0\}$, discussed in Example 5.3.1, is a continuous-time Markov process, with a discrete state pace $S = \{0, 1\}$.

5.17. Prove that in a Poisson process with transition rate λt, the probability mass function $p_x(t)$ is Poisson with parameter λt, that is,

$$p_x(t) = \frac{(\lambda t)^x}{x!} e^{-\lambda t}, \quad x = 0, 1, 2, \dots .$$

5.18. Show that for $n \geq i$, (5.5.1.16) reduces to (5.5.1.17).

5.19. In proof of Theorem 5.5.2.1, show that $\sum_{k=0}^{\infty} \rho^k p_{ki}(t) = 0$.

5.20. For stationary $M/M/1$, find the variance of the number of tasks in the waiting line.

5.21. Show that from (5.5.4.6) and (5.5.4.10), for $\rho - (\lambda/\rho) < 1$, we have:

$$E(B) = \frac{1}{\mu(1-\rho)} \quad \text{and} \quad Var(B) = \frac{1+\rho}{\mu^2(1-\rho)^3}.$$

5.22. For $M/M/1$ queue, show that the probability distribution, mean, and variance of the length of the busy period are, respectively, as follows:

$$P_0'(t) = \frac{i\sqrt{\mu/\lambda}}{t} e^{-(\lambda+\mu)t} I_i\left(2\sqrt{\lambda\mu t}\right),$$

where $I_v(z)$, $v = 0, \pm1, \pm2, \dots$, is the modified Bessel function of order v defined by (5.5.1.2), and

$$E(B_i) = \frac{i}{\mu(1-\rho)}, \quad \text{and} \quad Var(B_i) = \frac{i(1+\rho)}{\mu^2(1-\rho)^3}.$$

5.23. Solve the system

$$\begin{cases} \lambda a_0 P_0 = \mu P_1, \\ (\lambda a_n + \mu) P_n = \lambda a_{n-1} P_{n-1} + \mu P_{n+1}, \quad n \geq 1, \end{cases}$$

and show that

$$P_0 = \frac{1}{1 + \sum_{n=1}^{\infty} \left[\prod_{i=0}^{n-1} a_i \left(\frac{\lambda}{\mu}\right)^{i+1} \right]},$$

and

$$P_n = \left[\prod_{i=0}^{n-1} a_i \left(\frac{\lambda}{\mu}\right)^{i+1}\right] P_0, \quad n = 1, 2, \dots .$$

5.24. Develop the process $M/M/2$ and discuss, similar to the case $M/M/1$ in Section 5.5.

5.25. Show that the Laplace transform of the generating function of the queue length for system (5.9.1.9) is:

$$G^*(z, s) = \frac{z^{i+1} - \mu(1-z)\sum_{n=0}^{m-1}\left(m-n\right)z^n P_n^*(s)}{sz - (1-z)(m\mu - \lambda z)}.$$

5.26. Give the details of the iterative method and completion of the proof of Theorem 5.9.2.1.

5.27. Show that letting the traffic intensity be denoted by $\rho = \lambda/\mu$, the distribution of the queue size in an $M/M/2$ *loss system* is as follows:

$$P_{0,0} = \frac{2}{\rho^2 + 2\rho + 2}, \quad P_{0,1} = \frac{\rho^2}{(\rho^2 + 2\rho + 2)(1+\rho)},$$

$$P_{1,0} = \frac{\rho(2+\rho)}{(\rho^2 + 2\rho + 2)(1+\rho)}, \quad P_{1,1} = \frac{\rho^2}{\rho^2 + 2\rho + 2}.$$

5.28. Find the general solution of the system (5.11.1.1).

5.29. Prove the relation (5.11.2.3f).

APPENDIX

THE POISSON PROBABILITY DISTRIBUTION

$$F(x) = P(X \le x) = \sum_{k=0}^{x} \frac{\lambda^k e^{-\lambda}}{k!}$$

	$\lambda = E(X)$									
x	0.1	0.2	0.3	0.4	0.5	0.6	0.7	0.8	0.9	1.0
0	0.904837	0.818731	0.740818	0.670320	0.606531	0.548812	0.496585	0.449329	0.406570	0.367879
1	0.995321	0.982477	0.963064	0.938448	0.909796	0.878099	0.844195	0.808792	0.772482	0.735759
2	0.999845	0.998852	0.996400	0.992074	0.985612	0.976885	0.965858	0.952577	0.937143	0.919699
3	0.999996	0.999943	0.999734	0.999224	0.998248	0.996642	0.994247	0.990920	0.986541	0.981012
4	1.000000	0.999998	0.999984	0.999939	0.999828	0.999606	0.999214	0.998589	0.997656	0.996340
5	1.000000	1.000000	0.999999	0.999996	0.999986	0.999961	0.999910	0.999816	0.999656	0.999406
6	1.000000	1.000000	1.000000	1.000000	0.999999	0.999997	0.999991	0.999979	0.999957	0.999917
x	1.1	1.2	1.3	1.4	1.5	1.6	1.7	1.8	1.9	2.0
0	0.332871	0.301194	0.272532	0.246597	0.223130	0.201897	0.182684	0.165299	0.149569	0.135335
1	0.699029	0.662627	0.626823	0.591833	0.557825	0.524931	0.493246	0.462837	0.433749	0.406006
2	0.900416	0.879487	0.857112	0.833498	0.808847	0.783358	0.757223	0.730621	0.703720	0.676676
3	0.974258	0.966231	0.956905	0.946275	0.934358	0.921187	0.906811	0.891292	0.874702	0.857123
4	0.994565	0.992254	0.989337	0.985747	0.981424	0.976318	0.970385	0.963593	0.955919	0.947347
5	0.999032	0.998500	0.997769	0.996799	0.995544	0.993960	0.992001	0.989622	0.986781	0.983436
6	0.999851	0.999749	0.999596	0.999378	0.999074	0.998664	0.998125	0.997431	0.996554	0.995466
7	0.999980	0.999963	0.999936	0.999893	0.999830	0.999740	0.999612	0.999438	0.999207	0.998903
8	0.999998	0.999995	0.999991	0.999984	0.999972	0.999955	0.999928	0.999890	0.999837	0.999763

Difference and Differential Equations with Applications in Queueing Theory, First Edition.
Aliakbar Montazer Haghighi and Dimitar P. Mishev.
© 2013 John Wiley & Sons, Inc. Published 2013 by John Wiley & Sons, Inc.

	2.2	2.4	2.6	2.8	3.0	3.2	3.4	3.6	3.8	4.0
0	0.110803	0.090718	0.074274	0.060810	0.049787	0.040762	0.033373	0.027324	0.022371	0.018316
1	0.354570	0.308441	0.267385	0.231078	0.199148	0.171201	0.146842	0.125689	0.107380	0.091578
2	0.622714	0.569709	0.518430	0.469454	0.423190	0.379904	0.339740	0.302747	0.268897	0.238103
3	0.819352	0.778723	0.736002	0.691937	0.647232	0.602520	0.558357	0.515216	0.473485	0.433470
4	0.927504	0.904131	0.877423	0.847676	0.815263	0.780613	0.744182	0.706438	0.667844	0.628837
5	0.975090	0.964327	0.950963	0.934890	0.916082	0.894592	0.870542	0.844119	0.815556	0.785130
6	0.992539	0.988406	0.982830	0.975589	0.966491	0.955381	0.942147	0.926727	0.909108	0.889326
7	0.998022	0.996661	0.994666	0.991869	0.988095	0.983170	0.976926	0.969211	0.959893	0.948866
8	0.999530	0.999138	0.998513	0.997567	0.996197	0.994286	0.991707	0.988329	0.984016	0.978637
9	0.999899	0.999798	0.999624	0.999340	0.998898	0.998238	0.997291	0.995976	0.994201	0.991868
10	0.999980	0.999957	0.999913	0.999836	0.999708	0.999503	0.999190	0.998729	0.998071	0.997160
11	0.999996	0.999992	0.999982	0.999963	0.999929	0.999871	0.999777	0.999630	0.999408	0.999085
12	0.999999	0.999998	0.999996	0.999992	0.999984	0.999969	0.999943	0.999900	0.999832	0.999726

	4.2	4.4	4.6	4.8	5.0	5.2	5.4	5.6	5.8	6.0
0	0.014996	0.012277	0.010052	0.008230	0.006738	0.005517	0.004517	0.003698	0.003028	0.002479
1	0.077977	0.066298	0.056290	0.047733	0.040428	0.034203	0.028906	0.024406	0.020587	0.017351
2	0.210238	0.185142	0.162639	0.142539	0.124652	0.108787	0.094758	0.082388	0.071511	0.061969
3	0.395403	0.359448	0.325706	0.294230	0.265026	0.238065	0.213291	0.190622	0.169963	0.151204
4	0.589827	0.551184	0.513234	0.476259	0.440493	0.406128	0.373311	0.342150	0.312718	0.285056
5	0.753143	0.719912	0.685760	0.651006	0.615961	0.580913	0.546132	0.511861	0.478315	0.445680
6	0.867464	0.843645	0.818029	0.790805	0.762183	0.732393	0.701671	0.670258	0.638391	0.606303
7	0.936057	0.921421	0.904949	0.886666	0.866628	0.844922	0.821659	0.796975	0.771026	0.743980
8	0.972068	0.964197	0.954928	0.944183	0.931906	0.918065	0.902650	0.885678	0.867186	0.847237
9	0.988873	0.985110	0.980473	0.974859	0.968172	0.960326	0.951245	0.940870	0.929156	0.916076
10	0.995931	0.994312	0.992223	0.989583	0.986305	0.982301	0.977486	0.971778	0.965099	0.957379
11	0.998626	0.997992	0.997137	0.996008	0.994547	0.992690	0.990368	0.987513	0.984050	0.979908
12	0.999569	0.999342	0.999021	0.998578	0.997981	0.997191	0.996165	0.994856	0.993210	0.991173
13	0.999874	0.999799	0.999688	0.999527	0.999302	0.998992	0.998573	0.998019	0.997297	0.996372
14	0.999966	0.999942	0.999907	0.999853	0.999774	0.999661	0.999502	0.999284	0.998990	0.998600
15	0.999991	0.999984	0.999974	0.999957	0.999931	0.999892	0.999836	0.999756	0.999644	0.999491
16	0.999998	0.999996	0.999993	0.999988	0.999980	0.999968	0.999949	0.999922	0.999882	0.999825

	6.5	7.0	7.5	8.0	8.5	9.0	9.5	10.0	10.5	11.0
0	0.001503	0.000912	0.000553	0.000335	0.000203	0.000123	0.000075	0.000045	0.000028	0.000017
1	0.011276	0.007295	0.004701	0.003019	0.001933	0.001234	0.000786	0.000499	0.000317	0.000200
2	0.043036	0.029636	0.020257	0.013754	0.009283	0.006232	0.004164	0.002769	0.001835	0.001211
3	0.111850	0.081765	0.059145	0.042380	0.030109	0.021226	0.014860	0.010336	0.007147	0.004916
4	0.223672	0.172992	0.132062	0.099632	0.074364	0.054964	0.040263	0.029253	0.021094	0.015105
5	0.369041	0.300708	0.241436	0.191236	0.149597	0.115691	0.088528	0.067086	0.050380	0.037520
6	0.526524	0.449711	0.378155	0.313374	0.256178	0.206781	0.164949	0.130141	0.101632	0.078614
7	0.672758	0.598714	0.524639	0.452961	0.385597	0.323897	0.268663	0.220221	0.178511	0.143192
8	0.791573	0.729091	0.661967	0.592547	0.523105	0.455653	0.391823	0.332820	0.279413	0.231985
9	0.877384	0.830496	0.776408	0.716624	0.652974	0.587408	0.521826	0.457930	0.397133	0.340511
10	0.933161	0.901479	0.862238	0.815886	0.763362	0.705988	0.645328	0.583040	0.520738	0.459889
11	0.966120	0.946650	0.920759	0.888076	0.848662	0.803008	0.751990	0.696776	0.638725	0.579267
12	0.983973	0.973000	0.957334	0.936203	0.909083	0.875773	0.836430	0.791556	0.741964	0.688697
13	0.992900	0.987189	0.978435	0.965819	0.948589	0.926149	0.898136	0.864464	0.825349	0.781291
14	0.997044	0.994283	0.989740	0.982743	0.972575	0.958534	0.940008	0.916542	0.887888	0.854044

(Continued)

x	6.5	7.0	7.5	8.0	8.5	9.0	9.5	10.0	10.5	11.0
15	0.998840	0.997593	0.995392	0.991769	0.986167	0.977964	0.966527	0.951260	0.931665	0.907396
16	0.999570	0.999042	0.998041	0.996282	0.993387	0.988894	0.982273	0.972958	0.960394	0.944076
17	0.999849	0.999638	0.999210	0.998406	0.996998	0.994680	0.991072	0.985722	0.978138	0.967809
18	0.999949	0.999870	0.999697	0.999350	0.998703	0.997574	0.995716	0.992813	0.988489	0.982313
19	0.999984	0.999956	0.999889	0.999747	0.999465	0.998944	0.998038	0.996546	0.994209	0.990711
20	0.999995	0.999986	0.999961	0.999906	0.999789	0.999561	0.999141	0.998412	0.997212	0.995329
21	0.999999	0.999995	0.999987	0.999967	0.999921	0.999825	0.999639	0.999300	0.998714	0.997748
22	1.000000	0.999999	0.999996	0.999989	0.999971	0.999933	0.999855	0.999704	0.999430	0.998958
23	1.000000	1.000000	0.999999	0.999996	0.999990	0.999975	0.999944	0.999880	0.999758	0.999536

x	11.5	12.0	12.5	13.0	13.5	14.0	14.5	15.0	15.5	16.0
0	0.000010	0.000006	0.000004	0.000002	0.000001	0.000001	0.000000	0.000000	0.000000	0.000000
1	0.000127	0.000080	0.000050	0.000032	0.000020	0.000012	0.000008	0.000005	0.000003	0.000002
2	0.000796	0.000522	0.000341	0.000223	0.000145	0.000094	0.000061	0.000039	0.000025	0.000016
3	0.003364	0.002292	0.001555	0.001050	0.000707	0.000474	0.000317	0.000211	0.000140	0.000093
4	0.010747	0.007600	0.005346	0.003740	0.002604	0.001805	0.001246	0.000857	0.000587	0.00040C
5	0.027726	0.020341	0.014823	0.010734	0.007727	0.005532	0.003940	0.002792	0.001970	0.001384
6	0.060270	0.045822	0.034567	0.025887	0.019254	0.014228	0.010450	0.007632	0.005544	0.004006
7	0.113734	0.089504	0.069825	0.054028	0.041483	0.031620	0.023936	0.018002	0.013456	0.01000C
8	0.190590	0.155028	0.124916	0.099758	0.078995	0.062055	0.048379	0.037446	0.028787	0.021987
9	0.288795	0.242392	0.201431	0.165812	0.135264	0.109399	0.087759	0.069854	0.055190	0.043298
10	0.401730	0.347229	0.297075	0.251682	0.211226	0.175681	0.144861	0.118464	0.096116	0.077396
11	0.519798	0.461597	0.405761	0.353165	0.304453	0.260040	0.220131	0.184752	0.153783	0.126993
12	0.632947	0.575965	0.518975	0.463105	0.409333	0.358458	0.311082	0.267611	0.228269	0.193122
13	0.733040	0.681536	0.627835	0.573045	0.518247	0.464448	0.412528	0.363218	0.317081	0.274511
14	0.815260	0.772025	0.725032	0.675132	0.623271	0.570437	0.517597	0.465654	0.415407	0.367527
15	0.878295	0.844416	0.806029	0.763607	0.717793	0.669360	0.619163	0.568090	0.517011	0.466745
16	0.923601	0.898709	0.869308	0.835493	0.797545	0.755918	0.711208	0.664123	0.615440	0.565962
17	0.954250	0.937034	0.915837	0.890465	0.860878	0.827201	0.789716	0.748859	0.705184	0.659340
18	0.973831	0.962584	0.948148	0.930167	0.908378	0.882643	0.852960	0.819472	0.782464	0.742349
19	0.985682	0.978720	0.969406	0.957331	0.942128	0.923495	0.901224	0.875219	0.845508	0.812249
20	0.992497	0.988402	0.982692	0.974988	0.964909	0.952092	0.936216	0.917029	0.894367	0.868168
21	0.996229	0.993935	0.990600	0.985919	0.979554	0.971156	0.960377	0.946894	0.930430	0.910772
22	0.998179	0.996953	0.995094	0.992378	0.988541	0.983288	0.976301	0.967256	0.955837	0.941759
23	0.999155	0.998527	0.997536	0.996028	0.993816	0.990672	0.986340	0.980535	0.972960	0.963314
24	0.999622	0.999314	0.998808	0.998006	0.996783	0.994980	0.992406	0.988835	0.984018	0.977683
25	0.999837	0.999692	0.999444	0.999034	0.998385	0.997392	0.995923	0.993815	0.990874	0.986883
26	0.999932	0.999867	0.999749	0.999548	0.999217	0.998691	0.997885	0.996688	0.994962	0.992543
27	0.999973	0.999944	0.999891	0.999796	0.999633	0.999365	0.998939	0.998284	0.997308	0.995897
28	0.999989	0.999977	0.999954	0.999911	0.999833	0.999702	0.999485	0.999139	0.998607	0.997817
29	0.999996	0.999991	0.999981	0.999962	0.999927	0.999864	0.999757	0.999582	0.999301	0.998867
30	0.999999	0.999997	0.999993	0.999984	0.999969	0.999940	0.999889	0.999803	0.999660	0.999437
31	0.999999	0.999999	0.999997	0.999994	0.999987	0.999974	0.999951	0.999910	0.999839	0.999724
32	1.000000	1.000000	0.999999	0.999998	0.999995	0.999989	0.999979	0.999960	0.999926	0.999869

THE CHI-SQUARE DISTRIBUTION

$$F(x) = P(X \le x) = \int_0^x \frac{w^{r/2-1}e^{-w/2}}{\Gamma(r/2)2^{r/2}}\,dw,$$

$$0.005 \le P(X \le x) \le 0.50.$$

					$P(X \le x)$				
	0.005	0.01	0.025	0.05	0.10	0.20	0.30	0.40	0.50
1	0.000039	0.000157	0.000982	0.003932	0.015791	0.064185	0.148472	0.274996	0.454936
2	0.010025	0.020101	0.050636	0.102587	0.210721	0.446287	0.713350	1.021651	1.386294
3	0.071722	0.114832	0.215795	0.351846	0.584374	1.005174	1.423652	1.869168	2.365974
4	0.206989	0.297109	0.484419	0.710723	1.063623	1.648777	2.194698	2.752843	3.356694
5	0.411742	0.554298	0.831212	1.145476	1.610308	2.342534	2.999908	3.655500	4.351460
6	0.675727	0.872090	1.237344	1.635383	2.204131	3.070088	3.827552	4.570154	5.348121
7	0.989256	1.239042	1.689869	2.167350	2.833107	3.822322	4.671330	5.493235	6.345811
8	1.344413	1.646497	2.179731	2.732637	3.489539	4.593574	5.527422	6.422646	7.344121
9	1.734933	2.087901	2.700390	3.325113	4.168159	5.380053	6.393306	7.357034	8.342833
10	2.155856	2.558212	3.246973	3.940299	4.865182	6.179079	7.267218	8.295472	9.341818
11	2.603222	3.053484	3.815748	4.574813	5.577785	6.988674	8.147868	9.237285	10.340998
12	3.073824	3.570569	4.403788	5.226029	6.303796	7.807328	9.034277	10.181971	11.340322
13	3.565035	4.106915	5.008751	5.891864	7.041505	8.633861	9.925682	11.129140	12.339756
14	4.074675	4.660425	5.628726	6.570631	7.789534	9.467328	10.821478	12.078482	13.339274
15	4.600916	5.229349	6.262138	7.260944	8.546756	10.306959	11.721169	13.029750	14.338860
16	5.142205	5.812212	6.907664	7.961646	9.312236	11.152116	12.624349	13.982736	15.338499
17	5.697217	6.407760	7.564186	8.671760	10.085186	12.002266	13.530676	14.937272	16.338182
18	6.264805	7.014911	8.230746	9.390455	10.864936	12.856953	14.439862	15.893212	17.337902
19	6.843971	7.632730	8.906516	10.117013	11.650910	13.715790	15.351660	16.850433	18.337653
20	7.433844	8.260398	9.590777	10.850811	12.442609	14.578439	16.265856	17.808829	19.337429
21	8.033653	8.897198	10.282898	11.591305	13.239598	15.444608	17.182265	18.768309	20.337228
22	8.642716	9.542492	10.982321	12.338015	14.041493	16.314040	18.100723	19.728791	21.337045
23	9.260425	10.195716	11.688552	13.090514	14.847956	17.186506	19.021087	20.690204	22.336878
24	9.886234	10.856361	12.401150	13.848425	15.658684	18.061804	19.943229	21.652486	23.336726
25	10.519652	11.523975	13.119720	14.611408	16.473408	18.939754	20.867034	22.615579	24.336587
26	11.160237	12.198147	13.843905	15.379157	17.291885	19.820194	21.792401	23.579434	25.336458
27	11.807587	12.878504	14.573383	16.151396	18.113896	20.702976	22.719236	24.544005	26.336339
28	12.461336	13.564710	15.307861	16.927875	18.939242	21.587969	23.647457	25.509251	27.336229
29	13.121149	14.256455	16.047072	17.708366	19.767744	22.475052	24.576988	26.475134	28.336127
30	13.786720	14.953457	16.790772	18.492661	20.599235	23.364115	25.507759	27.441622	29.336032
35	17.191820	18.508926	20.569377	22.465015	24.796655	27.835874	30.178172	32.282116	34.335638
40	20.706535	22.164261	24.433039	26.509303	29.050523	32.344953	34.871939	37.133959	39.335345
45	24.311014	25.901269	28.366152	30.612259	33.350381	36.884407	39.584701	41.995025	44.335118
50	27.990749	29.706683	32.357364	34.764252	37.688648	41.449211	44.313307	46.863776	49.334937
60	35.534491	37.484852	40.481748	43.187958	46.458888	50.640618	53.809126	56.619995	59.334666
70	43.275180	45.441717	48.757565	51.739278	55.328940	59.897809	63.346024	66.396114	69.334474
80	51.171932	53.540077	57.153173	60.391478	64.277844	69.206939	72.915342	76.187932	79.334330
90	59.196304	61.754079	65.646618	69.126030	73.291090	78.558432	82.511097	85.992545	89.334218
100	67.327563	70.064895	74.221927	77.929465	82.358136	87.945336	92.128944	95.807848	99.334129
120	83.851572	86.923280	91.572642	95.704637	100.623631	106.805606	111.418574	115.464544	119.333996

THE CHI-SQUARE DISTRIBUTION (CONTINUED)

$$F(x) = P(X \le x) = \int_0^x \frac{w^{r/2-1}e^{-w/2}}{\Gamma(r/2)2^{r/2}}\,dw,$$

$$0.60 \le P(X \le x) \le 0.9995.$$

					$P(X \le x)$				
r	0.60	0.70	0.80	0.90	0.95	0.975	0.99	0.995	0.9995
1	0.708326	1.074194	1.642374	2.705543	3.841459	5.023886	6.634897	7.879439	12.11566
2	1.832581	2.407946	3.218876	4.605170	5.991465	7.377759	9.210340	10.596635	15.20180
3	2.946166	3.664871	4.641628	6.251389	7.814728	9.348404	11.344867	12.838156	17.72999
4	4.044626	4.878433	5.988617	7.779440	9.487729	11.143287	13.276704	14.860259	19.99735
5	5.131867	6.064430	7.289276	9.236357	11.070498	12.832502	15.086272	16.749602	22.10532
6	6.210757	7.231135	8.558060	10.644641	12.591587	14.449375	16.811894	18.547584	24.10279
7	7.283208	8.383431	9.803250	12.017037	14.067140	16.012764	18.475307	20.277740	26.01776
8	8.350525	9.524458	11.030091	13.361566	15.507313	17.534546	20.090235	21.954955	27.86804
9	9.413640	10.656372	12.242145	14.683657	16.918978	19.022768	21.665994	23.589351	29.66580
10	10.473236	11.780723	13.441958	15.987179	18.307038	20.483177	23.209251	25.188180	31.41981
11	11.529834	12.898668	14.631421	17.275009	19.675138	21.920049	24.724970	26.756849	33.13661
12	12.583838	14.011100	15.811986	18.549348	21.026070	23.336664	26.216967	28.299519	34.82127
13	13.635571	15.118722	16.984797	19.811929	22.362032	24.735605	27.688250	29.819471	36.47779
14	14.685294	16.222099	18.150771	21.064144	23.684791	26.118948	29.141238	31.319350	38.10940
15	15.733223	17.321694	19.310657	22.307130	24.995790	27.488393	30.577914	32.801321	39.71876
16	16.779537	18.417894	20.465079	23.541829	26.296228	28.845351	31.999927	34.267187	41.30807
17	17.824387	19.511022	21.614561	24.769035	27.587112	30.191009	33.408664	35.718466	42.87921
18	18.867904	20.601354	22.759546	25.989423	28.869299	31.526378	34.805306	37.156451	44.43377
19	19.910199	21.689127	23.900417	27.203571	30.143527	32.852327	36.190869	38.582257	45.97312
20	20.951368	22.774545	25.037506	28.411981	31.410433	34.169607	37.566235	39.996846	47.49845
21	21.991497	23.857789	26.171100	29.615089	32.670573	35.478876	38.932173	41.401065	49.01081
22	23.030661	24.939016	27.301454	30.813282	33.924438	36.780712	40.289360	42.795655	50.51111
23	24.068925	26.018365	28.428793	32.006900	35.172462	38.075627	41.638398	44.181275	52.00018
24	25.106348	27.095961	29.553315	33.196244	36.415028	39.364077	42.979820	45.558512	53.47875
25	26.142984	28.171915	30.675201	34.381587	37.652484	40.646469	44.314105	46.927890	54.94745
26	27.178880	29.246327	31.794610	35.563171	38.885139	41.923170	45.641683	48.289882	56.40689
27	28.214078	30.319286	32.911688	36.741217	40.113272	43.194511	46.962942	49.644915	57.85758
28	29.248618	31.390875	34.026565	37.915923	41.337138	44.460792	48.278236	50.993376	59.30002
29	30.282536	32.461168	35.139362	39.087470	42.556968	45.722286	49.587884	52.335618	60.73464
30	31.315863	33.530233	36.250187	40.256024	43.772972	46.979242	50.892181	53.671962	62.16185
35	36.474606	38.859140	41.777963	46.058788	49.801850	53.203349	57.342073	60.274771	69.19855
40	41.622193	44.164867	47.268538	51.805057	55.758479	59.341707	63.690740	66.765962	76.09460
45	46.760687	49.451713	52.728815	57.505305	61.656233	65.410159	69.956832	73.166061	82.87568
50	51.891584	54.722794	58.163797	63.167121	67.504807	71.420195	76.153891	79.489978	89.56051
60	62.134840	65.226507	68.972069	74.397006	79.081944	83.297675	88.379419	91.951698	102.6947
70	72.358347	75.689277	79.714650	85.527043	90.531225	95.023184	100.425184	104.214899	115.5775
80	82.566250	86.119710	90.405349	96.578204	101.879474	106.628568	112.328792	116.321056	128.2613
90	92.761420	96.523762	101.053723	107.565008	113.145270	118.135893	124.116319	128.298944	140.7822
100	102.945944	106.905761	111.666713	118.498004	124.342113	129.561197	135.806723	140.169489	153.1669
120	123.288988	127.615901	132.806284	140.232569	146.567358	152.211403	158.950166	163.648184	177.6029

THE STANDARD NORMAL PROBABILITY DISTRIBUTION

$$F(z) = P(X \leq z) = \frac{1}{\sqrt{2\pi}} \int_{-\infty}^{z} e^{-t^2/2} dt,$$

$$0 \leq P(Z \leq z) \leq 0.09.$$

					$P(Z \leq z)$					
z	0.00	0.01	0.02	0.03	0.04	0.05	0.06	0.07	0.08	0.09
-3.4	0.000337	0.000325	0.000313	0.000302	0.000291	0.000280	0.000270	0.000260	0.000251	0.000242
-3.3	0.000483	0.000466	0.000450	0.000434	0.000419	0.000404	0.000390	0.000376	0.000362	0.000349
-3.2	0.000687	0.000664	0.000641	0.000619	0.000598	0.000577	0.000557	0.000538	0.000519	0.000501
-3.1	0.000968	0.000935	0.000904	0.000874	0.000845	0.000816	0.000789	0.000762	0.000736	0.000711
-3.0	0.001350	0.001306	0.001264	0.001223	0.001183	0.001144	0.001107	0.001070	0.001035	0.001001
-2.9	0.001866	0.001807	0.001750	0.001695	0.001641	0.001589	0.001538	0.001489	0.001441	0.001395
-2.8	0.002555	0.002477	0.002401	0.002327	0.002256	0.002186	0.002118	0.002052	0.001988	0.001926
-2.7	0.003467	0.003364	0.003264	0.003167	0.003072	0.002980	0.002890	0.002803	0.002718	0.002635
-2.6	0.004661	0.004527	0.004396	0.004269	0.004145	0.004025	0.003907	0.003793	0.003681	0.003573
-2.5	0.006210	0.006037	0.005868	0.005703	0.005543	0.005386	0.005234	0.005085	0.004940	0.004799
-2.4	0.008198	0.007976	0.007760	0.007549	0.007344	0.007143	0.006947	0.006756	0.006569	0.006387
-2.3	0.010724	0.010444	0.010170	0.009903	0.009642	0.009387	0.009137	0.008894	0.008656	0.008424
-2.2	0.013903	0.013553	0.013209	0.012874	0.012545	0.012224	0.011911	0.011604	0.011304	0.011011
-2.1	0.017864	0.017429	0.017003	0.016586	0.016177	0.015778	0.015386	0.015003	0.014629	0.014262
-2.0	0.022750	0.022216	0.021692	0.021178	0.020675	0.020182	0.019699	0.019226	0.018763	0.018309
-1.9	0.028717	0.028067	0.027429	0.026803	0.026190	0.025588	0.024998	0.024419	0.023852	0.023295
-1.8	0.035930	0.035148	0.034380	0.033625	0.032884	0.032157	0.031443	0.030742	0.030054	0.029379
-1.7	0.044565	0.043633	0.042716	0.041815	0.040930	0.040059	0.039204	0.038364	0.037538	0.036727
-1.6	0.054799	0.053699	0.052616	0.051551	0.050503	0.049471	0.048457	0.047460	0.046479	0.045514
-1.5	0.066807	0.065522	0.064255	0.063008	0.061780	0.060571	0.059380	0.058208	0.057053	0.055917
-1.4	0.080757	0.079270	0.077804	0.076358	0.074934	0.073529	0.072145	0.070781	0.069437	0.068112
-1.3	0.096800	0.095098	0.093418	0.091759	0.090123	0.088508	0.086915	0.085343	0.083793	0.082264
-1.2	0.115070	0.113139	0.111232	0.109349	0.107488	0.105650	0.103835	0.102042	0.100273	0.098525
-1.1	0.135666	0.133500	0.131357	0.129238	0.127143	0.125072	0.123024	0.121000	0.119000	0.117023
-1.0	0.158655	0.156248	0.153864	0.151505	0.149170	0.146859	0.144572	0.142310	0.140071	0.137857
-0.9	0.184060	0.181411	0.178786	0.176186	0.173609	0.171056	0.168528	0.166023	0.163543	0.161087
-0.8	0.211855	0.208970	0.206108	0.203269	0.200454	0.197663	0.194895	0.192150	0.189430	0.186733
-0.7	0.241964	0.238852	0.235762	0.232695	0.229650	0.226627	0.223627	0.220650	0.217695	0.214764
-0.6	0.274253	0.270931	0.267629	0.264347	0.261086	0.257846	0.254627	0.251429	0.248252	0.245097
-0.5	0.308538	0.305026	0.301532	0.298056	0.294599	0.291160	0.287740	0.284339	0.280957	0.277595
-0.4	0.344578	0.340903	0.337243	0.333598	0.329969	0.326355	0.322758	0.319178	0.315614	0.312067
-0.3	0.382089	0.378280	0.374484	0.370700	0.366928	0.363169	0.359424	0.355691	0.351973	0.348268
-0.2	0.420740	0.416834	0.412936	0.409046	0.405165	0.401294	0.397432	0.393580	0.389739	0.385908
-0.1	0.460172	0.456205	0.452242	0.448283	0.444330	0.440382	0.436441	0.432505	0.428576	0.424655

THE STANDARD NORMAL PROBABILITY DISTRIBUTION (CONTINUED)

$$F(z) = P(Z \leq z) = \frac{1}{\sqrt{2\pi}} \int_{-\infty}^{z} e^{-t^2/2} dt,$$

$$0 \leq P(Z \leq z) \leq 0.09.$$

					$P(Z \leq z)$					
z	0.00	0.01	0.02	0.03	0.04	0.05	0.06	0.07	0.08	0.09
0.0	0.500000	0.496011	0.492022	0.488034	0.484047	0.480061	0.476078	0.472097	0.468119	0.464144
0.1	0.539828	0.535856	0.531881	0.527903	0.523922	0.519939	0.515953	0.511966	0.507978	0.503989
0.2	0.579260	0.575345	0.571424	0.567495	0.563559	0.559618	0.555670	0.551717	0.547758	0.543795
0.3	0.617911	0.614092	0.610261	0.606420	0.602568	0.598706	0.594835	0.590954	0.587064	0.583166
0.4	0.655422	0.651732	0.648027	0.644309	0.640576	0.636831	0.633072	0.629300	0.625516	0.621720
0.5	0.691462	0.687933	0.684386	0.680822	0.677242	0.673645	0.670031	0.666402	0.662757	0.659097
0.6	0.725747	0.722405	0.719043	0.715661	0.712260	0.708840	0.705401	0.701944	0.698468	0.694974
0.7	0.758036	0.754903	0.751748	0.748571	0.745373	0.742154	0.738914	0.735653	0.732371	0.729069
0.8	0.788145	0.785236	0.782305	0.779350	0.776373	0.773373	0.770350	0.767305	0.764238	0.761148
0.9	0.815940	0.813267	0.810570	0.807850	0.805105	0.802337	0.799546	0.796731	0.793892	0.791030
1.0	0.841345	0.838913	0.836457	0.833977	0.831472	0.828944	0.826391	0.823814	0.821214	0.818589
1.1	0.864334	0.862143	0.859929	0.857690	0.855428	0.853141	0.850830	0.848495	0.846136	0.843752
1.2	0.884930	0.882977	0.881000	0.879000	0.876976	0.874928	0.872857	0.870762	0.868643	0.866500
1.3	0.903200	0.901475	0.899727	0.897958	0.896165	0.894350	0.892512	0.890651	0.888768	0.886861
1.4	0.919243	0.917736	0.916207	0.914657	0.913085	0.911492	0.909877	0.908241	0.906582	0.904902
1.5	0.933193	0.931888	0.930563	0.929219	0.927855	0.926471	0.925066	0.923641	0.922196	0.920730
1.6	0.945201	0.944083	0.942947	0.941792	0.940620	0.939429	0.938220	0.936992	0.935745	0.934478
1.7	0.955435	0.954486	0.953521	0.952540	0.951543	0.950529	0.949497	0.948449	0.947384	0.946301
1.8	0.964070	0.963273	0.962462	0.961636	0.960796	0.959941	0.959070	0.958185	0.957284	0.956367
1.9	0.971283	0.970621	0.969946	0.969258	0.968557	0.967843	0.967116	0.966375	0.965620	0.964852
2.0	0.977250	0.976705	0.976148	0.975581	0.975002	0.974412	0.973810	0.973197	0.972571	0.971933
2.1	0.982136	0.981691	0.981237	0.980774	0.980301	0.979818	0.979325	0.978822	0.978308	0.977784
2.2	0.986097	0.985738	0.985371	0.984997	0.984614	0.984222	0.983823	0.983414	0.982997	0.982571
2.3	0.989276	0.988989	0.988696	0.988396	0.988089	0.987776	0.987455	0.987126	0.986791	0.986447
2.4	0.991802	0.991576	0.991344	0.991106	0.990863	0.990613	0.990358	0.990097	0.989830	0.989556
2.5	0.993790	0.993613	0.993431	0.993244	0.993053	0.992857	0.992656	0.992451	0.992240	0.992024
2.6	0.995339	0.995201	0.995060	0.994915	0.994766	0.994614	0.994457	0.994297	0.994132	0.993963
2.7	0.996533	0.996427	0.996319	0.996207	0.996093	0.995975	0.995855	0.995731	0.995604	0.995473
2.8	0.997445	0.997365	0.997282	0.997197	0.997110	0.997020	0.996928	0.996833	0.996736	0.996636
2.9	0.998134	0.998074	0.998012	0.997948	0.997882	0.997814	0.997744	0.997673	0.997599	0.997523
3.0	0.998650	0.998605	0.998559	0.998511	0.998462	0.998411	0.998359	0.998305	0.998250	0.998193
3.1	0.999032	0.998999	0.998965	0.998930	0.998893	0.998856	0.998817	0.998777	0.998736	0.998694
3.2	0.999313	0.999289	0.999264	0.999238	0.999211	0.999184	0.999155	0.999126	0.999096	0.999065
3.3	0.999517	0.999499	0.999481	0.999462	0.999443	0.999423	0.999402	0.999381	0.999359	0.999336
3.4	0.999663	0.999651	0.999638	0.999624	0.999610	0.999596	0.999581	0.999566	0.999550	0.999534

THE (STUDENT'S) *t* PROBABILITY DISTRIBUTION

$$F(x) = P(X \le x) = \frac{\Gamma\left(\dfrac{d+1}{2}\right)}{\Gamma\left(\dfrac{d}{2}\right)\sqrt{d\pi}} \int_{-\infty}^{x} \left(1 + \frac{t^2}{d}\right)^{-(d+1)/2} dt,$$

$$0.0005 \le P(X \le x) \le 0.30.$$

				$P(X \le x)$				
0.30	0.20	0.10	0.05	0.025	0.01	0.005	0.001	0.0005
0.726543	1.376382	3.077684	6.313752	12.706205	31.820516	63.656741	318.308839	636.619249
0.617213	1.060660	1.885618	2.919986	4.302653	6.964557	9.924843	22.327125	31.599055
0.584390	0.978472	1.637744	2.353363	3.182446	4.540703	5.840909	10.214532	12.923979
0.568649	0.940965	1.533206	2.131847	2.776445	3.746947	4.604095	7.173182	8.610302
0.559430	0.919544	1.475884	2.015048	2.570582	3.364930	4.032143	5.893430	6.868827
0.553381	0.905703	1.439756	1.943180	2.446912	3.142668	3.707428	5.207626	5.958816
0.549110	0.896030	1.414924	1.894579	2.364624	2.997952	3.499483	4.785290	5.407883
0.545934	0.888890	1.396815	1.859548	2.306004	2.896459	3.355387	4.500791	5.041305
0.543480	0.883404	1.383029	1.833113	2.262157	2.821438	3.249836	4.296806	4.780913
0.541528	0.879058	1.372184	1.812461	2.228139	2.763769	3.169273	4.143700	4.586894
0.539938	0.875530	1.363430	1.795885	2.200985	2.718079	3.105807	4.024701	4.436979
0.538618	0.872609	1.356217	1.782288	2.178813	2.680998	3.054540	3.929633	4.317791
0.537504	0.870152	1.350171	1.770933	2.160369	2.650309	3.012276	3.851982	4.220832
0.536552	0.868055	1.345030	1.761310	2.144787	2.624494	2.976843	3.787390	4.140454
0.535729	0.866245	1.340606	1.753050	2.131450	2.602480	2.946713	3.732834	4.072765
0.535010	0.864667	1.336757	1.745884	2.119905	2.583487	2.920782	3.686155	4.014996
0.534377	0.863279	1.333379	1.739607	2.109816	2.566934	2.898231	3.645767	3.965126
0.533816	0.862049	1.330391	1.734064	2.100922	2.552380	2.878440	3.610485	3.921646
0.533314	0.860951	1.327728	1.729133	2.093024	2.539483	2.860935	3.579400	3.883406
0.532863	0.859964	1.325341	1.724718	2.085963	2.527977	2.845340	3.551808	3.849516
0.532455	0.859074	1.323188	1.720743	2.079614	2.517648	2.831360	3.527154	3.819277
0.532085	0.858266	1.321237	1.717144	2.073873	2.508325	2.818756	3.504992	3.792131
0.531747	0.857530	1.319460	1.713872	2.068658	2.499867	2.807336	3.484964	3.767627
0.531438	0.856855	1.317836	1.710882	2.063899	2.492159	2.796940	3.466777	3.745399
0.531154	0.856236	1.316345	1.708141	2.059539	2.485107	2.787436	3.450189	3.725144
0.530892	0.855665	1.314972	1.705618	2.055529	2.478630	2.778715	3.434997	3.706612
0.530649	0.855137	1.313703	1.703288	2.051831	2.472660	2.770683	3.421034	3.689592
0.530424	0.854647	1.312527	1.701131	2.048407	2.467140	2.763262	3.408155	3.673906
0.530214	0.854192	1.311434	1.699127	2.045230	2.462021	2.756386	3.396240	3.659405
0.530019	0.853767	1.310415	1.697261	2.042272	2.457262	2.749996	3.385185	3.645959
0.529665	0.852998	1.308573	1.693889	2.036933	2.448678	2.738481	3.365306	3.621802
0.529353	0.852321	1.306952	1.690924	2.032244	2.441150	2.728394	3.347934	3.600716
0.529076	0.851720	1.305514	1.688298	2.028094	2.434494	2.719485	3.332624	3.582150
0.528828	0.851183	1.304230	1.685954	2.024394	2.428568	2.711558	3.319030	3.565678
0.528606	0.850700	1.303077	1.683851	2.021075	2.423257	2.704459	3.306878	3.550966
0.527760	0.848869	1.298714	1.675905	2.008559	2.403272	2.677793	3.261409	3.496013
0.527198	0.847653	1.295821	1.670649	2.000298	2.390119	2.660283	3.231709	3.460200
0.526797	0.846786	1.293763	1.666914	1.994437	2.380807	2.647905	3.210789	3.435015

REFERENCES AND FURTHER READINGS

Barlow, R. E., and Proschan, F. (1975). *Statistical Theory of Reliability and Life Testing Probability Models*. Holt, Rinehart and Winston, New York.

Bergland, G. D. (1969). A guided tour of the fast Fourier transform. *IEEE Spectrum*, 6, 41–52.

Bhat, U. N. (2008). *An Introduction to Queueing Theory: Modeling and Analysis in Applications*. Birkhäuser, Boston.

Birkhoff, G., and Rota, G.-C. (1989). *Ordinary Differential Equations*, 4th edition. John Wiley & Sons, New York.

Bojadjiev, L., and Kamenov, O. (2000). *Higher Mathematics*, Vol. 3. Ciela, Sofia, Bulgaria. (In Bulgarian.)

Brémaud, P. (1978). Streams of a *M/M/*1 feedback queue in statistical equilibrium. *Z. Wahrscheinkeitstheorie verw. Gebiete*, 45, 21–33.

Chan, W. C., and Lin, Y.-B. (2003). The waiting time distribution for the *M/M/m* queue. *IEE Proceedings—Communications*, 150(3), 159–162.

Cohen, J. W., and Boxma, O. J. (1983). *Boundary Value Problems in Queueing System Analysis*. North-Holland, New York.

Conrad, B. P. (2003). *Differential Equations with Boundary Value Problems*. Pearson Education, Upper Saddle River, NJ.

Cook, L. M. (1965). Oscillation in the simple logistic growth model. *Nature*, 207, 316.

Devore, J. L. (2000). *Probability and Statistics for Engineering and the Sciences*, 5th edition. Brooks/Cole, Belmont, CA.

Disney, R. L., and Kiessler, P. C. (1987). *Traffic Processes in Queueing Networks: A Markov Renewal Approach*. Johns Hopkins University Press, Baltimore, MD.

Duhamel, P., and Vetterli, M. (1990). Fast Fourier transforms: a tutorial review. *Signal Processing*, 19, 259–299.

Durham, S., Haghighi, A. M., and Goddard, P. (1991). Differential Markov chains: an introduction to applied probability. Lecture notes, Department of Statistics, University of South Carolina.

Difference and Differential Equations with Applications in Queueing Theory, First Edition.
Aliakbar Montazer Haghighi and Dimitar P. Mishev.
© 2013 John Wiley & Sons, Inc. Published 2013 by John Wiley & Sons, Inc.

Durham, S., Flournoy, N., Goddard, C., and Haghighi, A. M. (1995). An introduction to applied probability. Lecture notes, Department of Statistics, University of South Carolina.

Durham, S. D., Flournoy, N., and Rosenberger, W. F. (1997). A random walk rule for phase I clinical trials. *Biometrics*, 53, 745–760.

Dym, H., and McKean, H. P. (1972). *Fourier Series and Integrals*, Academic Press, New York.

Erdelyi, A. (1954). *Tables of Integral Transforms*, Vol. 1. McGraw-Hill, New York.

Fayolle, G. (1979). Méthodes analytiques pour les filess d'attente couplées. Thesis, Université Paris VI.

Feller, W. (1968). *An Introduction to Probability Theory and Its Applications*, Vol. 1, 3rd edition, revised printing. John Wiley & Sons, New York.

Finizio, N., and Ladas, G. (1982). *An Introduction to Differential Equations with Difference Equations, Fourier Series, and Partial Differential Equations*. Wadsworth, Belmont, CA.

Fitzpatrick, P. M. (2009). *Advanced Calculus*, 2nd edition. AMS, Providence, RI.

Foley, R. D., and Disney, R. L. (1983). Queues with delayed feedback. *Advances in Applied Probability*, 15, 162–182.

Gakhov, F. D. (1966). *Boundary Value Problems* (I. N. Sneddon, translation ed.). Dover, New York. Originally published by Pergamon Press, Oxford, UK.

Goldberg, S. (1986). *Introduction to Difference Equations with Illustrative Examples from Economics, Psychology, and Sociology*. Dover, New York. Originally published by John Wiley & Sons, New York, 1958.

Gross, D., Shortle, J. F., Thompson, J. M., and Harris, C. M. (2011). *Fundamentals of Queueing Theory*, 4th edition. John Wiley & Sons, New York.

Gyori, I., and Ladas, G. (1991). *Oscillation Theory of Delay Differential Equations: With Applications*. Oxford University Press, New York.

Haar, A. (1910). Zur Theorie der orthogonalen Funktionensysteme. *Mathematische Annalen*, 69, 331–371.

Haghighi, A. M., and Mishev, D. P. (2007). A tandem queueing system with task-splitting, feedback, and blocking. *International Journal of Operational Research*, 2(2), 208–230.

Haghighi, A. M., and Mishev, D. P. (2009). Analysis of a two-node task-splitting feedback tandem queue with infinite buffers by functional equation. *International Journal of Mathematics in Operational Research*, 1(1/2), 246–277.

Haghighi, A. M., and Mishev, D. P. (2013). *Queueing Models in Industry and Business*, 2nd edition. (First edition published 2008.) Nova Science Publishers, New York.

Haghighi, A. M., and Shayib, M. (2010). Reliability computation using logistic and extreme-value distributions. *International Journal of Statistics and Economics*, 4(S10), 55–73.

Haghighi, A. M., Mishev, D., and Chukova, S. S. (2008). A single-server Poisson queueing system with delayed-service. *International Journal of Operational Research*, 3(4), 363–383.

Haghighi, A. M., Lian, J., and Mishev, D. P. (2011a). *Advanced Mathematics for Engineers with Applications in Stochastic Processes*, revised edition. Nova Science Publishers, New York.

Haghighi, A. M., Chukova, S. S., and Mishev, D. (2011b). Single-server Poisson queueing system with splitting and delayed-feedback: part I. *International Journal of Mathematics in Operational Research*, 3(1), 1–21.

Hannibalsson, I., and Disney, R. L. (1977). An *M/M/*1 queues with delayed feedback. *Naval Research Logistics Quarterly*, 24, 281–291.

Hartman, P. (2002). *Ordinary Differential Equations*, 2nd edition. SIAM, Philadelphia.

Hassell, M. P. (1975). Density-dependence in single species populations. *Journal of Animal Ecology*, 44(1), 283–295.

Higgins, J. J., and Keller-McNulty, S. (1995). *Concepts in Probability and Stochastic Modeling*. Duxbury Press, Belmont, CA.

Hogg, R. V., and Tanis, E. A. (1993). *Probability and Statistical Inference*, 4th edition. Macmillan, New York.

Hurewicz, W. (1947). Filters and servosystems with pulsed data. In *Theory of Servomechanism* (H. J. James, N. B. Nichols, and R. S. Phillips, eds.). Massachusetts Institute of Technology, Radiation Laboratory Series, vol. 25. McGraw-Hill, New York, pp. 231–261.

Hutchinson, G. E. (1948). Circular causal systems in ecology. *Annals of the New York Academy of Sciences*, 50, 221–246.

Ivanova, A., Montazer-Haghighi, A., Mohanty, S. G., and Durham, S. D. (2003). Improved up-and-down designs for phase I trials. *Statistics in Medicine*, 22, 69–82.

Jackson, J. R. (1957). Networks of waiting lines. *Operations Research*, 5, 518–521.

Jagerman, D. L. (2000). *Difference Equations with Applications to Queues*. Marcel Dekker, New York.

Jain, J. L., Mohanty, S. G., and Böhm, W. (2007). *A Course on Queueing Models*. Chapman & Hall/CRC, Taylor & Francis Group, Boca Raton, FL.

Jordan, C. (1979). *Calculus of Finite Differences*. Chelsea Publishing Company, New York. Originally published in 1939.

Kaplan, J. L., and Yorke, J. A. (1975). On the stability of a periodic solution of a differential delay equation. *SIAM Journal of Mathematical Analysis*, 6, 268–282.

Karlin, S., and Taylor, H. (1975). *A First Course in Stochastic Processes*, 2nd edition. Academic Press, New York.

Kelley, W. G., and Peterson, A. (2001). *Difference Equations: An Introduction with Applications*. Academic Press, New York.

Kemeny, J. G., and Snell, J. L. (1960). *Finite Markov Chains*. Van Nostrand, Princeton, NJ.

Kendall, M. (1953). The analysis of time series, part 1: prices. *Journal of the Royal Statistical Society, Series A (General)*, 116 (1), 11–34.

Kleinrock, L. (1975a). *Queueing Systems, Problems and Solutions*, Vol. 1: *Theory*. John Wiley & Sons, New York.

Kleinrock, L. (1975b). *Queueing Systems, Problems and Solutions*, Vol. 2: *Theory*. John Wiley & Sons, New York.

Kleinrock, L., and Gail, R. (1996). *Queueing Systems, Problems and Solutions*. John Wiley & Sons, New York.

Kotz, S., Lumelskii, Y., and Pensky, M. (2003). *The Stress-Strength Model and Its Generalizations, Theory and Applications*. World Scientific, Singapore.

Krebs, J. R. (1972). *Ecology: The Experimental Analysis of Distribution and Abundance*. Harper and Row, New York.

Lehmann, E. L. (1983). *Theory of Point Estimation*. John Wiley & Sons, New York.

Leslie, P. H. (1957). An analysis of the data for experiments carried out by Gause with populations of the protozoa, *Paramecium aurelia* and *Paramecium caudatum*. *Biometrika*, 44(3–4), 314–327.

Levine, S. H., Scudo, F. M., and Plunkett, D. J. (1977). Persistence and convergence of ecosystems: an analysis of some second order difference equations. *Journal of Mathematical Biology*, 4(2), 171–182.

Li, T. Y., and Yorke, J. A. (1975). Period three implies chaos. *American Mathematical Monthly*, 82(10), 985–992.

Lipsky, L. (2009). *Queueing Theory*, 2nd edition. Springer, New York.

Little, J. D. C. (1961). A proof of a queueing formula: $L = \lambda W$. *Operations Research*, 9, 383–387.

Lorenz, E. N. (1963). The mechanics of vacillations. *Journal of Atmospheric Science*, 20, 448–464.

Lorenz, E. N. (1964). The problem of deducing the climate from the governing equations. *Tellus*, 16, 1–11.

Macfadyen, A. (1963). *Animal Ecology*. Sir Isaac Pitman & Sons, London.

May, R. M. (1973a). On relationships among various types of population model. *American Naturalist*, 107, 46–57.

May, R. M. (1973b). *Stability and Complexity in Model Ecosystems*. Princeton University Press, Princeton, NJ.

May, R. M. (1974a). Biological populations with nonoverlapping generations: stable points, stable cycles, and chaos. *Science (New Series)*, 186(4164), 645–647.

May, R. M. (1974b). Ecosystem patterns in randomly fluctuating environments. In *Progress in Theoretical Biology*, Vol. 3 (R. Rosen and F. Snell, eds.). Academic Press, New York, pp. 1–50.

May, R. M. (1975). Biological populations obeying difference equations: stable points, stable cycles, and chaos. *Journal of Theoretical Biology*, 51, 511–524.

May, R. M. (1976). Simple mathematical models with very complicated dynamics. *Nature*, 261(5560), 459–467.

May, R. M., Conway, G. R., Hassell, M. P., and Southwood, T. R. E. (2015). Time delays, density-dependence and single-species oscillations. *The Journal of Animal Ecology*, 43, 747–770.

Maynard Smith, J. (1968). *Mathematical Ideas in Biology*. Cambridge University Press, Cambridge, UK.

Maynard Smith, J. (1974). *Models in Ecology*. Cambridge University Press, Cambridge, UK.

McMuritrie, R. (1978). Persistence and stability of single-species and prey-predator systems in spatially heterogeneous environments. *Mathematical Biosciences*, 39, 11–51.

Medhi, J. (2003). *Stochastic Models in Queueing Theory*, 2nd edition. Academic Press, San Diego, CA.

Mendenhall, W., Beaver, R. J., and Beaver, B. M. (2009). *Introduction to Probability and Statistics*, 13th edition. Brooks/Cole, Cengage Learning, Belmont, CA.

Mickens, R. E. (1987). *Difference Equations: Theory and Applications*, 2nd edition. Van Nostrand Reinhold, New York.

Miller, K. (1968). *Linear Difference Equations*. W. A. Benjamin, New York.

Montazer-Haghighi, A. (1976). Many-server queueing system with feedback. PhD dissertation, Case Western Reserve University.

Montazer-Haghighi, A. (1981). A many-server queueing system with feedback. *Bulletin of Iranian Mathematical Society*, 9(1), 65–74. Serial no. 16.

Montazer-Haghighi, A., Medhi, J., and Mohanty, S. G. (1986). On a multiserver Markovian queueing system with balking and reneging. *Computers and Operations Research*, 13(4), 421–425.

Muskhelishvili, N. I. (1992). *Singular Integral Equations, Boundary Problems of Function Theory and Their Application to Mathematical Physics*, 2nd edition. Dover Press, New York. Translated from Russian by J. R. M. Radok. Second Russian edition, Moscow, 1946.

Nagle, K., Saff, E. B., and Snider, D. (2012). *Fundamentals of Differential Equations and Boundary Value Problems*, 6th edition. Pearson Education, Boston.

Nakamura, G. (1971). A feedback queueing model for an interactive computer system. *AFIPS Proceedings of the Fall Joint Conference*.

Neuts, M. F. (1973). *Probability*. Allyn and Bacon, Boston.

Oppenheim, A. V., Willsky, A. S., and Nawab, S. H. (1997). *Signals and Systems*, 2nd edition, Prentice Hall Signal Processing Series. Prentice Hall, Upper Saddle River, NJ.

Pennycuik, C. J., Compton, R. M., and Beckingham, L. (1968). A computer model for simulating the growth of a population or of two interacting populations. *Journal of Theoretical Biology*, 18, 316–329.

Pielou, E. C. (1969). *An Introduction to Mathematical Ecology*. John Wiley & Sons, New York.

Prabhu, N. U. (2007). *Stochastic Processes*. World Scientific, Singapore.

Ragazzini, J. R., and Zadeh, L. A. (1952). The analysis of sampled-data systems. Transactions of the American Institute of Electrical Engineers, Part II (Applications and Industry), 1(3), 225–234.

Rhee, H., Aris, R., and Amundson, N. R. (2001). *First Order Partial Differential Equations*. Dover, Mineola, NY.

Robbins, M. (2003). A NIMITZ-Class Carrier, *Popular Science*, December.

Robertson, T., Wright, F. T., and Dykstra, R. L. (1988). *Ordered Restricted Statistical Inference*. John Wiley & Sons, New York.

Ross, S. (2010). *Introduction to Probability Models*, 10th edition. Academic Press/Elsevier, Burlington, MA.

Ross, S. M. (1996). *Introductory Statistics*. McGraw-Hill, New York.

Ryan, T. (1989). *Statistical Methods for Quality Improvement*. John Wiley & Sons, New York.

Saaty, T. L. (1983). *Elements of Queueing Theory with Application*. Dover Press, New York. Originally published by McGraw-Hill, 1961.

Skellam, J. G. (1952). Studies in statistical biology, I. Spatial pattern. *Biometrika*, 39, 346–362.

Smirnov, V. I. (1964). *A Course of Higher Mathematics*, Vol. 3. Pergamon Press, Oxford, UK.

Snell, J. L. (1988). *Introduction to Probability*, Random House/Birkhäuser Mathematics Series. Random House, New York.

Spiegel, M. (1971). *Theory and Problems of Calculus of Finite Differences and Difference Equations*. McGraw-Hill, New York.

Stewart, J. (2012). *Calculus*, 7th edition. Brooks/Cole, Belmont, CA.

Takács, L. (1962). *Introduction to the Theory of Queues*. Oxford University Press, New York.

Takács, L. (1963). A single server queue with feedback. *The Bell System Technical Journal*, 42, 505–519.

Trench, W. F. (2000). *Elementary Differential Equations*. Brooks/Cole, Thomson Learning, Pacific Grove, CA.

Usher, M. B. (1972). Developments in the Leslie matrix model. In *Mathematical Models in Ecology* (J. N. R. Jeffers, ed.). Blackwell, Oxford, UK, pp. 29–60.

Utida, S. (1967). Damped oscillation of population density at equilibrium. *Researches on Population Ecology (Kyoto)*, 9, 1–9.

Van Der Vaart, H. R. (1973). A comparative investigation of certain difference equation and related differential equations: implications for model-building. *Bulletin of Mathematical Biology*, 35, 195–211.

Van Loan, C. (1992). *Computational Frameworks for the Fast Fourier Transform*. SIAM, Philadelphia.

Varley, G. C., Gradwell, G. R., and Hassell, M. P. (1973). *Insect Population Ecology*. Oxford, Blackwell, UK, p. 20.

Williamson, M. (1974). In *Ecological Stability* (M. B. Usher and M. Williamson, eds.). Chapman and Hall, London, pp. 17–34.

Zill, D. G. (2009). *A First Course in Differential Equations with Modeling Applications*, 9th edition. Brooks/Cole, Belmont, CA.

1.54. a. There are 27 possible samples of size 3 as in the following table.
b. 1/27.

Note that M_m denotes median.

Possible Samples of $n = 3$ Measures	$\bar{X}$	M_m	Probability
0, 0, 0	0	0	1/27
0, 0, 3	1	0	1/27
0, 0, 12	4	0	1/27
0, 3, 0	1	0	1/27
0, 3, 3	2	3	1/27
0, 3, 12	5	3	1/27
0, 12, 0	4	0	1/27
0, 12, 3	5	3	1/27
0, 12, 12	8	12	1/27
3, 0, 0	1	0	1/27
3, 0, 3	2	3	1/27
3, 0, 12	5	3	1/27
3, 3, 0	2	3	1/27
3, 3, 3	3	3	1/27
3, 3, 12	6	3	1/27
3, 12, 0	5	3	1/27
3, 12, 3	6	3	1/27
3, 12, 12	9	12	1/27
12, 0, 0	4	0	1/27
12, 0, 3	5	3	1/27
12, 0, 12	8	12	1/27
12, 3, 0	5	3	1/27
12, 3, 3	6	3	1/27
12, 3, 12	9	12	1/27
12, 12, 0	8	12	1/27
12, 12, 3	9	12	1/27
12, 12, 12	12	12	1/27

Difference and Differential Equations with Applications in Queueing Theory, First Edition.
Aliakbar Montazer Haghighi and Dimitar P. Mishev.
© 2013 John Wiley & Sons, Inc. Published 2013 by John Wiley & Sons, Inc.

From the table, we can see that the sample mean $\bar{X}$ takes the following values: $0, 1, 2, 3, 4, 5, 6, 8, 9,$ and 12. Since $\bar{X} = 0$ occurs only in one sample, we will have $P(\bar{X} = 0) = 1/27$. On the other hand, since $\bar{X} = 1$ occurs in 3 samples, $(0, 0, 3), (0, 3, 0),$ and $(3, 0, 0),$ we will have $P(\bar{X} = 1) = 3/27 = 1/9$. We can calculate other probabilities from the table and summarize the results in the following distribution form:

$\bar{X}$	0	1	2	3	4	5	6	8	9	12
$P_{\bar{X}}$	1/27	3/27	3/27	1/27	3/27	6/27	3/27	3/27	3/27	1/27

1.55. In this case:

$$E(X) = m = (0)(1/3) + (3)(1/3) + (12)(1/3) = 5,$$

$$E(\bar{X}) = (0)(1/27) + (1)(3/27) + (2)(3/27) + \cdots + (12)(3/27) = 5.$$

Hence, $\mu = 5 = E(\bar{X})$ implies that $\bar{X}$ is an unbiased estimator of μ.

1.56. Note that θ is a point in E^2 and not a real number, but MLE is the same as for a real parameter. Now, we have:

$$f_\theta(x) = \frac{1}{(2\pi\sigma^2)^{\pi/2}} e^{-\frac{1}{2\sigma^2} \sum_{i=1}^{n} (x_i - \mu)^2},$$

from which we have:

$$\ln f_\theta(x) = -\frac{n}{2} \ln(2\pi) - n \ln \sigma - \frac{1}{2\sigma^2} \sum_{i=1}^{n} (x_i - \mu)^2.$$

Hence,

$$\frac{\partial}{\partial \mu} \ln f_\theta(x) = \frac{1}{\sigma^2} \sum_{i=1}^{n} (x_i - \mu)^2 = \frac{n}{\sigma^2} - (\bar{x} - \mu)$$

and

$$\frac{\partial}{\partial \mu} \ln f_\theta(x) = -\frac{n}{\sigma} + \frac{1}{\sigma^3} \sum_{i=1}^{n} (x_i - \mu)^2 = \frac{n}{\sigma^3} \left(-\sigma^2 + \frac{1}{n} \sum_{i=1}^{n} (x_i - \mu)^2 \right).$$

Setting the partial derivative equal to 0, we obtain $\hat{\theta} = (\bar{x}, s^2)$, where:

$$s^2 = \frac{1}{n} \sum_{i=1}^{n} (x_i - \bar{x})^2.$$

Thus,

$$\hat{\theta}(X_1, \ldots, X_n) = (\bar{X}, S^2).$$

2.16. a.

$$\mathcal{L}^{-1}\left\{\frac{2s+1}{s^2+2s+4}\right\} = \mathcal{L}^{-1}\left\{\frac{2s+1}{(s+1)^2+3}\right\} = \mathcal{L}^{-1}\left\{\frac{2(s+1)-1}{(s+1)^2+3}\right\}$$

$$= 2\mathcal{L}^{-1}\left\{\frac{s+1}{(s+1)^2+3}\right\} - \mathcal{L}^{-1}\left\{\frac{1}{(s+1)^2+3}\right\}$$

$$= 2e^{-t}\cos\sqrt{3}t - e^{-t}\frac{\sin\sqrt{3}t}{\sqrt{3}t}.$$

2.22. a.

$$G(z) = \sum_{n=0}^{\infty}\left(\frac{1}{5}\right)^n z^n = \sum_{n=0}^{\infty}\left(\frac{z}{5}\right)^n = \frac{1}{1-\frac{z}{5}} = \frac{5}{5-z}.$$

2.23. b. $G(z) = (z/(1-2z)) = z(1/(1-2z)) = z\sum_{n=0}^{\infty}(2z)^n = \sum_{n=0}^{\infty}2^n z^n$.

3.1. a. Second-order linear
   b. Second-order nonlinear
   c. Third-order linear
   d. Second-order nonlinear

3.3. a. $y = 125(x^2+1)^{-2}$.
   b. $\ln|\tan y/4| + 2\sin x/2 = 0$.
   c. $y = 3e^{\sqrt{1-y^2}}$.
   d. $\sqrt{1-x^2} = \sin^{-1}x$.

3.5. a. $y = 1/x^2(\sin x - x\cos x)$.
   b. $y = (\ln|x|)^2 - \ln|x|$.

3.9. a. $x^2 - y^2 = (5/27)y^3$.
   b. $e^x \sin y + y^2 = C$.

3.11. a. $\ln x + e^{y/x} = C$.
   b. $(x^2 - 4y^2)^3 = 27x^{-2}$.
   c. $|y - x| = (1/16)|y + 3x|$.
   d. $3(x+y)^2 = 16x^2y$.

3.13. a. $x + y + 1 = \tan(x + (\pi/3))$
   b. $y = 2x - 3 + 1/4(x + 3)^2$.

3.17. a. $y = (C_1 + C_2)e^x + x^2 e^x$.

b. $y = \cos x + 2 \sin x - 4/9 \cos 2x - 1/3\, x \sin 2x$.

c. $y = C_1 e^x + (C_2 - x^2 + 2x)e^{-x}$.

d. $y = (8x - 2)e^x + \cos 2x - 2 \sin 2x$.

3.19. a. $y = C_1 \cos(2\ln x) + C_2 \sin(2\ln x)$.

b. $y = x + 3x \ln x + 1/2x\, (\ln x)^2$.

3.28. a.

$$y(x) = a_0 \sum_{n=0}^{\infty} \frac{(-1)^n}{(2n)!} x^{2n} + a_0 \sum_{n=0}^{\infty} \frac{(-1)^n}{(2n+1)!} x^{2n+1}.$$

b.

$$a_{n+2} = \frac{a_n}{(n+1)(n+2)},\ y_1(x) = \sum_{n=0}^{\infty} \frac{x^{2x}}{(2n)!} = \cosh x,\ y_2(x) = \sum_{n=0}^{\infty} \frac{x^{2x+1}}{(2n+1)!} = \sinh x.$$

c.

$$a_{n+2} = \frac{a_n}{n+2},\ y_1(x) = \sum_{n=0}^{\infty} \frac{x^{2n}}{2^n n!},\ y_2(x) = \sum_{n=0}^{\infty} \frac{2^n n! x^{2x+1}}{(2n+1)!}.$$

3.29. a. $z(x, y) = y - x + F(x^2 - y^2)$.

b. $z(x, y) = xF(x^2 + y^2)$.

4.1. b. The characteristic equation is $r^2 + 4r + 1 = 0$. Thus, the roots are $r_1 = -2 - \sqrt{3}$ and $r_2 = -2 + \sqrt{3}$. The general solution of this equation is $y_n = c_1(-2 - \sqrt{3})^n + c_2(-2 - \sqrt{3})^n$.

4.2. b. From the general solution of 4.1.b and the initial conditions are $y_0 = 0$ and $y_1 = 1$, we have:

$$\begin{cases} y_0 = c_1(-2 - \sqrt{3})^0 + c_2(-2 - \sqrt{3})^0, \\ y_1 = c_1(-2 - \sqrt{3}) + c_2(-2 - \sqrt{3}). \end{cases}$$

Thus,

$$\begin{cases} c_1 + c_2 = 0, \\ c_1(-2 - \sqrt{3}) + c_2(-2 - \sqrt{3}) = 1. \end{cases}$$

Hence, $c_1 = -(\sqrt{3}/6)$ and $c_2 = \sqrt{3}/6$. Therefore, the particular solution is:

$$y_n = -\frac{\sqrt{3}}{6}(-2 - \sqrt{3})^n + \frac{\sqrt{3}}{6}(-2 - \sqrt{3})^n.$$

4.3. c. The homogeneous part of the given equation is $y_{n+2} - y_{n+1} - 2y_n = 0$. The characteristic equation is $r^2 - r - 2 = 0$. Thus, the roots are $r_1 = 2$ and $r_2 = -1$. The general solution of the homogenous part is $\varphi_n = A(2)^n + B(-1)^n$. For the particular solution, assume $\varphi_p = an^2 + bn + c$. Substituting this in the original equation and using equivalence of the polynomials, we obtain:

$$\begin{cases} -2a = 1, \\ 2a - 2b = 0, \\ 3a - b - 2c = 0. \end{cases}$$

Hence, $a = b = c - 1/2$. Hence, the particular solution is $\varphi_p = -(1/2)n^2 - (1/2)n - (1/2)$. The general solution, therefore, is $yn = A(2)^n + B(-1)^n - (1/2)n^2 - (1/2)n - (1/2)$.

f. Dividing both sides of the equation, we will have:

$$y_{n+2} - \frac{3}{4}y_{n+1} + \frac{1}{8}y_n = \frac{1}{8}\sin\left(\frac{n\pi}{2}\right).$$

The characteristic equation of the homogenous part is $r^2 - 3/4r + 1/8 = 0$. Thus, the roots are $r_1 = 1/2$ and $r_2 = 1/4$. The general solution of the homogenous part is $\varphi_n = A(1/2)^n + B(1/4)^n$. For the particular solution, assume:

$$\phi_p = a\cos\left(\frac{n\pi}{2}\right) + b\sin\left(\frac{n\pi}{2}\right).$$

Substituting this in the original equation and using equivalence of the polynomials, we obtain:

$$\begin{cases} -\dfrac{7}{8}a - \dfrac{6}{8}b = 0, \\ \dfrac{7}{8}b - \dfrac{6}{8}a = \dfrac{5}{8}. \end{cases}$$

Hence, the general solution of the original problem is:

$$y_n = A\left(\frac{1}{2}\right)^n + B\left(\frac{1}{4}\right)^n + \frac{6}{17}\cos\left(\frac{n\pi}{2}\right) - \frac{7}{17}\sin\left(\frac{n\pi}{2}\right).$$

4.5. a.

$$G(z) = \sum_{n=0}^{\infty}(1-3n)z^n = \frac{1}{1-z} - 3z\sum_{n=0}^{\infty}nz^{n-1}$$

$$= \frac{1}{1-z} - 3z\sum_{n=0}^{\infty}\frac{d}{dz}(z)^n = \frac{1-4z}{1-2z+z^2}.$$

d.

$$G(z) = \sum_{n=0}^{\infty} \frac{n(n+1)}{2} z^n = \frac{z^2}{2} \sum_{n=0}^{\infty} n(n-1) z^{n-2} + z \sum_{n=0}^{\infty} n z^{n-1}$$

$$= \frac{z^2}{2} \frac{d^2}{dz^2} \left(\frac{1}{1-z} \right) + z \frac{d}{dz} \left(\frac{1}{1-z} \right)$$

$$= \frac{z}{1 - 3z + 3z^2 - z^3}.$$

4.6. c.

$$G(z) = \frac{z^2}{1-z} = z^2 \sum_{n=0}^{\infty} z^n = \sum_{n=0}^{\infty} z^{n+2} = \sum_{n=0}^{\infty} a_n z^{n+2}, \quad a_n = 1.$$

4.11.

$$P_{1,0} = \rho P_{0,0}$$
$$P_{2,0} = \rho P_{1,0} = \rho^2 P_{0,0}$$
$$P_{3,0} = \rho P_{2,0} = \rho^3 P_{0,0}$$
$$\vdots$$
$$P_{k,0} = \rho P^{k-1} = \rho^k P_{0,0}$$

$$P_{k,0} = \begin{cases} \rho^k P_{0,0}, & 0 \le k \le n \\ \theta^{k-n} \rho^n P_{0,0}, & k \ge n+1. \end{cases}$$

Now, we have:

$$P_{2,1} = (1+\rho) P_{1,1} - \rho P_{0,1},$$
$$P_{3,1} = (1+\rho) P_{2,1} - \rho P_{1,1} = (1+\rho)[(1+\rho) P_{1,1} - \rho P_{0,1}] - \rho P_{1,1}$$
$$= (1+\rho)^2 P_{1,1} - \rho(1+\rho) P_{0,1} - \rho P_{1,1} = [(1+\rho)^2 - \rho] P_{1,1} - \rho(1+\rho) P_{0,1},$$
$$P_{4,1} = (1+\rho) P_{3,1} - \rho P_{2,1} = (1+\rho) \{ [(1+\rho)^2 - \rho] P_{1,1} - \rho(1+\rho) P_{0,1} \}$$
$$\quad - \rho[(1+\rho) P_{1,1} - \rho P_{0,1}],$$
$$\vdots$$
$$P_{n,1} = (1+\rho) P_{n-1,1} - \rho P_{n-2,1}.$$

4.12.

$$\begin{cases} P_{1,1} = \rho P_{0,0}, & \\ P_{i,0} = \rho P_{i-1,0} & 1 \le i \le n, \\ P_{i,0} = \theta P_{i-1,0} & n \le i, \\ P_{i,1} = (1+\rho) P_{i-1,1} - \rho P_{i-2,1} & 2 \le i \le n, \\ P_{i,1} = (1+\rho) P_{i-1,1} - \rho P_{i-2,1} - w P_{i-1,0} & n+1 \le i. \end{cases}$$

$$P_{1,0} = \rho P_{0,0}$$
$$P_{2,0} = \rho P_{1,0} = \rho^2 P_{0,0}$$
$$P_{3,0} = \rho P_{2,0} = \rho^3 P_{0,0}$$
$$\vdots$$
$$P_{k,0} = \rho P^{k-1} = \rho^k P_{0,0}$$
$$P_{k,0} = \begin{cases} \rho^k P_{0,0}, & 0 \le k \le n \\ \theta^{k-n} \rho^n P_{0,0}, & k \ge n+1. \end{cases}$$

We have:

$$P_{2,1} = (1+\rho) P_{1,1} - \rho P_{0,1},$$
$$P_{3,1} = (1+\rho) P_{2,1} - \rho P_{1,1} = (1+\rho)[(1+\rho) P_{1,1} - \rho P_{0,1}] - \rho P_{1,1},$$
$$= (1+\rho)^2 P_{1,1} - \rho(1+\rho) P_{0,1} - \rho P_{1,1} = \left[(1+\rho)^2 - \rho\right] P_{1,1} - \rho(1+\rho) P_{0,1}$$
$$P_{4,1} = (1+\rho) P_{3,1} - \rho P_{2,1}$$
$$= (1+\rho)\left[\left[(1+\rho)^2 - \rho\right] P_{1,1} - \rho(1+\rho) P_{0,1}\right] - \rho[(1+\rho) P_{1,1} - \rho P_{0,1}],$$
$$\vdots$$
$$P_{n,1} = (1+\rho) P_{n-1,1} - \rho P_{n-2,1}$$

$$P_{n+1,0} = \theta P_{n,0} = \theta \rho^n P_{0,0},$$
$$P_{n+2,0} = \theta P_{n+1,0} = \theta^2 \rho^n P_{0,0},$$
$$P_{n+3,0} = \theta P_{n+2,0} = \theta^3 \rho^n P_{0,0},$$
$$\vdots$$
$$P_{k,0} = \theta^{k-n} \rho^n P_{0,0}, \ k \ge n+1.$$

5.3.

$$\mathbf{P} = \begin{array}{c} \\ 0 \\ 1 \\ 2 \\ 3 \\ \vdots \end{array} \begin{array}{ccccc} 0 & 1 & 2 & 3 & \cdots \\ \left(1-p \right. & p & 0 & 0 & \cdots \\ 0 & 0 & 1-p & p & \cdots \\ 0 & 1-p & p & 0 & \cdots \\ 0 & 0 & 0 & 1-p & \cdots \\ \vdots & \vdots & \vdots & \vdots & \left. \cdots \right) \end{array}$$

Difference and Differential Equations with Applications in Queueing Theory, First Edition.
Aliakbar Montazer Haghighi and Dimitar P. Mishev.
© 2013 John Wiley & Sons, Inc. Published 2013 by John Wiley & Sons, Inc.

Printed and bound by CPI Group (UK) Ltd, Croydon, CR0 4YY

16/04/2025

14658521-0004